Mechanical Engineering Series

Frederick F. Ling
Series Editor

Peter A. Engel

Structural Analysis of Printed Circuit Board Systems

With 192 Figures

Springer-Verlag
New York Berlin Heidelberg London Paris
Tokyo Hong Kong Barcelona Budapest

Peter A. Engel
Department of Mechanical and Industrial
  Engineering
State University of New York at Binghamton
Binghamton, NY 13902-6000 USA

Series Editor
Frederick F. Ling
Ernest F. Gloyna Regents Chair in Engineering
Department of Mechanical Engineering
The University of Texas at Austin
Austin, TX 78712-1063 USA
  and
William Howard Hart Professor Emeritus
Department of Mechanical Engineering,
  Aeronautical Engineering and Mechanics
Rensselaer Polytechnic Institute
Troy, NY 12180-3590 USA

Library of Congress Cataloging-in-Publication Data
Engel, Peter A.
    Structural analysis of printed circuit board systems/Peter A.
  Engel.
      p.   cm. — (Mechanical engineering series)
    Includes bibliographical references and indexes.
    ISBN 0-387-97939-5   ISBN 3-540-97939-5
    1. Printed circuits – Design and construction. 2. Structural
  analysis (Engineering) I. Title. II. Series: Mechanical
engineering series (Berlin, Germany)
TK7868.P7E53   1993
621.3815′31 – dc20   92-21531

Printed on acid-free paper.

Production managed by Hal Henglein; manufacturing supervised by Vincent R. Scelta.
Typeset by Macmillan India Ltd., Bangalore, India.
Printed and bound by Edwards Brothers, Inc., Ann Arbor, MI.
Printed in the United States of America.

9 8 7 6 5 4 3 2 1

ISBN 0-387-97939-5 Springer-Verlag New York Berlin Heidelberg
ISBN 3-540-97939-5 Springer-Verlag Berlin Heidelberg New York

To Fanya

Preface

Electronics, avionics, and opto-electronics packaging demand a great deal of mechanical design skill for achieving sound and reliable products. This mechanical engineering activity has thus become a crucial contributor to the computer, telecommunications, aerospace, and other industries.

Traditionally, printed circuit cards are categorized as "second level packaging" since they support the chip carrier (the "first level package"). The cards are, most often, further connected to larger "boards" and the latter to frames, constituting the higher levels (third and fourth) of packaging. These manifold interactions make the realm of printed circuit card and board systems one of the most structurally fundamental and fascinating areas in the electronics hardware hierarchy. A multiplicity of thermal stress, handling, and vibration problems arise here in transient, repetitive, or steady state form. On their successful solution depends the manufacturing, testing, and operational life of the product.

This book was written with a focus on the mechanical principles involved in the system components and their assembly. The author's thinking in this field was nurtured by his many years of work on industrial design, development, testing, analysis, and consulting in electronics and avionics packaging. Much of the book involves his own research.

Some words must be said right at the outset about terminology, which varies widely in the area to be treated. Printed circuit boards are also often called "printed wiring boards" – the acronym PWB occurs in many a journal article title. The size distinction between "cards" and the larger, thicker (more multilayered) "boards" also enters the semantics syndrome.

Module, component, chip carrier and package are some of the names appended to first level structures attached to cards and boards. In this work, the term module is adopted in most cases, although exceptions may be made, especially when a quoted work employs a different terminology. One must realize that terminology is also a function of locality, company, or industry and, as such, a matter of professional traditions. The author hopes that his readers will find his choices reasonable.

A special note is due regarding the roles of finite element method versus classical theory and experiment. There is no clear "winner" here in general. Finite element solutions must have confirmation by simplified or approximate analytical methods. Such "classical" methods have always been needed in engineering for basic understanding to fall back on. The author emphasizes the use of the latter, and attempts to show the direction to develop them further as the need arises. Because finite element problems can also be extremely demanding in memory for complex multileaded, multimodule systems, computational procedures simplifying the tasks of finding lead force distributions, structural stiffness, and solder joint stress are required. Experimental stress analysis in these highly miniaturized components, using ingenious applications of strain gauging, holography, and Moire techniques, for example, is on the rise. There is no more convincing, albeit far from cheap, proof of understanding a mechanical system than through experiments. In general, it is one of the author's basic views that electronics packaging structural design must contain all three phases: analysis, finite elements, and experiment.

This book is intended as a statement and discussion of structural principles, and the ensuing solution methods are applied to various circuit board systems. It is for use in the research laboratory, design office, by testing agencies, and in university curricula. Fundamental mechanical engineering concepts are stressed, with the mathematical handling corresponding to a senior or first year graduate level.

The subject matter is introduced in Chapter 1, dedicated to classical structural analysis concepts and methods. Beams (for example, those supported on elastic foundations), plates, thermal stresses, plasticity, and other topics are discussed, in connection with typical circuit-board problems. This comes in handy for reference in later chapters on that type of component.

Chapter 2 is a concise wrapup of finite element structural analysis. Theoretical fundamentals and examples are shown, referred to the circuit board subject matter.

The physical properties of various components, cards, modules, leads, solder joints, and their testing methods are described in Chapter 3. Some mechanical data (e.g., for modulus, creep, and fatigue) are graphed and tabulated, ready to furnish input for later analytical treatment.

Chapter 4 is devoted to the fundamentally important leadless chip carrier. Hall's analysis of solder forces and his constitutive solder equations are described.

Thermal stresses in pin-grid arrays are analyzed in Chapter 5. The "primary" system of forces arises on a single pin (the corner pin is highest stressed) subjected to thermal mismatch between the module and card. A basic ingredient is the role of solder as an elastic foundation. In Chapter 6, the finite flexibilities of the module and card are also included, furnishing a "secondary" force system, and tending to relieve some of the stresses generated by the primary mechanism.

Flexural treatment of simple compliant leaded, surface mounted systems, called "local assemblies," is the subject of Chapter 7. The latter are the "building blocks" of populated circuit cards, and their solution facilitates computation of more complex assemblies (module clusters) discussed in Chapter 8. Such technological entities as double sided, stacked, and hybrid configurations are also treated here.

Chapter 9 introduces a simplified analysis of a circuit card subjected to twist. At first, a single module attachment is treated. Next, an approximate engineering computational procedure for highly populated circuit cards is given. An analytical torsional stiffness and lead force computation method is expounded in Chapter 10, for elastic systems. Experimental work on torsional testers, and the fatigue of interconnecting leads resulting from this type of loading, is shown.

Thermal stress analysis in compliant leaded modules is the subject of Chapter 11. Structural analysis methods are given, and finite element treatment is exemplified by Lau's modeling of J- and gullwing leads. The cyclic thermal fatigue analysis procedures of Engelmaier and his co-workers and their "figure of merit" calculation are introduced.

Chapter 12 is concerned with dynamic analysis. This is crucial to avionics packaging problems. For vibration analysis, module-populated systems can be represented by "smearing" or "lumping" techniques. Random vibration testing and the fragility (damage boundary) method are described. The analysis of a mixed thermal/vibrational fatigue problem is exhibited.

Plated holes (PTH and vias) are featured in Chapter 13. PTHs have two major failure mechanisms. The "global" (module-to-card) stressing tends to cause longitudinal cracks in the solder; their self-strain relieving tendencies are discussed. The second failure mechanism is the "z-directional mismatch" between board and copper barrel. Experience on the failure phenomena, and work on through-holes and on vias by experimental, analytical, and finite element methods, are discussed.

The book concludes with a description, in Chapter 14, of the assembly structural analysis work for the IBM 9370 card enclosure system. The design of the frame, boards, and their stiffeners required excessive rigidity to promote a sizable wipe during insertion of cards into a zero insertion force connector. The actuation of a card is analyzed for its role on the other contacts; tribological analysis of the multilayer plated contacts is also featured.

This book initiated from the author's work and consulting experience at the IBM Endicott Development Laboratory. He fondly recalls the fun of sharing work with R.G. Bayer, W.L. Brodsky, D.V. Caletka, W.T. Chen, E.Y. Hsue, C.K. Lim, M.R. Palmer, N.G. Payne, D.L. Questad, D.H. Strope, M.D. Toda, A.K. Trivedi, T.E. Wray, and many others. He has also learned much from discussions with many eminent colleagues, such as J.H. Lau, B.G. Sammakia, and E. Suhir, just to name a few.

The author taught some of the material in graduate courses at the State University of New York at Binghamton; he warmly recalls the interest of his students. Special thanks go to students who worked with the author on some of the topics involved in the book. They include T. Albert, B. Banerjee, A.R. Chitsaz, Y. Ling, T.M. Miller, R. Prasanna, A. Sahay, J.T. Vogelmann, J.R. Webb, K.R. Wu, and Q. Yang. The author thanks his colleagues J.M. Pitarresi and V. Prakash for their useful comments on the manuscript.

A word of indebtedness is due to several professional societies and publishers who permitted reproduction of some illustrations in this book. The American Society of Mechanical Engineers (ASME) has spearheaded efforts to promote electronics packaging advances in the mechanical engineering area; their conferences and their Transaction Journal of Electronic Packaging rallied much initiative. The Institute of Electrical and Electronic Engineers (IEEE) has sponsored conferences and a great deal of research, such as the task force on compliant leaded structures. The International Electronic Packaging Society (IEPS) and International Society of Hybrid Microelectronics (ISHM) have done much for the discipline. Quotations in this book were also taken from conferences and publications of the American Society of Metals, the Institute of Environmental Research, National Electronic Packaging Conferences (NEPCON), and the journals Circuit World, Connection Technology, the IBM Journal of Research and Development, Solid State Technology, and Soldering and Surface Mount Technology.

Last, but not least, the author thanks his wife for the inspiration she was always ready to give.

Binghamton, New York   Peter A. Engel
November, 1992

Contents

Nomenclature

a_i = polynomial coefficient
a = plate dimension (square side)
A = cross sectional area
b = plate dimension; hole radius
$[B]$ = derivative matrix of shape function
c = radius of circular lead
    distance of extreme fiber
    damping coefficient
    subscript for card
$[C]$ = damping matrix
$\{d\}, \{D\}$ = nodal displacement
D = plate rigidity
E = modulus of elasticity
$f(t)$ = forcing function
f = system reduction factor
f_1 = fundamental natural frequency
F = (axial) force; friction force
$g(y), g$ = moment development function
G = shear modulus of elasticity
h = thickness
H = thickness, hardness
I = moment of inertia, current
J = Jacobian matrix
k = foundation modulus, stiffness
k_T = torsional stiffness
K = stiffness
$[K]$ = stiffness matrix
l, L = length
m = mass, subscript for module
M = bending moment
$[M]$ = mass matrix
M_y = elastic limit moment

M_p = plastic moment
n, n' = number of leads
N = number of cycles; normal force
$N(x, y)$ = shape function
N_x, N_y, N_{xy} = in-plane stress resultants
N_f^* = mean cyclic fatigue life
P = force
q = distributed load
Q = plate shear; corner force; heat
r = radial coordinate
$\{r\}, \{R\}$ = nodal force column matrix
R = beam rigidity, EI; radius of curvature; electrical resistance
s = lead spacing
S = structural stiffness, lead force sum
S_e = endurance limit
S_u = ultimate strength
t = time
T = temperature, torque
u = displacement
U = strain energy; nondimensional function
v = displacement
V = shear force, potential energy
w = transverse displacement (of a card, a plate or a beam)
W = transverse displacement (of a module); wear
ΔW = cyclic viscoplastic strain energy
x, y, z = coordinates
y_0 = plasticity distance
Z_i = modal participation

Greek symbols

α = coefficient of thermal expansion, CTE
γ = shear strain (angle)
δ_{ij} = Kronecker delta
Δ = deflection, change
ΔT = temperature change
ε = strain
$\eta = y_0/c;\ W - w$
θ = rotation
λ = foundation constant; eigenvalue
$\Lambda(r: r')$ = displacement influence function
μ = nondimensional moment; friction coefficient
v = Poisson's ratio

ξ = distance
ξ_i = damping factor
$\rho = \Delta/L$; λL; density
σ = normal stress component
σ_y = normal yield stress
τ = shear stress component
τ_y = shear yield stress
ϕ = twist angle
Φ = body force; twist angle
ω = natural circular frequency
Ω = potential energy of external loads
∇ = gradient operator
∇^2 = Laplace operator

Chapter 1

Elements of Structural Analysis

Circuit cards and boards, modules, and leads can often be individually analyzed acting as simple structural elements, such as rods, beams, or plates. These may allow closed form or series solutions under thermal or mechanical loading; even as parts of assemblies, their behavior is often expediently explained through simplified analysis. Structural theory is nurtured by the mechanics of materials, the theories of elasticity and plasticity, of plates and shells, vibrations, thermal stresses, and probably half a dozen other engineering endeavors.

1. Rods

The simplest of all structural elements is acted upon by an axial load F and is stretched longitudinally by an amount u; a linearly elastic elongation has the relation involving the length of the rod L, its area A, and the modulus of elasticity E:

$$u = FL/AE . \tag{1.1}$$

2. Beams

2.1. Flexure

A hierarchy of static differential equations applies between distributed load $q(x)$, shear force $V(x)$, and bending moment $M(x)$. These are further related to the displacement $w(x)$, slope dw/dx, and curvature d^2w/dx^2. The sign conventions of Fig. 1.1a are adopted:

$$q(x) = dV/dx \tag{1.2}$$

$$V(x) = dM/dx \tag{1.3}$$

$$M(x) = EI\, d^2w/dx^2 . \tag{1.4}$$

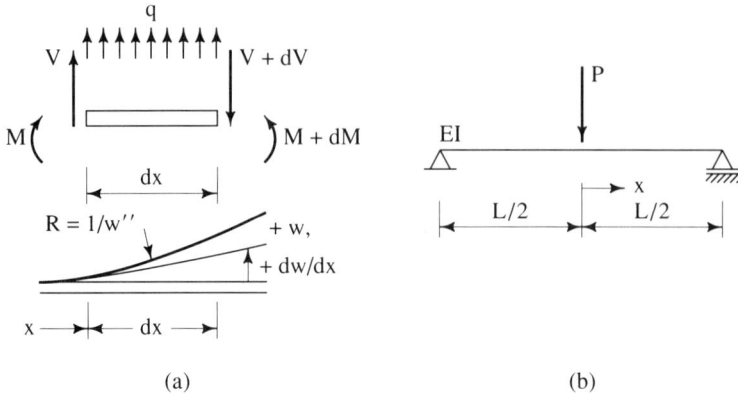

(a)   (b)

FIGURE 1.1. Schematic for beams. (a) Sign convention; (b) simple beam with concentrated load.

The term EI (modulus × moment of inertia) is the (bending-) rigidity of the beam; connecting the force and deformation terms, EI enters the beam equation:

$$d^4w/dx^4 = q(x)/EI . \qquad (1.5)$$

This fourth-order differential equation has the general solution: $w(x) = \sum_1^4 C_i x^i$, which must be solved for a given problem, using four boundary conditions. For example, Fig. 1.1b shows a simple beam under concentrated central load P, modeling the "three-point bending test" of a module. The boundary conditions are:

$$x = 0: \ dw/dx = 0 \qquad (1.6)$$

$$x = L/2: w = 0, \qquad M(= EId^2w/dx^2) = 0,$$

$$V(= EId^3w/dx^3) = -P/2 , \qquad (1.7)$$

yielding the solution

$$w = \frac{P}{48EI}(-L^3 + 6Lx^2 - 4x^3) . \qquad (1.8)$$

2.2. Beams on Elastic Foundation

In Chapter 5 we shall model a cylindrical lead soldered into a circuit card hole; the solder acts as an elastic foundation of modulus $k\,(N/mm^2)$. An elastic foundation exerts a distributed reaction load q on the beam, proportional to the negative of the displacement. Thus, substituting $q = -kw$ into Eq. (1.5), the differential equation of the beam in Fig. 1.2a becomes

$$d^4w/dx^4 + kw/EI = 0 . \qquad (1.9)$$

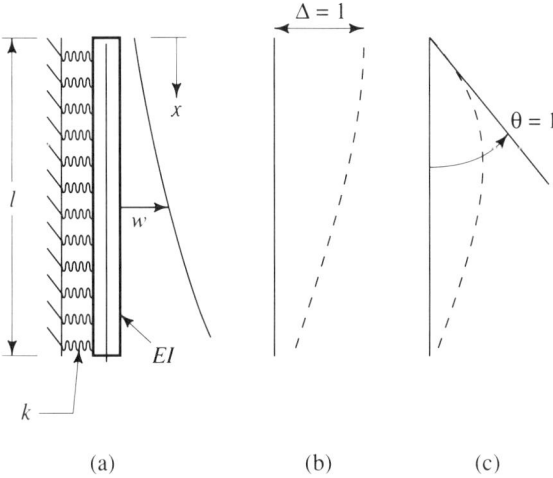

FIGURE 1.2. Elastically supported beam. (a) Schematic; (b) the first degree of freedom; (c) the second degree of freedom.

Introducing the "foundation constant" $\lambda = (k/4EI)^{1/4}$, we get

$$d^4w/dx^4 + 4\lambda^4 w = 0 , \tag{1.10}$$

the general solution of which, in a form preferred for infinite beams [Eq. (1.11)] and finite beams [Eq. (1.12)], is

$$w(x) = e^{\lambda x}(C_1 \cos \lambda x + C_2 \sin \lambda x) + e^{-\lambda x}(C_3 \cos \lambda x + C_4 \sin \lambda x) \tag{1.11}$$

$$w(x) = \cosh \lambda x(C_1 \cos \lambda x + C_2 \sin \lambda x)$$
$$+ \sinh \lambda x(C_3 \cos \lambda x + C_4 \sin \lambda x) . \tag{1.12}$$

As an example relevant to soldered lead analysis, we shall evaluate a 2×2 reaction force matrix $[k]$ of a beam segment embedded through its length l; $[k]$ consists of the force-elements (V, M) at the top $x = 0$, sustaining two alternate degrees of freedom (dof) of displacement: $\Delta = w(x = 0)$ and $\theta = dw/dx(x = 0)$. Figure 1.2b illustrates the beam segment deformed into its first dof; it consists of $w(0) = \Delta = 1$ and $dw/dx(0) = 0$. The second dof (Fig. 1.2c) will be θ: it consists of $w(0) = 0$, and $dw/dx(0) = \theta = 1$. For both dofs, the shear and moment at $x = l$ vanish. The k_{ij} element of $[k]$ is defined as "the force at i when dof j is unity, all other dofs being zero", and will be computed by using two sets of four boundary conditions (Table 1.1). Each dof supplies a column of $[k]$.

For the calculation of the coefficients (C_1, C_2, C_3, C_4) in both dofs, Eq. (1.12) must be evaluated with columns 2 and 3 of Table 1.1. The

TABLE 1.1. Boundary conditions for two degrees of freedom of elastically supported (embedded) pin portion.

Location, x	Displacement dof $k_{DD}, k_{RD}; C_1, C_2, C_3, C_4$		Rotation dof $k_{DR}, k_{RR}; C_1, C_2, C_3, C_4$	
0	$w = 1$	1	$w = 0$	0
0	$w' = 0$	1	$w' = 1$	$1/\lambda$
l	$w'' = 0$	-1	$w'' = 0$	0
l	$w''' = 0$	-1	$w''' = 0$	$-1/\lambda$

expression resulting for dof 1 is

$$\begin{Bmatrix} C_2 \\ C_4 \end{Bmatrix} = \frac{\begin{bmatrix} \text{ch}\cdot\text{c} & -\text{sh}\cdot\text{c} + \text{ch}\cdot\text{s} \\ -\text{sh}\cdot\text{c} - \text{ch}\cdot\text{s} & 2\text{ch}\cdot\text{c} \end{bmatrix}}{2\,\text{ch}^2\cdot\text{c}^2 + \text{ch}^2\cdot\text{s}^2 - \text{sh}^2\cdot\text{c}^2} \begin{Bmatrix} \text{ch}\cdot\text{s} + \text{sh}\cdot\text{c} \\ \text{sh}\cdot\text{s} \end{Bmatrix} \tag{1.13}$$

with $C_1 = 1$, $C_3 = -C_2$. We used the notation: $\rho = \lambda l$, ch $= \cosh\rho$, sh $= \sinh\rho$, c $= \cos\rho$, and s $= \sin\rho$.

As for dof 2 (a unit rotation at $x = 0$), we get

$$\begin{Bmatrix} C_2 \\ C_4 \end{Bmatrix} = \frac{\begin{bmatrix} \text{sh}\cdot\text{c} - \text{ch}\cdot\text{s} & -\text{ch}\cdot\text{c} \\ -2\text{ch}\cdot\text{c} & \text{sh}\cdot\text{c} + \text{ch}\cdot\text{s} \end{bmatrix}}{(\text{sh}^2\cdot\text{c}^2 - \text{ch}^2\cdot\text{s}^2 - 2\text{ch}^2\cdot\text{c}^2)\lambda} \begin{Bmatrix} \text{ch}\cdot\text{s} \\ \text{sh}\cdot\text{s} + \text{ch}\cdot\text{c} \end{Bmatrix} \tag{1.14}$$

with $C_1 = 0$ and $C_4 = -C_2 = 1/\lambda$.

If the foundation length l is large enough, $\rho = \lambda l > 3$ and a semi-infinite foundation is approached. Now important simplifications occur in Eqs. (1.13) and (1.14), owing to $\sinh\rho \approx \cosh\rho$. In that case, often satisfied by lead geometries, Eq. (1.13) becomes

$$C_1 = 1, \qquad C_2 = 1, \qquad C_3 = -1, \qquad C_4 = -1 \tag{1.15}$$

whereas Eq. (1.14) simplifies to

$$C_1 = 0, \qquad C_2 = 1/\lambda, \qquad C_3 = 0, \qquad C_4 = -1/\lambda. \tag{1.16}$$

The force coefficients $V(x = 0)$ and $M(x = 0)$ from both dofs are evaluated from Eqs. (1.3) and (1.4) by differentiations of w: twice for M and once more for V. For dof 1, we get

$$k_{DD} = 4EI\lambda^3, \qquad k_{RD} = -2EI\lambda^2 \tag{1.17}$$

and from dof 2,

$$k_{DR} = 2EI\lambda^2, \qquad k_{RR} = -2EI\lambda, \tag{1.18}$$

where the subscript D means "displacement" or Δ and R refers to "rotation" or θ. In matrix form, finally,

$$\begin{Bmatrix} V \\ M \end{Bmatrix} = 2EI\lambda \begin{bmatrix} 2\lambda^2 & \lambda \\ -\lambda & -1 \end{bmatrix} \begin{Bmatrix} \Delta \\ \theta \end{Bmatrix} = \begin{bmatrix} k_{DD} & k_{DR} \\ k_{RD} & k_{RR} \end{bmatrix} \begin{Bmatrix} \Delta \\ \theta \end{Bmatrix}. \tag{1.19}$$

The negative sign here is explained by the opposite senses for $+M$ and $+\theta$ stemming from the sign convention. Maxwell's reciprocal relation is valid, as $|1 \cdot k_{DR}| = |1 \cdot k_{RD}|$.

2.3. Torsion

The torsion of elastic beams (or plates) has two constituents, in general: St. Venant's torsion and warping torsion. The former is characterized by the following equation connecting torque T and twist angle ϕ along the beam:

$$d\phi/dx = T(x)/GJ . \tag{1.20}$$

This may characterize a circuit card subjected to twist (see Chapters 9 and 10). We shall not discuss warping torsion [1], which is significant in flanged members.

2.4. Frames

Steinberg [2] modeled compliant lead-supported modules soldered to circuit boards as plane frames. In the pre-finite-element era, elastic plane frames were preferentially solved by the method of slope-deflection equations [3]. For each segment of the frame, two such end-moment equations may be written, containing three unknowns: the two end rotations θ_A and θ_B, and the end-to-end rotation $\rho = \Delta/L$ (Fig. 1.3a):

$$M_{AB} = M_{AB}^F + \frac{2EI}{L}(2\theta_A + \theta_B - 3\rho) \tag{1.21}$$

$$M_{BA} = M_{BA}^F + \frac{2EI}{L}(2\theta_B + \theta_A - 3\rho) \tag{1.22}$$

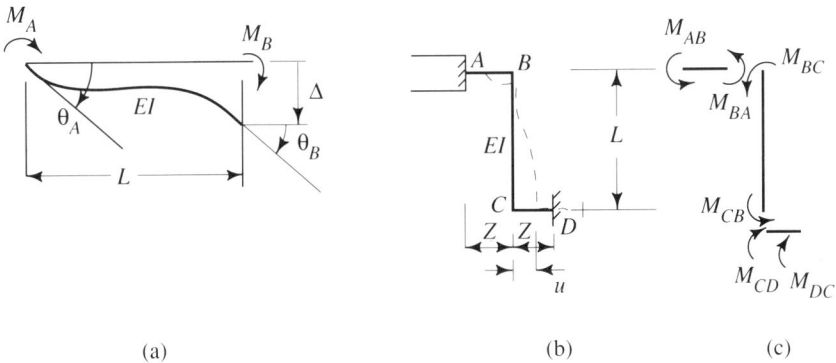

(a)   (b)   (c)

FIGURE 1.3. Slope-deflection analysis. (a) Beam element; (b) schematic of gullwing lead subjected to thermal displacement; (c) moments at ends of beam segments.

where M^F_{AB} and M^F_{BA} are imposed fixed-end moments, as from intermediate loading or side-sway.

Figure 1.3 shows the sign convention generally adopted, in connection with the example of a compliant gullwing lead. For simplicity we assume fixity at the module and at the solder joint, equal and straight horizontal segments z, and a constant cross section and material EI. The load is a thermal displacement u of the base. There are three beam segments, so that the number of unknowns includes 3×3 displacements and 2×3 end moments (Fig. 1.3c), i.e., 15. We proceed to write two equations [(1.21) and (1.22)] for each of the three-beam segments; two additional equations result from the equilibrium of the joints B and C:

$$M_{BA} + M_{BC} = 0$$

$$M_{CB} + M_{CD} = 0 .$$

Two equations result from $\theta_{BA} = \theta_{BC}$ and $\theta_{CB} = \theta_{CD}$ due to rigid joints. Two fixed ends A and D yield two equations $(\theta_{A,D} = 0)$. In addition, three static external equilibrium equations are available, bringing the number of equations to that of the unknowns, 15. The corner moment is $M_{BA} = -M_{BC} = 12EIu/(2L^2 + 3Lz)$; at the fixity $M_{AB} = (3/2)M_{BA}$.

3. Plates

3.1. Cylindrical Bending

Beams are loaded in a two-dimensional (x, y) state of stress called "plane stress" [4] characterized by $\sigma_z = \tau_{xz} = \tau_{yz} = 0$. Cylindrically bent plates also exhibit a two-dimensional state of stress, with the difference that now $\varepsilon_z = 0$, while $\sigma_z \neq 0$; this is called "plane strain." An example is the action on a circuit card simply supported along opposite connectors parallel to the z-axis, and free on the remaining sides, with some flexure in the x, y plane occurring during insertion (Fig. 1.4a). The Hooke relation of stress and strain components in three dimensions includes E and the Poisson ratio v:

$$\varepsilon_x = \frac{1}{E}[\sigma_x - v(\sigma_y + \sigma_z)] \tag{1.23}$$

$$\gamma_{xy} = \frac{\tau_{xy}}{G} . \tag{1.24}$$

Both Eqs. (1.23) and (1.24) express three equations with the permutation of the subscripts x, y, and z. From these equations, the special two-dimensional states yield Eqs. (1.25a) and (1.25b).

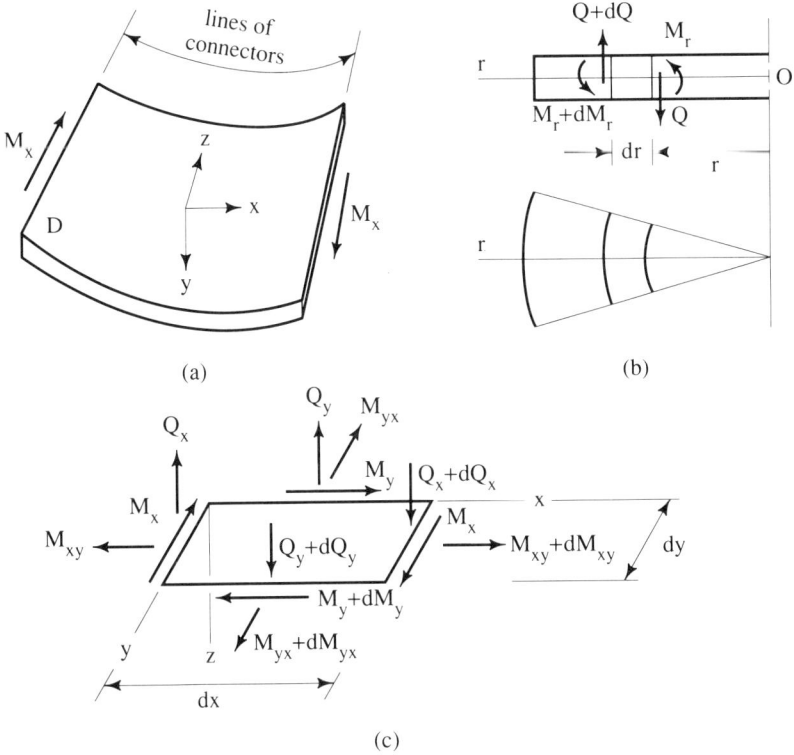

FIGURE 1.4. Flexure of plates. (a) Cylindrical bending; (b) circular plate; (c) rectangular plate. Moments designated by straight arrows rotate according to the right-hand rule, i.e., counter clockwise as seen from the tip of the arrow.

Plane stress:

$$\begin{Bmatrix} \sigma_x \\ \sigma_y \\ \tau_{xy} \end{Bmatrix} = \frac{E}{1-v^2} \begin{bmatrix} 1 & v & 0 \\ v & 1 & 0 \\ 0 & 0 & \dfrac{1-v}{2} \end{bmatrix} \begin{Bmatrix} \varepsilon_x \\ \varepsilon_y \\ \gamma_{xy} \end{Bmatrix} \qquad (1.25a)$$

Plane strain:

$$\frac{E}{(1+v)(1-2v)} \begin{bmatrix} 1-v & v & 0 \\ v & 1-v & 0 \\ 0 & 0 & \dfrac{1-2v}{2} \end{bmatrix} \begin{Bmatrix} \varepsilon_x \\ \varepsilon_y \\ \gamma_{xy} \end{Bmatrix}. \qquad (1.25b)$$

Since in bending we always assume $\sigma_y = 0$, cylindrical bending with plane strain ($\varepsilon_z = 0$) has, from Eq. (1.23), $\sigma_z = v\sigma_x$.

The bending strain in the x-direction, resubstituting into Eq. (1.23) is

$$\varepsilon_x = \frac{\sigma_x}{E}(1 - v^2),$$ (1.26)

showing an increase, $(1 - v^2)^{-1}$-fold, of the bending rigidity. For this reason, the beam equations [(1.4) and (1.5)] are applicable to unit widths of cylindrically bent plates, with EI replaced by the plate rigidity

$$D = Eh^3/12(1 - v^2).$$ (1.27)

3.2. Pure Bending

Neglecting shear forces, the relationship between moments and curvatures is established analogously to beam theory [5]:

$$M_x = -D\left(\frac{\partial^2 w}{\partial x^2} + v\frac{\partial^2 w}{\partial y^2}\right)$$ (1.28)

$$M_y = -D\left(\frac{\partial^2 w}{\partial y^2} + v\frac{\partial^2 w}{\partial x^2}\right)$$ (1.29)

$$M_{xy} = D(1 - v)\frac{\partial^2 w}{\partial x\partial y} = -M_{yx}$$ (1.30)

where M_{xy} is the twisting moment.

We shall solve the problem of a square plate twisted by corner forces, which is practical for the testing of circuit cards. Let us begin by considering pure moments M_1 and M_2 $(M_1 = -M_2)$ acting on the sides of a square plate $L \times L$ (Fig. 1.5). Solving for the curvatures from Eqs. (1.28) and (1.29), we get

$$\partial^2 w/\partial x^2 = -\frac{M_1}{D(1 - v)} = -\partial^2 w/\partial y^2.$$ (1.31)

This "anticlastic" (saddle) curvature is integrated to

$$w = -\frac{M_1}{2D(1 - v)}(x^2 - y^2).$$ (1.32)

An *inscribed* square has dimensions $L^* = \sqrt{2}L/2$, and its corners are deflected to

$$w_i = \pm\frac{M_1 L^2}{8D(1 - v)}.$$ (1.33)

By a coordinate transformation $x = \frac{1}{\sqrt{2}}(x' - y')$, $y = \frac{1}{\sqrt{2}}(x' + y')$, the displacements of the inscribed square are written, by Eq. (1.31)

$$w = \frac{M_1}{D(1 - v)}x'y'.$$ (1.34)

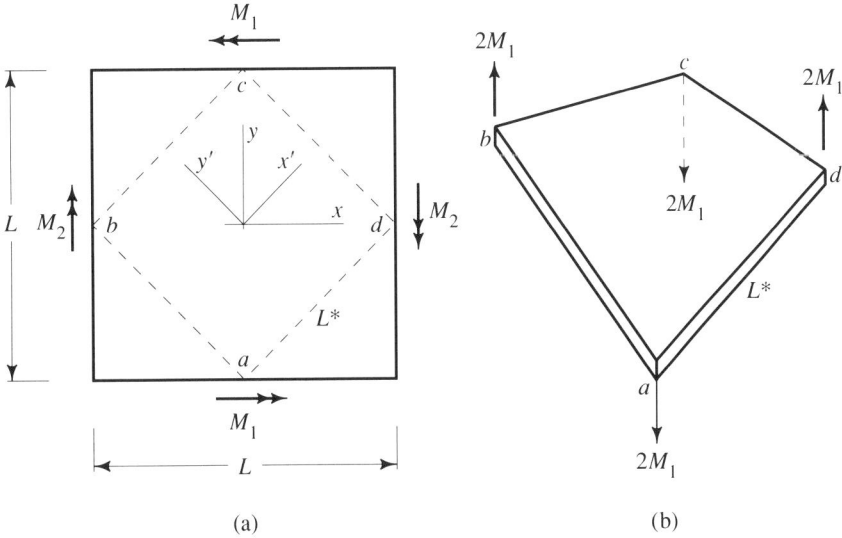

FIGURE 1.5. Pure bending of a square plate. (a) Moments M_1 and $M_2 = -M_1$ acting on sides L; (b) corner forces on inscribed square of side L^*.

A typical phenomenon present in rectangular plate bending is recalled now: the tendency for corners to curl up. This tendency must be overcome by anchoring forces $R = 2M_{xy}$ applied at the corners [5]. The twisting moment $M_{xy} = -M_{yx}$ is evaluated for Eq. (1.34) from Eq. (1.30), and we get $M_{x'y'} = M_1$. Thus the forces P that must be exerted at the corners of the inscribed square L^* are alternatively $+2M_1$ and $-2M_1$. By Eq. (1.33), this yields $w_i = PL^{*2}/8(1-v)D$ for the corner displacement of a square plate L^* under the twisting corner forces P. (Compare this with Eq. (9.3).)

The reader may verify from Eq. (1.20) that the same result would ensue from applying a torque PL^* at the opposite sides of the inscribed square having a torsion constant $J = L^* h^3/3$ and shear modulus $G = E/2(1+v)$.

3.3. Circular Plates

Linear elastic thin plate theory is valid for displacements w not exceeding a third of the plate thickness [5]. At greater displacements the in-plane (membrane) forces add substantive stiffness, as seen, e.g., in Chapter 10. The partial differential equation governing thin plate bending with polar symmetry is

$$\nabla^4 w = q(r)/D . \tag{1.35}$$

The biharmonic operator, the square of the Laplacian V^2, may be written out as

$$V^4 \equiv (V^2)^2 = \frac{1}{r}\frac{d}{dr}\left\{ r\frac{d}{dr}\left[\frac{1}{r}\frac{d}{dr}\left(r\frac{dw}{dr} \right) \right] \right\}. \qquad (1.36)$$

The nomenclature we shall follow is Timoshenko's [5]; see Fig. 1.4b. Figure 1.6 shows the geometric relationship between the meridional and tangential radii of curvature (R and R_t, respectively):

$$1/R_t = -(1/r)(dw/dr), \qquad 1/R = -d^2w/dr^2 , \qquad (1.37)$$

allowing the following statements for the stress resultants M_r, M_t, and Q_r (radial and tangential plate moments, and radial shear force, respectively), based on Eqs. (1.28) and (1.29):

$$M_r = -D\left(\frac{d^2w}{dr^2} + \frac{v}{r}\frac{dw}{dr} \right) \qquad (1.38)$$

$$M_t = -D\left(v\frac{d^2w}{dr^2} + \frac{1}{r}\frac{dw}{dr} \right) \qquad (1.39)$$

$$Q_r = D\left(\frac{d^3w}{dr^3} + \frac{1}{r}\frac{d^2w}{dr^2} - \frac{1}{r^2}\frac{dw}{dr} \right). \qquad (1.40)$$

The bending problem of modules arising due to differential thermal expansion with the supporting card is often simplified by assuming a circular plate of radius a for the module. Let us calculate the radius of curvature R induced

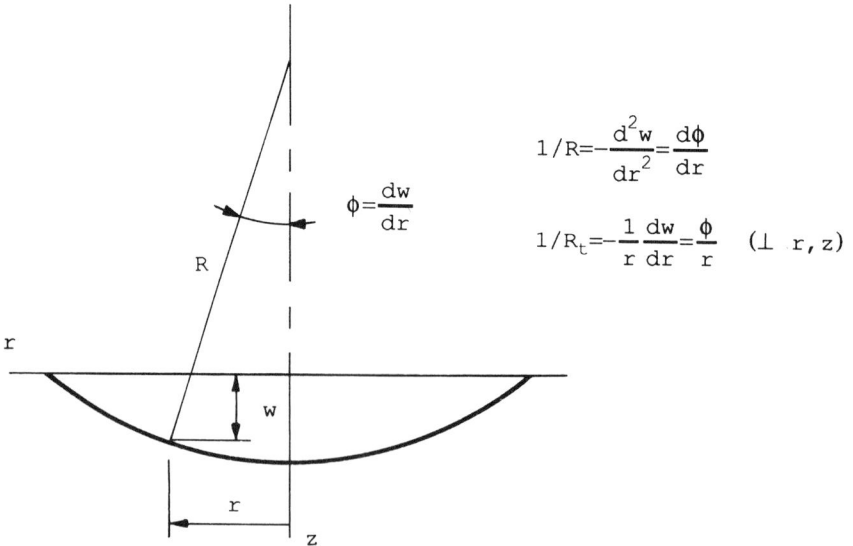

$$1/R = -\frac{d^2w}{dr^2} = \frac{d\phi}{dr}$$

$$1/R_t = -\frac{1}{r}\frac{dw}{dr} = \frac{\phi}{r} \qquad (\perp r, z)$$

$$\phi = \frac{dw}{dr}$$

FIGURE 1.6. Geometry of axisymmetrical deformation of circular plate.

in the module, its leads spaced at a distance s. Considering the radial plate moment M_r at $r = a$ as added up by the frame moments M_A (Figs. 1.3b and 3c) of $1/s$ equally spaced soldered lead resistances, $M_r = M_A/s$, we shall find the displacements of a circular plate bent by edge moments M_r. Using the assumption of parabolic plate displacement

$$w(r) = w_0 - r^2/2R \tag{1.41}$$

and substituting into Eq. (1.38), we get

$$R = D(1 + v)/M_r . \tag{1.42}$$

As for the effect of significant membrane forces N_r, Timoshenko [1] gives the differential equation

$$\zeta^2 d^2\phi/d\zeta^2 + \zeta \, d\phi/d\zeta - (\zeta^2 + 1) \phi = 0 \tag{1.43}$$

where the changes of variable $\zeta = \alpha r$, $\alpha^2 = N_r/D$, and $\phi = dw/dr$ were made. The solution, in terms of Bessel functions of the first and second order, J_1 and Y_1, is

$$\phi = A_1 \cdot J_1(\zeta) + A_2 \cdot Y_1(\zeta) . \tag{1.44}$$

3.4. Rectangular Plates in Flexure

In x, y coordinates (Fig. 1.4c) the plate equation (1.35) is $\nabla^4 w = q(x, y)/D$. Rectangular $(a \times b)$ simply supported thin plates can be solved in the Navier form [5] by assuming the deflection in a double trigonometric series

$$w = \frac{1}{\pi^4 D} \sum_{m=1}^{\infty} \sum_{n=1}^{\infty} \frac{a_{mn}}{\left(\dfrac{m^2}{a^2} + \dfrac{n^2}{b^2}\right)^2} \sin \frac{m\pi x}{a} \sin \frac{n\pi y}{b} \tag{1.45}$$

where the load

$$q(x, y) = \sum_{m=1}^{\infty} \sum_{n=1}^{\infty} a_{mn} \sin \frac{m\pi x}{a} \sin \frac{n\pi y}{b} . \tag{1.46}$$

Note that the moments M_x, M_y, etc., shown as arrows in Fig. 1. 4c are understood to be rotating counterclockwise when viewed from the tip of the arrow.

Example 1

A module, peripherally supported on surface soldered leads, is manually pressed against the circuit card, resulting in a uniformly distributed load q_0. Assuming that the leads constitute a simple support all around the module, the Fourier coefficients of $q(x, y) = q_0$ are then evaluated:

$$a_{mn} = \frac{4q_0}{ab} \int_0^a \int_0^b \sin \frac{m\pi x}{a} \sin \frac{n\pi y}{b} \, dx dy = \frac{16q_0}{\pi^2 mn} . \tag{1.47}$$

The plate equation may further be solved in the single-series Levy form [5]. By superposition, various boundary conditions on alternative sides of the rectangular plate (fixed, free) may be accounted for; Roark [6] offers many loading cases.

4. Thermal Stress

4.1. One-Dimensional Treatment: Bimaterial Rods

The linear coefficient of thermal expansion (CTE or α) is a key parameter in the behavior of electronics packaging materials. We may write the strain in each of two parallel rods of a bimaterial structure, short enough to minimize bending and thus subject to axial stress only. Subjecting rod 1 to a temperature rise ΔT_1 and rod 2 to ΔT_2 (Fig. 1.7), the respective strains are

$$\varepsilon_1 = \alpha_1 \Delta T_1 + \frac{F}{A_1 E_1} \tag{1.48}$$

$$\varepsilon_2 = \alpha_2 \Delta T_2 - \frac{F}{A_2 E_2} . \tag{1.49}$$

Invoking compatibility, $\varepsilon_1 = \varepsilon_2$, we obtain the axial force:

$$F = \frac{\alpha_2 \Delta T_2 - \alpha_1 \Delta T_1}{\dfrac{1}{A_1 E_1} + \dfrac{1}{A_2 E_2}} . \tag{1.50}$$

A special consideration arises in soldering a chip to a substrate by tiny solder bumps of height h in the "flip-chip" arrangement. Heating the assembly, the elastic strains F/AE are now negligible in both connected bodies as the chip and substrate play the role of the rods. The solder, of relatively small rigidity AE, is sheared by the differential thermal displacement $\Delta u = (\varepsilon_1 - \varepsilon_2)r$ incurred on the length r between the solder joint and the middle of the assembly; for a general nonuniform temperature rise $\Delta T_1 \neq \Delta T_2$:

$$\Delta u = (\alpha_1 \Delta T_1 - \alpha_2 \Delta T_2)r , \tag{1.51}$$

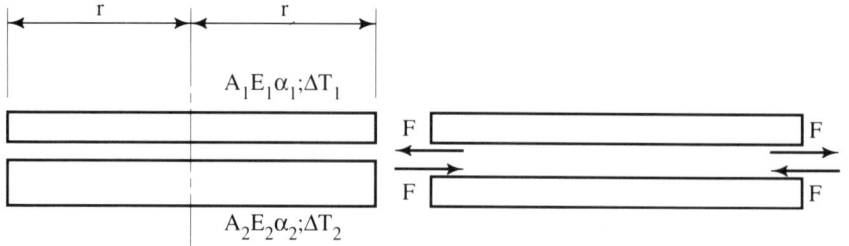

FIGURE 1.7. Thermal stress in rods bonded at their ends; bending is neglected.

and in the case of uniform temperature rise, $\Delta T_1 = \Delta T_2 = \Delta T$:

$$\Delta u = (\alpha_1 - \alpha_2)\Delta T \cdot r . \tag{1.52}$$

The shear strain on the solder joint is consequently

$$\gamma = \Delta u / h . \tag{1.53}$$

4.2. Timoshenko's Formula for Thermal Bending of Bimaterial Circular Plates

The original derivation [7], made for beams, is redone here for thin plates. Figure 1.8 shows the cross section of two thin plates of different plate rigidities, bonded together. Upon a temperature rise ΔT, common to both parts, internal forces (P_1 and P_2) and moments (M_1 and M_2) arise. By equilibrium,

$$P_1 = P_2 = P \tag{1.54}$$

$$-Ph/2 = M_1 + M_2 . \tag{1.55}$$

By invoking the plate equation (1.38), we get

$$M_1 = -D_1\left(\frac{d^2w}{dr^2} + \frac{v}{r}\frac{dw}{dr}\right) \tag{1.56}$$

$$M_2 = -D_2\left(\frac{d^2w}{dr^2} + \frac{v}{r}\frac{dw}{dr}\right) \tag{1.57}$$

and, by Eq. (1.55),

$$Ph/2 = (D_1 + D_2)\left(\frac{d^2w}{dr^2} + \frac{v}{r}\frac{dw}{dr}\right) . \tag{1.58}$$

Now w may be approximated as

$$w = w_0 - r^2/2R . \tag{1.59}$$

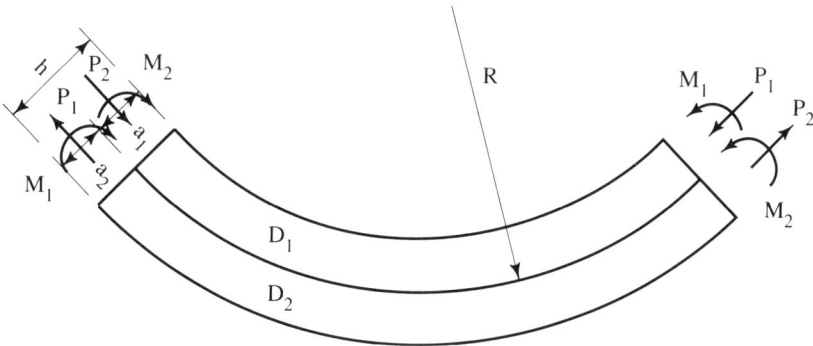

FIGURE 1.8. Bimaterial circular plate (or beam) bent by temperature. Note: the bond forces a common curvature $1/R$.

Thus

$$dw/dr = -r/R, \qquad d^2w/dr^2 = -1/R ,$$

and

$$Ph/2 = (D_1 + D_2)(1 + v)(-1/R) . \qquad (1.60)$$

Bending, axial force, and temperature result in the following strains at the interfaces:

$$\varepsilon_1 = \alpha_1 \Delta T - P_1/a_1 D_1 + a_1/2R \qquad (1.61)$$

$$\varepsilon_2 = \alpha_2 \Delta T + P_2/a_2 D_2 - a_2/2R . \qquad (1.62)$$

Equating the two strains yields the common curvature

$$1/R = \frac{6}{h} \left\{ \frac{(\alpha_2 - \alpha_1)\Delta T(1 + a_1/a_2)^2}{3(1 + a_1/a_2)^2 + \left[1 + \left(\dfrac{a_1}{a_2}\right)\left(\dfrac{D_1}{D_2}\right)\right](a_1/a_2)^2 + \dfrac{1}{(a_1/a_2)(D_1/D_2)}} \right\}.$$

$$(1.63)$$

Many examples exist in substrate and circuit card structures for bi- and multi-material sandwiches. The peeling and shear stress concentrations near the outer circumference of the bond, not addressed by Timoshenko, have been an especially lively topic for study in recent years, e.g., [8, 9]. Their order-of-magnitude values are $E \propto \Delta T$, reached within a length of the layer thickness.

5. Plastic Beam Deformation

Thermal stress may plastically deform the leads in a pin-grid array, a subject to be treated in Chapter 5. We begin the analysis with the differential equation of a beam under the "planes remain plane" assumption, which can be written on a purely geometric ground:

$$d^2w/dx^2 = \varepsilon_{max}(x)/c \qquad (1.64)$$

where ε_{max} is the bending strain in the extreme fiber at a distance $y = c$ away from the neutral axis. Equation (1.64) includes Eq. (1.4) as its special elastic case, when $\varepsilon_{max} = Mc/EI$.

It is desirable to express $\varepsilon_{max}(x)$ in terms of the bending moment $M(x)$ for integration of Eq. (1.64). $M(x)$ can be calculated from the internal normal stress distribution $\sigma(y)$ on a cross section. Let us idealize the stress-strain curve as an elastic–perfectly plastic one with yield stress σ_y. Then, for a circular cross section, Fig. 1.9 shows the strain and stress, which is elastic up to $y = y_0$ and plastic above that, up to the extreme fiber. We have, by definition:

$$M = \int_A \sigma(y) \cdot y \cdot dA \qquad (1.65)$$

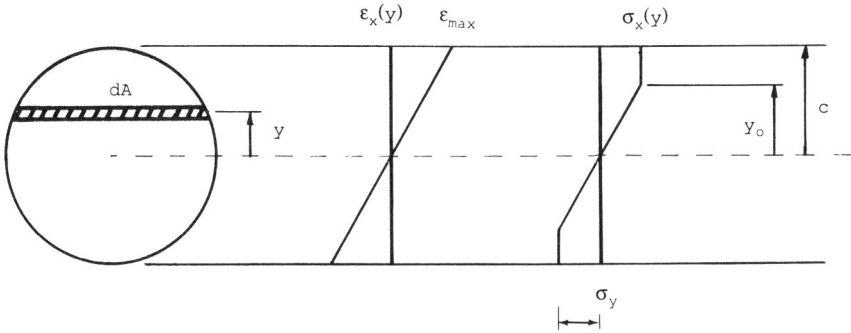

FIGURE 1.9. Elasto–plastic strain and stress distribution for a circular pin.

which, integrated [10], yields a three-term expression in y_0/c:

$$M = \sigma_y c^3 \left\{ \frac{[1 - (y_0/c)^2]^{3/2}}{3} + \frac{\sqrt{1 - (y_0/c)^2}}{2} + \frac{\sin^{-1}(y_0/c)}{2(y_0/c)} \right\}. \quad (1.66)$$

For a square pin of size a the corresponding equation is

$$M = \frac{\sigma_y}{4} a^3 \left[1 - \frac{(2y_0/a)^2}{3} \right]. \quad (1.67)$$

The elastic limit moment M_y (making $y_0 = c$ or $a/2$) and fully plastic moment M_p (making $y_0 = 0$) are listed in Table 1.2. Now we define two nondimensional quantities: firstly, the plasticity distance $\eta = y_0/c$ for circular and $\eta = 2y_0/a$ for square cross sections; secondly, the nondimensional moment $\mu = M/M_y$. From Eqs. (1.66) and (1.67) we get their respective nondimensional forms:

$$\mu = \frac{4}{\pi} \left[\frac{(1 - \eta^2)^{3/2}}{3} + \frac{\sqrt{1 - \eta^2}}{2} + \frac{\sin^{-1}\eta}{2\eta} \right], \quad 0 \leqslant \eta \leqslant 1 \quad (1.68)$$

$$\mu = \frac{3}{2} [1 - \eta^2/3], \quad 0 \leqslant \eta \leqslant 1. \quad (1.69)$$

TABLE 1.2. Plasticity properties of circular and square cross sections.

Shape	Fully plastic moment M_p	Elastic limit moment M_y	η	A	B
Circular, radius c	$4\sigma_y c^3/3$	$\pi\sigma_y c^3/4$	y_0/c	2.433	1.433
Square, side a	$\sigma_y a^3/4$	$\sigma_y a^3/6$	$Y_0/(a/2)$	3	2

These expressions may be inverted and solved for η in the form [11]

$$\eta = (A - B\mu)^{1/2} \tag{1.70}$$

where A and B are constants, characteristic of the shape of the cross section, and as such are listed in Table 1.2.

Realizing that $\varepsilon_{max} = \sigma_y c / E y_0 = \sigma_y / E \eta$, the differential equation (1.64) is finally rewritten:

$$d^2 w / dx^2 = (\sigma_y / Ec) \cdot [A - B\mu(x)]^{-1/2} . \tag{1.71}$$

If the moment $M(x)$ is given along the beam [10] (which is the case for a fixed–fixed pin under thermal mismatch load), then $\mu(x)$ becomes available and Eq. (1.71) may be integrated.

6. Energy Methods in Structural Analysis

The principle of stationary potential energy requires the variation of the latter to vanish for an elastic body in equilibrium:

$$\delta V = 0 . \tag{1.72}$$

Under static loading the total potential energy $V = U + \Omega$, where U is the strain energy and Ω the potential energy of external loads; both are integral expressions over the body. Inserting these into Eq. (1.72) the differential equation of the problem together with the natural boundary conditions can be extracted [12]. Alternatively, with some insight into the boundary conditions and the displacement function $w(c_i; x, y, z)$, the potential energy may be written in terms of the unknown parameters c_i of w. A thin plate has the expression [5]

$$U = \frac{D}{2} \int \int \left\{ \left(\frac{\partial^2 w}{\partial x^2} + \frac{\partial^2 w}{\partial y^2} \right)^2 - 2(1 - v) \left[\frac{\partial^2 w}{\partial x^2} \frac{\partial^2 w}{\partial y^2} - \left(\frac{\partial^2 w}{\partial x \partial y} \right)^2 \right] \right\} dx dy . \tag{1.73}$$

In the preceding Example of Section 3.4, a rectangular plate had trigonometric functions as "base functions"; by Eq. (1.45), we may write one of these as $w_i = c_i \sin(i\pi x/a) \cdot \sin(i\pi y/b)$. Since $w = \sum w_i$, it is possible to integrate the base functions to get U over the plate by Eq. (1.73). Likewise, for a distributed load $q(x, y)$, integration of the base functions yields $\Omega = -\int \int q \sum^N w_i \, dS = -\sum^N \int \int q w_i \, dS$. The Ritz method evaluates the N unknown coefficients c_i by requiring, based on Eq. (1.72), that the partial derivatives $\partial V / \partial c_i = 0$. This yields N linear simultaneous equations in N unknowns.

For example, in the uniform loading of a simply supported plate, a single base function $w_1(x, y)$ would yield, by Eq. (1.73):

$$U = \frac{Dab\pi^2}{8} \left(\frac{1}{a^2} + \frac{1}{b^2} \right)^2 c_1^2 \tag{1.74}$$

and

$$\Omega = -\frac{4qab}{\pi^2} c_1 .$$

(1.75)

Thus $\partial(U + \Omega)/\partial c_1 = 0$ results in $c_1 = \dfrac{16q}{\pi^6 D} \dfrac{1}{\left(\dfrac{1}{a^2} + \dfrac{1}{b^2}\right)^2}$, w_{max} being

slightly (2.5%) higher than the series solution. Such overestimation is always present in applications of energy methods where a displacement function, no matter how plausible, is artificially imposed, creating greater than real constraint.

For beams the strain energy expression is comparable, but much simpler than Eq. (1.73) for a plate; neglecting shear contributions, we have

$$U = \int \frac{EI}{2} \left(\frac{d^2 w}{dx^2}\right)^2 dx .$$

(1.76)

There are several other energy methods [12] (such as Galerkin's and Reissner's) that are related to Eq. (1.72). Variational formulations abound in finite element methods.

7. Experimental Methods of Analysis

The static experimental methods enumerated below have been found useful and eminently complement various theoretical and numerical analysis methods used in circuit board systems. Some dynamic measurement methods are described in Chapter 12.

7.1. Load Testers

The Instron and MTS systems have been widely used. A multidegree-of-freedom microstructural tester has been devised by Niu et al. [13].

7.2. Strain Gauges

A classical application by Hall and co-workers [14] investigated thermal mismatch stresses between circuit cards and ceramic substrates.

7.3. Capacitance Measurement

Very accurate displacement measurements (25 nm) are based on the dependence of the capacitance in an electrical circuit upon distance across a gap. Cornell university research [15] involving solder joints and nano-indentation has used capacitive devices.

7.4. Fiberoptic Probe/Photodiode Measurement

The relationship between the distance from an illuminated object and the photodiode voltage generated is calibrated for use of this device. An example is the experimental measurement of thermal displacements of a silicon chip and the supporting ceramic substrate in a thermal grease-enhanced module [16].

7.5. Photoelasticity

Stresses in birefringent materials (such as some polymers used in electronic packaging) under stress can be analyzed by photoelastic principles [17].

7.6. Holographic Interferometry

The out-of-plane thermal displacements due to power dissipation in a module-card system were studied by holography [18]. In that work, traditional holographic systems were replaced by optical fibers.

7.7. Piezo-Electric Stress Sensors

Electronic packaging applications to chip stress distributions were described by Bittle et al. [19]. Electrical connector contact force measurements are described in Chapter 14.

7.8. Moire Interferometry

A cross-lined high-frequency grating is formed on the surface of a deformed specimen (Fig. 1.10). By the "geometric Moire" method, in-plane displacements are measured by counting the fringes resulting from the interference of light from the deformed grating with a grating placed above. Thermal strain measurements [20], those of electronic packages [21], and second-level applications [22] (to solder joints of pin-grid arrays and compliant leads) have been made successfully. "Shadow Moire" methods are capable of transverse displacement measurement. Here an oblique light source creates shadows through the upper undeformed grating which interfere with the grating formed on the transversely displaced surface. Both Moire methods, combined, were used in measuring the equibiaxial tensile stretch of hyperelastic (rubber-like) materials [23]. This test allowed a significant extension of the range of applicability of constitutive equations used in the structural analysis of elastomer connectors.

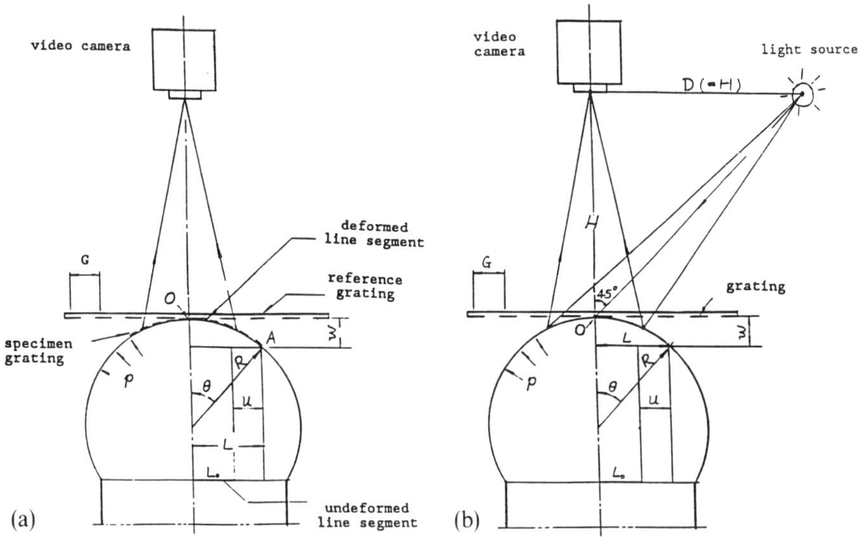

FIGURE 1.10. Moire analysis of an inflated elastomer membrane. (a) Geometric Moire method; (b) shadow Moire method.

7.9. Electrical Resistance Method

An electrical resistance method, based on the change of R in leaded interconnections subjected to sinusoidal strain, was introduced by Constable and Sahay [24]; it is intended for measuring lead forces and fatigue life.

References

1. Timoshenko, S.P., and Gere, J.M. (1961), *Theory of Elastic Stability*, 2d ed., McGraw-Hill, New York.
2. Steinberg, D.S. (1988), *Vibration Theory of Electronic Equipment*, 2d ed., Wiley and Sons, New York.
3. Flügge, W., Ed. (1962), *Handbook of Engineering Mechanics*, McGraw-Hill, New York.
4. Timoshenko, S.P., and Goodier, J.N. (1951), *Theory of Elasticity*, 2d ed., McGraw-Hill, New York.
5. Timoshenko, S.P., and Woinowsky-Krieger, S. (1959), *Theory of Plates and Shells*, 2d ed., McGraw-Hill, New York.
6. Roark, R.J., and Young, W.C. (1992), *Formulas for Stress and Strain.* 6th ed., McGraw-Hill, New York.
7. Timoshenko, S.P. (1953), *Collected Papers*, McGraw-Hill, New York.
8. Suhir, E. (1989), "Interfacial Stresses in Bimetal Thermostats," *J. Appl. Mech.*, **56**, 595–600.
9. Pao, Y-H., and Eisele, E. (1991), "Interfacial Shear and Peel Stresses in Multilayered Thin Stacks Subjected to Uniform Thermal Loading," *ASME J. Elec. Packag.*, **113**(2), 164–172.

10. Engel, P.A., Lim, C.K., Toda, M.D., and Gjone, R. (1984), "Thermal Stress Analysis of Soldered Pin Connectors for Complex Electronics Modules," *Comp. in Mech. Eng.*, **2**(6), 59–69.

11. Engel, P.A., and Lim, C.K. (1991), "Stress Analysis for Soldered Pin-Through-Hole Connectors for Electronic Packaging Structures," *Proceedings of the PACAM II Conference*, Valparaiso, Chile, pp. 118–121.

12. Langhaar, H.L. (1989), *Energy Methods in Applied Mechanics*, Krieger, Malabar, Fla.

13. Niu, T.M., Burke, E.J., Black, W.E., and Case, J.R. (1992), "6-Axis Submicron Fatigue Tester," *Proc. ASME/JSME Int'l Elec. Packaging Conference*, Milpitas, Calif., Vol. 1, pp. 937–946.

14. Hall, P.M., Dudderar, T.D., and Argyle, J.F. (1983), "Thermal Deformations Observed in Leadless Ceramic Chip Carriers Surface Mounted to Printed Wiring Boards," *IEEE Trans.*, **CHMT-6**(4), 544–557.

15. Wilcox, J.R. (1990), "Inelastic Deformation and Fatigue Damage in Materials at High Homologous Temperatures," Ph.D. Thesis, Cornell University.

16. Engel, P.A., Strope, D.H., and Wray, T.E. (1989), "Mechanical Analysis for Thermal Grease Enhanced Modules Enclosing a Silicon Chip," *ASME J. Elec. Packag.*, **111**, 90–96.

17. Burger, C.P., Kobayashi, A.S., Ed. (1987), "Photoelasticity," Chapter 5 in *Handbook of Experimental Mechanics*, Prentice-Hall, New York.

18. Dudderar, T.D., Hall, P.M., and Gilbert, J.A. (1985), "Holo-interferometric Measurement of the Thermal Deformation Response to Power Dissipation in Multilayer Printed Wiring Boards," *Exper. Mechanics*, **25**, 95–104.

19. Bittle, D.A., Suhling, J.C., Beaty, R.E., Jaeger, R.C., and Johnson, R.W. (1991), "Piezoresistive Stress Sensors for Structural Analysis of Electronic Packages," *ASME J. Elec. Packag.*, **113**(3), 203–215.

20. Post, D., and Wood, J. (1987), "Determination of Thermal Strains by Moire Interferometry," *Exper. Mechanics*, **29**, Sept.

21. Bastawros, A.F., and Voloshin, A.S. (1990), "Thermal Strain Measurements in Electronic Packages Through Fractional Fringe Moire Interferometry," *ASME J. Elec. Packag.*, **112**(4) 303–308.

22. Guo, Y., and Woychik, C.G. (1992), "Thermal Strain Measurements of Solder Joints in Second Level Interconnections Using Moire Interferometry," *ASME J. Elec. Packag.*, **114**(1), 88–92.

23. Ling. Y., Engel. P.A., Brodsky, W.L., and Guo, Y. (1992), "Finding the Constitutive Equation of a Specific Elastomer," ASME Paper No. 92-WA-EEP-22.

24. Constable, J.H., and Sahay, C. (1992), "Electrical Resistance as an Indicator of Fatigue," *IEEE* **CHMT** 15(6).

Chapter 2

Finite Element Analysis

1. Preliminaries

Finite elements (FE) is the only numerical method that can, in detail, analyze electronics packaging structures of complex geometric shapes, material combinations, and constitutive behaviors under variegated loading conditions. As a good example, the structure of Fig. 2.1 is a plastic quad flatpack chip carrier, its gull wing leads surface soldered to an epoxy-glass circuit card, the assembly deformed owing to temperature [1]. A brief introduction to the highly developed subject of FE will be given here.

Customary application codes use the "displacement method"; the objective is transformation of the partial differential equations of solid mechanics into algebraic equations in a restricted number of degrees of freedom (dof) that account for the displacements of the structure as a whole. The mathematical language, methodically keeping track of geometry, material, and loads, is

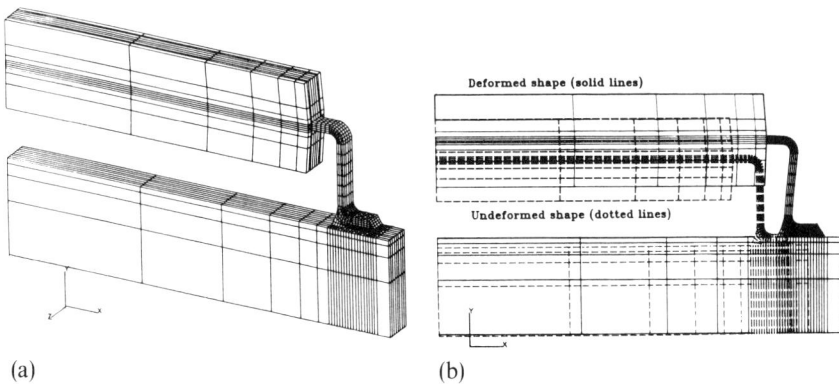

(a)   (b)

FIGURE 2.1. Finite element modeling of a plastic quad flat pack (from Lau [1]. (a) 3-D finite element model for the PQFP SMT assembly; (b) deformation of the PQFP SMT assembly.

matrix-algebra and calculus [2]. If the total structure has its n dofs written in a column matrix $\{D\}$ and the forces $\{R\}$ acting at these as another column matrix, then the structural stiffness matrix $[K]$ connecting the two is an $n \times n$ square matrix:

$$[K]\{D\} = \{R\} . \tag{2.1}$$

The transpose will be designated by a superscript T; a row matrix by $\lfloor \quad \rfloor$, and the inverse of a square matrix by $[\quad]^{-1}$. The latter can be calculated by $\mathrm{adj}[A]/\det[A]$. The product $[A][A]^{-1} = [I]$ is the unit matrix, a special diagonal matrix of elements 1. The product of an $m \times n$ matrix with an $n \times k$ is an $m \times k$. A useful rule is $(AB)^T = B^T A^T$. Often matrix symbols are mixed or interchanged with subscripted (tensor) symbols, e.g.,

$$\frac{\partial [A]}{\partial \{x\}} = \frac{\partial [A]}{\partial x_k} = \frac{\partial (a_{ij})}{\partial x_k} .$$

Matrices well express structural concepts. Work, a scalar quantity, is computed from a multiplication $\{D\}^T \{R\}$. Maxwell's law of reciprocal deflections is written $\{D_1\}^T \{R_2\} = \{D_2\}^T \{R_1\}$. Strain energy is, by Eq. (2.1)

$$U = \tfrac{1}{2}\{D\}^T\{R\} = \tfrac{1}{2}\{D\}^T[K]\{D\} . \tag{2.2}$$

U is a special example for a quadratic functional $F(x_1, \ldots, x_n) = 1/2 \lfloor x \rfloor [A]\{x\} - \lfloor x \rfloor \{b\}$. A more general one is the total potential energy $V = U + \Omega$, where $\Omega = -\{D\}^T\{R\}$ is the potential energy of external loads. The rules for taking derivatives (as applied, e.g., to the minimum potential energy principle) are as follows:

$$\frac{\partial \{x\}^T[A]\{x\}}{\partial \{x\}} = 2[A]\{x\} \tag{2.3}$$

$$\frac{\partial \{x\}^T\{b\}}{\partial \{x\}} = \{b\} \tag{2.4}$$

$$\frac{\partial F}{\partial \{x\}} = [A]\{x\} - \{b\} . \tag{2.5}$$

2. Direct Stiffness Matrix Approach

One may assemble the total stiffness matrix $[K]$ from element stiffness matrices $[k]$. Each element is defined by its dofs $\{d\}$, which are connected to its nodal forces $\{r^*\}$ by $[k]$:

$$[k]\{d\} = \{r^*\} . \tag{2.6}$$

Each matrix member k_{ij} is computed by finding the nodal force at dof i, ensuring equilibrium when the displacement at dof j is 1, all other dofs being zero (i.e., $d_j = \delta_{jk}$) .

Two examples follow. The first one is a uniaxial rod (E, A, L) having two nodes, i and j, and only 2 dofs, u_i and u_j.

$$[k] = AE/L \begin{bmatrix} 1 & -1 \\ -1 & 1 \end{bmatrix}; \quad \{d\} = \begin{Bmatrix} u_i \\ u_j \end{Bmatrix}; \quad \{r^*\} = \begin{Bmatrix} F_i \\ F_j \end{Bmatrix}. \quad (2.7)$$

When a total structure is built up of three such rods in series (Fig. 2.2), the element matrices are telescoped at the nodes; one gets Eq. (2.1) in the form

$$\begin{bmatrix} A_1 E_1/L_1 & -A_1 E_1/L_1 & 0 & 0 \\ -A_1 E_1/L_1 & A_1 E_1/L_1 + A_2 E_2/L_2 & -A_2 E_2/L_2 & 0 \\ 0 & -A_2 E_2/L_2 & A_2 E_2/L_2 + A_3 E_3/L_3 & -A_3 E_3/L_3 \\ \hline 0 & 0 & -A_3 E_3/L_3 & A_3 E_3/L_3 \end{bmatrix}$$

$$\begin{Bmatrix} u_1 \\ u_2 \\ u_3 \\ u_4 \end{Bmatrix} = \begin{Bmatrix} P_1 \\ P_2 \\ P_3 \\ P_4 \end{Bmatrix}. \quad (2.8)$$

The total structural matrix $[K] = \Sigma[k]$ relates the structural displacements $\{D\} = \Sigma\{d\}$ to the external nodal forces $\{R\}$. For an unsupported structure, however, Eq. (2.1) cannot be performed for arbitrary loads; one cannot "solve" for the displacements $\{D\}$ by taking $[K]^{-1}\{R\}$, because $[K]^{-1}$ does not exist. This can be ascertained by the fact that $\det[K] = 0$.

For the computation of a *supported* stable structure, $[K]$ is partitioned, as in Eq. (2.9), according to unknown displacement components $\{D_x\}$ and constrained ones $\{D_c\}$, these being related to unknown (reaction-) forces $\{R_x\}$ and the known ones $\{R_c\}$:

$$\begin{bmatrix} K_{11} & K_{12} \\ \hline K_{21} & K_{22} \end{bmatrix} \begin{Bmatrix} D_x \\ \hline D_c \end{Bmatrix} = \begin{Bmatrix} R_c \\ \hline R_x \end{Bmatrix}. \quad (2.9)$$

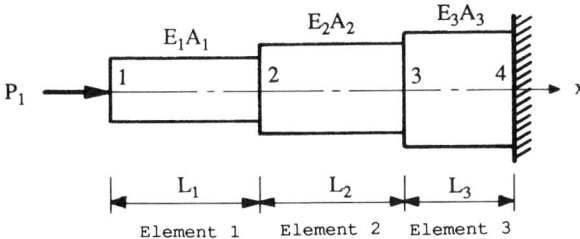

FIGURE 2.2. Assembly of elastic rods; the structure has four nodes and three elements.

For the load P applied to node 1 of Fig. 2.2, these partitions are then easily made. The unknown displacements are $\{D_x\} = [K_{11}]^{-1}(\{R_c\} - [K_{12}]\{D_c\})$, and the reactions $\{R_x\} = [K_{21}]\{D_x\} + [K_{22}]\{D_c\}$. In our example, (Eq. (2.8)):

$$\{D_x\} = \begin{Bmatrix} u_1 \\ u_2 \\ u_3 \end{Bmatrix}; \qquad \{D_c\} = \{0\} \tag{2.10}$$

$$\{R_c\} = \begin{Bmatrix} P_1 \\ 0 \\ 0 \end{Bmatrix}; \qquad \{R_x\} = \{P_4\}. \tag{2.11}$$

The second element-example is an elastic beam (E, I, L) with its four dofs (Figs. 2.3a, b): $d_1 = w_1$, $d_2 = \theta_1$, $d_3 = w_2$, $d_4 = \theta_2$. Any member of the $[k]$ matrix can easily be calculated, e.g., by slope-deflection equations $[(1.21)$ and $(1.22)]$. For example,

$$k_{23} = \frac{2EI}{L}\left(2\theta_i + \theta_j - \frac{3\Delta}{L}\right) = \frac{2EI}{L}\left(0 + 0 - \frac{3}{L}\right) = -\frac{6EI}{L^2}.$$

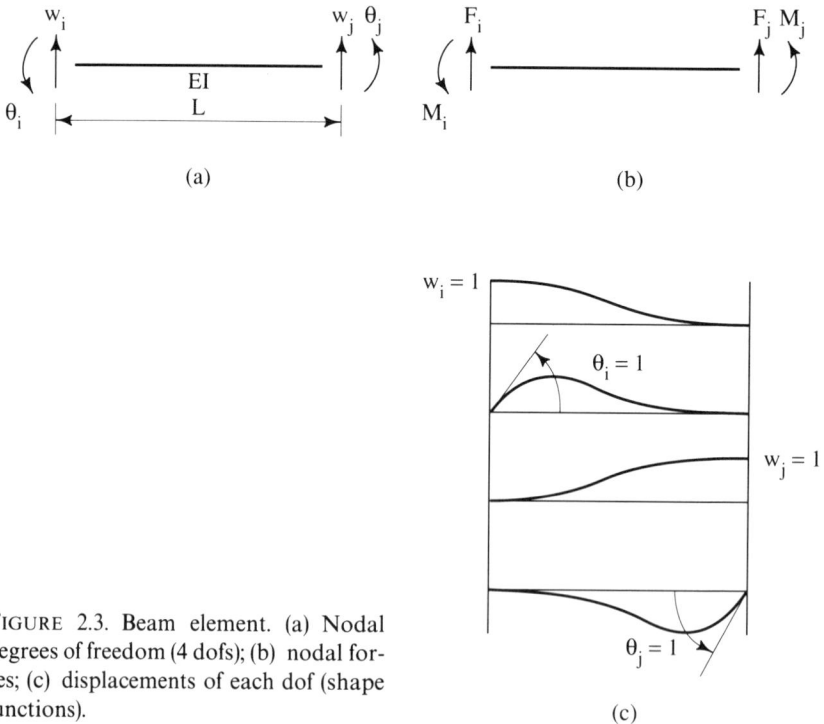

(a)   (b)

FIGURE 2.3. Beam element. (a) Nodal degrees of freedom (4 dofs); (b) nodal forces; (c) displacements of each dof (shape functions).

(c)

Thus, calculating column by column for consecutive dofs, like $\{k_{13}, k_{23}, k_{33}, k_{43}\} = \{k_{i3}\}$, we can assemble $[k]$ as in Eq. (2.12):

$$\frac{EI}{L^3}\begin{bmatrix} 12 & 6L & -12 & 6L \\ 6L & 4L^2 & -6L & 2L^2 \\ -12 & -6L & 12 & -6L \\ 6L & 2L^2 & -6L & 4L^2 \end{bmatrix}\begin{Bmatrix} w_i \\ \theta_i \\ w_j \\ \theta_j \end{Bmatrix} = \begin{Bmatrix} F_i \\ M_i \\ F_j \\ M_j \end{Bmatrix}. \qquad (2.12)$$

If a beam problem such as a simple beam under concentrated central load P is posed, one could cut the beam structure into two elements at the load P, telescope the two $[k]$ matrices into a $[K]$, and then partition the latter according to free dofs $\{D_x\}$ and constrained ones $\{D_c\}$. $\{D_x\}$ and $\{R_x\}$ are evaluated as in the previous example; see eqs. (2.10) and (2.11).

3. The Principle of Minimum Potential Energy

The principle $\delta V = 0$ is equivalent to the statement of equilibrium [Eq. (2.1)], since $\partial V/\partial\{D\} = \partial U/\partial\{D\} + \partial\Omega/\partial\{D\} = [K]\{D\} - \{R\}$. It also results in standard mathematical procedures for obtaining both $[k]$ and an equivalent element load matrix $\{r_e\}$ in more complex structures than rods or beams [3]. This is apparent by taking the derivatives of the strain energy, and of the energy of external loads:

$$\partial U/\partial\{d\} = [k]\{d\}, \qquad \partial\Omega/\partial\{d\} = -\{r_e\} . \qquad (2.13)$$

We start from the statement of V for an arbitrary solid:

$$V = \int_{vol} (\tfrac{1}{2}\{\varepsilon\}^T[E]\{\varepsilon\} - \{\varepsilon\}^T[E]\{\varepsilon_0\} + \{\varepsilon\}^T\{\sigma_0\})d\text{vol}$$

$$U$$

$$- \int_{vol} \{u\}^T\{F\}d\text{vol} - \int_S \{u\}^T\{\Phi\}dS - \{D\}^T\{P\} \qquad (2.14)$$

$$\Omega$$

where $\{\varepsilon\}$ = strain field; $\{\varepsilon_0\}$ = prestrain; $\{\sigma_0\}$ = prestress; $[E]$ = material stiffness matrix, for $\{\sigma\} = [E]\{\varepsilon\}$; $\{F\}$ = traction; $\{\Phi\}$ = body force; $\{u\}$ = displacement field; $\{D\}$ = nodal displacements; $\{P\}$ = nodal forces.

For example, Eq. (2.14) for a beam in general (Fig. 2.4a) yields Eq. (2.15), and for a more special loading (Fig. 2.4b), Eq. (2.16). Partial derivatives are designated by a comma. Using Eq. (1.76) we can write

$$V = \int_0^L \tfrac{1}{2}EIw_{,xx}^2 dx - \int_0^L wq\,dx - \{w\}^T\{F\} - \{\theta\}^T\{M\} \qquad (2.15)$$

$$V = \int_0^L \tfrac{1}{2}EIw_{,xx}^2 dx - \int_0^L wq\,dx . \qquad (2.16)$$

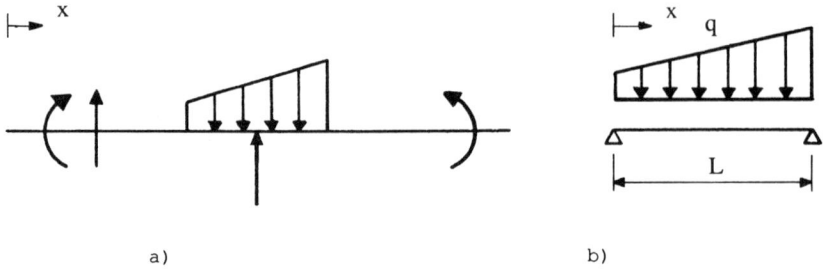

FIGURE 2.4. Beam Loading. (a) General loads; (b) distributed load.

This involves the elastic line $w(x)$, which may be formulated in two ways. The customary polynomial expression is $w(x) = \lfloor X \rfloor \{a\}$, where $\lfloor X \rfloor = \lfloor 1 \ x \ x^2 \ x^3 \rfloor$, and $\{a\}^T = \lfloor a_1 \ a_2 \ a_3 \ a_4 \rfloor$. (Coefficients $\lfloor a \rfloor$ are usually calculated by the Rayleigh-Ritz method, setting $\partial V/\partial \lfloor a \rfloor = 0$, for simple structures.) The other way is by using the beam dofs $\{d\} = \lfloor w_i, \theta_i, w_j, \theta_j \rfloor$ as shown in Fig. 2.3a as parameters, and a "shape function" row matrix $\lfloor N \rfloor$ associated with each d_i, so that $w(x) = \lfloor N \rfloor \{d\}$. Since from $w = \lfloor X \rfloor \{a\}$ we have

$$\begin{Bmatrix} w_1 \\ \theta_1 \\ w_2 \\ \theta_2 \end{Bmatrix} = \begin{bmatrix} 1 & 0 & 0 & 0 \\ 0 & 1 & 0 & 0 \\ 1 & L & L^2 & L^3 \\ 0 & 1 & 2L & 3L^2 \end{bmatrix} \begin{Bmatrix} a_1 \\ a_2 \\ a_3 \\ a_4 \end{Bmatrix} \qquad (2.17)$$

the relationship between $\{d\}$ and $\{a\}$ is determined by the 4×4 square matrix $[A]$ in Eq. (2.17); thus $\{d\} = [A]\{a\}$, and solving for $\lfloor N \rfloor$:

$$\lfloor N \rfloor = \lfloor X \rfloor [A]^{-1} . \qquad (2.18)$$

These four shape functions, not too surprisingly, turn out to be the cubic functions dictated by each dof in Fig. 2.3c:

$$N_1 = 1 - \frac{3x^2}{L^2} + \frac{2x^3}{L^3}$$

$$N_2 = x - \frac{2x^2}{L} + \frac{x^3}{L^2}$$

$$N_3 = \frac{3x^2}{L^2} - \frac{2x^3}{L^3} \qquad (2.19)$$

$$N_4 = -\frac{x^2}{L} + \frac{x^3}{L^2} .$$

The U expression for an element, by Eq. (2.16), is now written

$$U = \tfrac{1}{2} \int_0^L \{w,_{xx}\}^T EI \{w,_{xx}\} dx . \qquad (2.20)$$

Using $w = \lfloor N \rfloor \{d\}$ and $w'' = \lfloor B \rfloor \{d\}$, where $\lfloor B \rfloor = d^2/dx^2 \lfloor N \rfloor$,

$$U = \frac{1}{2} \int_0^L \{d\}^T \lfloor B \rfloor^T EI \lfloor B \rfloor \{d\} \, dx$$

$$= \frac{1}{2} \{d\}^T \left(\int_0^L \lfloor B \rfloor^T EI \lfloor B \rfloor \, dx \right) \{d\} \tag{2.21}$$

and so, based on Eq. (2.2), we conclude that

$$[k] = \int_0^L \lfloor B \rfloor^T EI \lfloor B \rfloor \, dx . \tag{2.22}$$

By the same token,

$$\Omega = - \int_0^L \{d\}^T \lfloor N \rfloor^T q \, dx , \tag{2.23}$$

so that the "consistent" nodal loads are, for distributed q, using Eq. (2.13b):

$$\{r_e\} = \int_0^L \lfloor N \rfloor^T q \, dx . \tag{2.24}$$

The shape functions or "interpolation functions" prescribe a plausible (or, as with the above third-order beam segments, exact) displacement for the element. For two- or three-dimensional elements (e.g., plate and body elements) it is equally easy to obtain $\lfloor N \rfloor$. For example, the Lagrange-interpolation formula gives n functions of $(n - 1)$ degree to interpolate between n points along a line. Considering three points, the three shape functions $N_i(x)$ for the three dofs ϕ_i are

$$N_1 = \frac{(x_2 - x)(x_3 - x)}{(x_2 - x_1)(x_3 - x_1)}$$

$$N_2 = \frac{(x_1 - x)(x_3 - x)}{(x_1 - x_2)(x_3 - x_2)} \tag{2.25}$$

$$N_3 = \frac{(x_1 - x)(x_2 - x)}{(x_1 - x_3)(x_2 - x_3)}$$

so that the displacements along x are described by $w(x) = \lfloor N \rfloor \{d\}$ $= \lfloor N \rfloor \{\phi\}$.

Linear interpolation for a two-dimensional rectangular plate (Fig. 2.5a) of size $2a \times 2b$ is useful for in-plane displacements $u(x, y)$, $v(x, y)$. The interpolation is now based on four corner point dofs ϕ_i; the functions $u(x, y)$, $v(x, y)$ throughout the plate are $\lfloor N \rfloor \{\phi\}$, where

$$N_1 = \frac{(a - x)(b - y)}{4ab} , \qquad N_2 = \frac{(a + x)(b - y)}{4ab}$$

$$N_3 = \frac{(a + x)(b + y)}{4ab} , \qquad N_4 = \frac{(a - x)(b + y)}{4ab} . \tag{2.26}$$

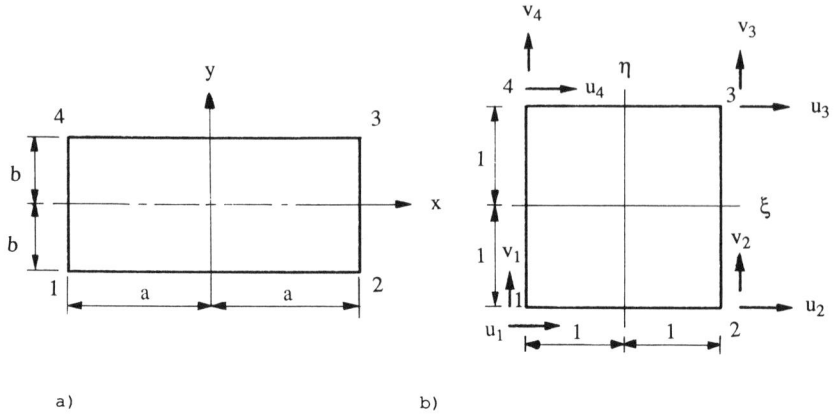

FIGURE 2.5. Rectangular plane element. (a) Global coordinates (x, y); (b) natural (element) coordinates (ξ, η) and four pairs of dofs (u_i, v_i) are shown.

4. Element Types

The type of element chosen should reflect not only the relevant structural action (rod, beam, plate, etc.), but also the shape of the structure and the shape of the displacement field. The shape functions described for the beam [Eq. (2.19)] were of the third order in x, $(n_s = 3)$; however, they were only first order in x in describing geometry $(n_g = 1)$, since elements were defined by their two endpoints. The in-plane rectangular plate element [Eq. (2.26)] had $n_s = 1$ and $n_g = 1$.

Elements for which $n_s = n_g$ are called isoparametric, and these have certain advantages, especially in the numerical calculation effort for $[k]$. This arises in most cases as "natural," i.e., element-associated coordinates (ξ_i) are used instead of a fixed-coordinate system x_i in computing shape functions. For a plane quadrilateral bilinear isoparametric element, for example, we can introduce the natural coordinates $\xi = x/a$ and $\eta = y/b$, both existing between -1 and $+1$, (Fig. 2.5b). This element has four u and four v displacement dofs $\{u^*\} = \lfloor u_1, u_2, u_3, u_4 \rfloor^T$ and $\{v^*\} = \lfloor v_1, v_2, v_3, v_4 \rfloor^T$ with bilinear shape functions N_1, N_2, N_3, N_4 as in Eq. (2.27):

$$N_1 = \tfrac{1}{4}(1 - \xi)(1 - \eta), \qquad N_2 = \tfrac{1}{4}(1 + \xi)(1 - \eta)$$
$$N_3 = \tfrac{1}{4}(1 + \xi)(1 + \eta), \qquad N_4 = \tfrac{1}{4}(1 - \xi)(1 + \eta) . \qquad (2.27)$$

In order to formulate the strain energy U of an element of thickness h, we shall need the integral $U = 1/2 \int \{\varepsilon\}^T [E] \{\varepsilon\} \, d\text{Vol}$, where the strain vector $\{\varepsilon\}$ is related by the elasticity matrix $[E]$ to the stress vector $\{\sigma\}$. For plane stress and plane strain [E] is written out in Eqs. (1.25). For plane stress we have:

$$\left\{\begin{matrix} \sigma_x \\ \sigma_y \\ \tau_{xy} \end{matrix}\right\} = [E]\left\{\begin{matrix} \varepsilon_x \\ \varepsilon_y \\ \gamma_{xy} \end{matrix}\right\}; \qquad [E] = \frac{E}{1-v^2}\begin{bmatrix} 1 & v & 0 \\ v & 1 & 0 \\ 0 & 0 & \dfrac{1-v}{2} \end{bmatrix}. \qquad (2.28)$$

By the strain-displacement relations we can write

$$\{\varepsilon\} = \left\{\begin{matrix} \varepsilon_x \\ \varepsilon_y \\ \gamma_{xy} \end{matrix}\right\} = \begin{bmatrix} 1 & 0 & 0 & 0 \\ 0 & 0 & 0 & 1 \\ 0 & 1 & 1 & 0 \end{bmatrix}\left\{\begin{matrix} u_{,x} \\ u_{,y} \\ v_{,x} \\ v_{,y} \end{matrix}\right\}. \qquad (2.29)$$

where the 3×4 rectangular matrix will be designated [H].

From $u = \lfloor N \rfloor\{u^*\}$ and $v = \lfloor N \rfloor\{v^*\}$ we have

$$\left\{\begin{matrix} u_{,x} \\ u_{,y} \\ v_{,x} \\ v_{,y} \end{matrix}\right\} = \begin{bmatrix} N_{1,x} & N_{2,x} & N_{3,x} & N_{4,x} & 0 & 0 & 0 & 0 \\ N_{1,y} & N_{2,y} & N_{3,y} & N_{4,y} & 0 & 0 & 0 & 0 \\ 0 & 0 & 0 & 0 & N_{1,x} & N_{2,x} & N_{3,x} & N_{4,x} \\ 0 & 0 & 0 & 0 & N_{1,y} & N_{2,y} & N_{3,y} & N_{4,y} \end{bmatrix}\left\{\begin{matrix} u^* \\ \cdots \\ v^* \end{matrix}\right\}, \qquad (2.30)$$

$\qquad 4\times 1 \qquad\qquad\qquad\qquad 4\times 8 \qquad\qquad\qquad\qquad 8\times 1$

where the rectangular 4×8 matrix will be referred to as $[L_{xy}]$. Now we have

$$\{\varepsilon\} = [H][L_{xy}]\{d\} = [B]\{d\}, \qquad (2.31)$$

so that we can write the 8×8 element stiffness matrix, similarly to Eq. (2.22), in the global coordinate form:

$$[k] = \iint [B]^T[E][B]\,h\,dx\,dy. \qquad (2.32)$$

We may also derive [k] by conversion to natural, nondimensional coordinates $\xi = x/a$, $\eta = y/b$. We start by expressing the derivatives entering Eq. (2.29):

$$\left\{\begin{matrix} u_{,\xi} \\ u_{,\eta} \end{matrix}\right\} = [J]\left\{\begin{matrix} u_{,x} \\ u_{,y} \end{matrix}\right\}; \qquad \left\{\begin{matrix} v_{,\xi} \\ v_{,\eta} \end{matrix}\right\} = [J]\left\{\begin{matrix} v_{,x} \\ v_{,y} \end{matrix}\right\}, \qquad (2.33)$$

where the Jacobian matrix is defined as

$$[J] = \begin{bmatrix} x_{,\xi} & y_{,\xi} \\ x_{,\eta} & y_{,\eta} \end{bmatrix} = \begin{bmatrix} a & 0 \\ 0 & b \end{bmatrix} \qquad (2.34)$$

and its inverse:

$$[J]^{-1} = [\Gamma]. \qquad (2.35)$$

Now we write

$$\left\{\begin{matrix} u_{,x} \\ u_{,y} \\ v_{,x} \\ v_{,y} \end{matrix}\right\} = \begin{bmatrix} [\Gamma] & 0 \\ \hline 0 & [\Gamma] \end{bmatrix}\left\{\begin{matrix} u_{,\xi} \\ u_{,\eta} \\ v_{,\xi} \\ v_{,\eta} \end{matrix}\right\} = [M]\left\{\begin{matrix} u_{,\xi} \\ u_{,\eta} \\ v_{,\xi} \\ v_{,\eta} \end{matrix}\right\}. \qquad (2.36)$$

Similarly to Eq. (2.30) we have

$$\begin{Bmatrix} u_{,\xi} \\ u_{,\eta} \\ v_{,\xi} \\ v_{,\eta} \end{Bmatrix} = \begin{bmatrix} N_{1,\xi} & N_{2,\xi} & N_{3,\xi} & N_{4,\xi} & 0 & 0 & 0 & 0 \\ N_{1,\eta} & N_{2,\eta} & N_{3,\eta} & N_{4,\eta} & 0 & 0 & 0 & 0 \\ 0 & 0 & 0 & 0 & N_{1,\xi} & N_{2,\xi} & N_{3,\xi} & N_{4,\xi} \\ 0 & 0 & 0 & 0 & N_{1,\eta} & N_{2,\eta} & N_{3,\eta} & N_{4,\eta} \end{bmatrix} \begin{Bmatrix} u^* \\ \cdots \\ v^* \end{Bmatrix}, \quad (2.37)$$

$$\underset{4 \times 8}{} \qquad\qquad\qquad\qquad\qquad\qquad\qquad\qquad \underset{8 \times 1}{}$$

where the 4×8 rectangular matrix is denoted $[L_{\xi\eta}]$.

We thus get

$$\{\varepsilon\} = [H][M][L_{\xi\eta}]\{d\} . \tag{2.38}$$

Substituting

$$[B'] = [H][M][L_{\xi\eta}] , \tag{2.39}$$

$$\underset{3 \times 8}{} \quad \underset{3 \times 4}{} \; \underset{4 \times 4}{} \; \underset{4 \times 8}{}$$

the element stiffness matrix resulting from nondimensional coordinates is finally computed from

$$[k] = \int_{-1}^{1} \int_{-1}^{1} [B']^{T}[E][B'] \cdot h \cdot J \, d\xi \, d\eta , \tag{2.40}$$

with $J = ab$ being the determinant of $[J]$, so that $dx \cdot dy = J \cdot d\xi \cdot d\eta$.

These concepts are valid to one-, two-, and three-dimensional elements (e.g., rods and beams, triangles and quadrilaterals, and tetrahedra and brick elements, respectively). Linear, quadratic, and cubic shape functions may be used to interpolate between two neighboring nodes. A sample of practical elements is shown in Fig. 2.6. Some of these various elements are known to converge better than others in diverse circumstances [3]; for the investigation of their stability and convergence properties, various tests such as the patch test have been devised.

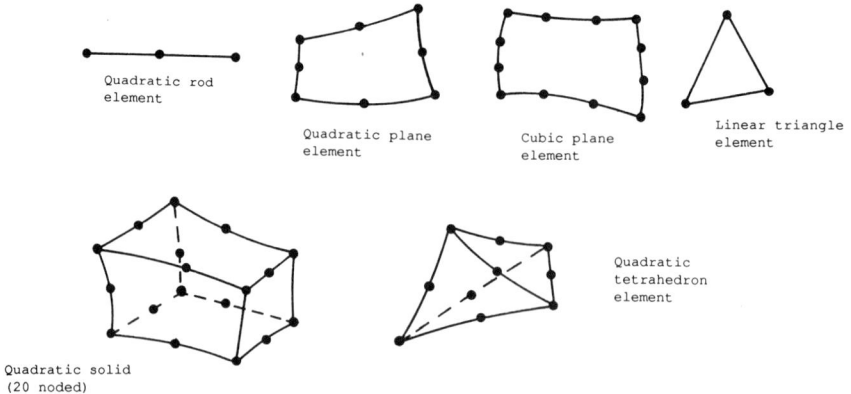

FIGURE 2.6. Some practical elements.

5. Finite Element Dynamic Analysis

Dynamic equilibrium of an element and of the total structure, respectively, may be formulated:

$$[m]\{\ddot{d}\} + [c]\{\dot{d}\} + [k]\{d\} - (\bar{r}\} \tag{2.41}$$

$$[M]\{\ddot{D}\} + [C]\{\dot{D}\} + [K]\{D\} = \{R\} . \tag{2.42}$$

The "consistent" mass matrix $[m] = \int\rho \lfloor N \rfloor^T \lfloor N \rfloor d\text{Vol}$ rather than a lumped diagonal matrix is often used. For damping, the Rayleigh-type or proportional damping is sometimes used, $[C] = \alpha[K] + \beta[M]$, so as to permit vibrational modes to be uncoupled.

The natural mode shapes are obtained from Eq. (2.42), by inserting $\{D\} = \{\bar{D}\} \sin\omega t$, $\omega = \lambda^{1/2}$ being a natural circular frequency of the structure. Thus we get

$$([K] - \lambda[M])\{\bar{D}\} = \{0\} . \tag{2.43}$$

Since Eq. (2.43) is a homogeneous system of algebraic equations, the eigenvalue $\lambda = \omega^2$ is computed from the condition $\det([K] - \lambda[M]) = 0$. The natural mode vectors $\{D\}_i$ are orthogonal to one another with respect to both the $[K]$ and $[M]$ matrices, meaning $\{D\}_i^T[K]\{D\}_j = 0$, $\{D\}_i^T[M]\{D\}_j = 0$, $i \neq j$. The normalized eigenvectors satisfy $\{\bar{D}\}_i^T[M]\{\bar{D}\}_i = 1$, $\{\bar{D}\}_i^T[K]\{\bar{D}\}_i = \omega_i^2$.

The first task of dynamic elastic response analysis is to find the natural frequencies and mode shapes. Having done that, the modal matrix $[u]$ is obtained: this is the matrix of normalized eigenvectors strung together as columns, so that $[u]^T[M][u] = [I]$ and $[u]^T[K][u] = [\omega^2]$, where $[\omega^2]$ is the $n \times n$ square diagonal matrix of the squares of natural frequencies ω_i. The forced vibration solution may then be stipulated in the form

$$\{D\} = [u]\{Z\} \tag{2.44}$$

where $\{Z(t)\}$ is the modal participation or amplitude vector. One solution for $\{Z\}$ is by premultiplying Eq. (2.42) by $[u]^T$, and substituting Eq. (2.44), getting $\{\ddot{Z}\} + [u]^T[C][u] + [\omega^2]\{Z\} = [u]^T\{R\}$. If $[C]$ is a Rayleigh-type damping, then the dynamic matrix equation decouples into n ordinary second-order differential equations

$$\ddot{Z}_i + 2\xi_i\omega_i\dot{Z}_i + \omega_i^2 Z_i = f_i; \qquad f_i = [u]^T\{R\} . \tag{2.45}$$

Since the forcing function $f_i(t)$ is known, premultiplying Eq. (2.44) by the matrix product $[u]^T[M]$ will yield the initial values $\{Z(0)\} = [u]^T[M]\{D(0)\}$ and $\{\dot{Z}(0)\} = [u]^T[M]\{\dot{D}(0)\}$. Further time-integration of Eq. (2.45) is possible either by exact or numerical methods.

In practice, most dynamic problems are reduced to calculations of only a few selected (low-frequency) "master" degrees of freedom, without losing appreciable accuracy. Guyan reduction [4] is one scheme redistributing the structure by appropriate coordinate transformations into "master" and "slave" dofs.

Transient problems such as impact and shock are often handled by directly integrating Eq. (2.41). The "implicit" and "explicit" methods are distinguished, the latter catering to impact problems requiring very small time increments [3].

6. Stress and Strain Calculations

Equation (2.14) was a general formulation of the total potential energy, which included thermal strains (e.g., due to linear expansion $\varepsilon_0 = \alpha \cdot \Delta T$) and initial stress σ_0. The structural solution is then performed, resulting in node displacements $\{D\}$. Once these have been solved, the stresses and strains within the body will obey the stress versus strain relationship; for elastic materials, in the xy plane we get for any element

$$\begin{Bmatrix} \sigma_x \\ \sigma_y \\ \tau_{xy} \end{Bmatrix} = [E]\left([B]\{d\} - \begin{Bmatrix} \alpha\Delta T \\ \alpha\Delta T \\ 0 \end{Bmatrix} \right). \tag{2.46}$$

Special care must be exercised to correctly calculate stresses along interfaces of two materials; in circuit boards this may correspond to a copper conductor bonded to epoxy glass, with greatly different moduli. Figure 2.7 shows the stresses along the juncture: σ_y and τ_{xy} must be continuous across the boundary; however, σ_{x1} should not be equal to σ_{x2}. If the structure is a strip in the x-direction, subjected to temperature rise, then the unequal expansion properties of the neighboring materials should, in fact, induce normal stresses σ_x of opposite sign on two sides of the interface. Lau [5] has

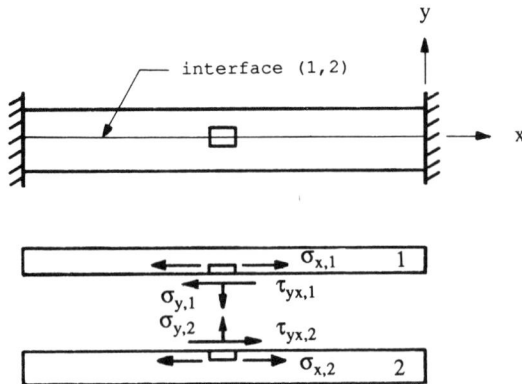

FIGURE 2.7. Thermal stress components at bimaterial strip interface. Note: the stress components σ_y and τ_{xy} match at the interface, but $\sigma_{x,1} \neq \sigma_{x,2}$.

called attention to the inadmissability of indiscriminate stress averaging along material interfaces, to be watched by code writers and users alike.

Plasticity or material nonlinearity is an important addition to modern codes. The material σ-ε relation is usually input as a bi- or multi-linear curve. Loads (or displacements) are added gradually, in steps. The structural matrix [K] must be recomputed at each stage of loading, so that the decremental stiffness is correctly included in each element. The structural solution therefore consists of a two-fold step-by-step approach: the load incrementation process and the iterative solution at each load increment.

Other nonlinearities included in FE analysis are those of 1) geometric nonlinearities (large displacements, stress stiffening, and buckling) and 2) local stresses. While in deflected plates membrane stresses tend to arise and stiffen the structure ([K] is increased), an axially compressed slender structure loses its bending stiffness under load. Both of these effects start as elastic, linear ones; however, [K] is not independent of {R}, hence the nonlinearity.

In contact problems (involving "local stress"), overall stiffness grows with displacements, because the loaded boundary is enlarged with progressive loading. One type of solution introduces load-resistive "gap elements" of stiffness k_c, which are interposed at nodes along the potential contact surface between two bodies; the gaps develop a reaction force $k_c(\Delta)$ (which is not, in general, linear) only when a positive interference Δ is indicated. This, therefore, requires an implicit (iterative) computation. Contact problems generally involve high local stresses and may induce plasticity and large displacements as well. Fracture mechanics uses crack-tip elements, another nonlinear entity.

With all these enumerated complexities of electronics packaging structural problems, finite element analysis has been a mainstay of computer-aided design [6, 7].

7. Structural Codes

All FE codes must process information in some form including 1) problem assignment (e.g., static, dynamic, buckling, heat transfer); 2) geometry definition (node or grid points and numbering); 3) element definition (e.g., plate, beam, numbering); 4) material properties (e.g., E, v, α); 5) loads (e.g., forces, pressures, moments, temperatures); and 6) constraints (e.g., boundary conditions, displacement constraints). Succinct descriptions are offered here of four specific codes with which the author is familiar through use. He has no intention, however, to rank any of the available codes; it is, in fact, the engineer's skill that most often makes the decisive difference.

Begun in 1964, NASTRAN (NASA Structural Analysis) exists in two versions: the U.S. government Cosmic Version and the private MSC version, which is costlier, but highly competitive. An "Executive control deck" identifies the problem solution, and the "Case control deck" refers to the numerical data, all of which are contained in the "Bulk deck" of input lines. (The name "deck" reminds us that program statements were cards in the old days).

ANSYS (Swanson Analysis, Houston, Pa.) is a general-purpose FE program, for practical structural, thermal, and flow problems. It has often been updated since its inception in 1970. ANSYS has numerous files, e.g., for pre- and post-processing, memory, and resuming computation. It is also often used with other preprocessors such as IDEAS or CAEDS (as are NASTRAN and ABAQUS).

ABAQUS (HKS Inc., Providence, R.I.) is a general-purpose FE program that has a batch program to assemble a deck. The data are organized in two groups: model data (elements, nodes, element properties, material definitions) and history data (sequence of events, loadings, steps). ABAQUS has complex contact (with friction) and plasticity routines.

DYNA2 and DYNA3, by Lawrence Livermore Labs, are especially geared to impact dynamics and champion the explicit solution method. These programs have proven to be valuable research tools.

The author has greatly benefited from attending some of the seminars the FE coding firms have offered.

8. Steps in the Design and Use of Finite Element Analysis

1. Discretization (nodes and elements).
2. Choice of shape (approximation) function.
3. Constitutive relation.
4. Element equation: $[k]\{d\} = \{r^*\}$.
5. Assembly of equations.
6. Boundary conditions (forces and displacements, temperature,)
7. Solve $[K]\{D\} = \{R\}$.
8. Calculate and plot stresses, strains, etc.

References

1. Lau, J.H.: (1989), "Thermal Stress Analysis of SMT PQFP Packages and Inter-connections," *ASME J. Elec. Packag.*, **111**(1), 2–8.
2. Wiley, C.R., Jr. (1960), *Advanced Engineering Mathematics*, 2d ed., McGraw-Hill, New York.
3. Cook, R.D., Malkus, D.S., and Plesha, M.E. (1989), *Concepts and Applications of Finite Element Analysis*, 3d ed., Wiley and Sons, New York.
4. Humar, J.L. (1990), *Dynamics of Structures*, Prentice-Hall, Englewood Cliffs, N.J.
5. Lau, J.H. (1989), "A Note on the Calculation of Thermal Stresses in Electronic Packaging by Finite Element Methods," *ASME J. Elec. Packag.*, **111**(4), 313–326.
6. Hsu, T.-R., and Sinha, D.K. (1992), *Computer-Aided Design: An Integrated Approach*, West Publishing Co., St. Paul, Minn.
7. Adkin, J.E. (1990), *Computer-Assisted Mechanical Design*, Prentice-Hall, Englewood Cliffs, N.J.

Chapter 3

Components, Data, and Testing

The three major structural elements making up a populated circuit card in electronics packaging are the module, leads, and the printed circuit card. "Module" as a generic name will be used in this book for all sorts of components attached to a circuit card. For their mechanical roles, featured in this book, modules are characterized by elastic, mass and strength properties. Some of these properties, along with the test procedures followed to obtain them, are considered in this chapter. Special attention is paid to solder, the ubiquitous material of interconnections. Plated through holes of multilayer circuit boards will be treated in Chapter 13. Mass considerations will be included in Chapter 12, in connection with dynamic effects.

1. Modules

For function, we distinguish memory and logic modules, resistors, and capacitors. As for material, various ceramics (e.g., alumina and glass ceramics) and plastics (e.g., PLCC or plastic leaded chip carrier) ordinarily constitute the basic structure [1, 2]. In the former type "first-level package," a chip is attached to a ceramic substrate by solder bumps, wire bonding, or beam leads. In the interior of plastic packages, a chip is attached to a metal lead frame and subsequently encapsulated. First-level packages also include the TAB structure, obtained by the tape automated bonding method [3]; there are challenging mechanics problems associated with each structure; however, these are beyond the scope of this book.

As for their method of attachment, we classify modules as leadless, pin-grid array, and compliant-leaded ones. Leadless chip carriers (e.g., CLCC – ceramic leadless chip carrier) are attached on their periphery by stubby solder posts to a printed circuit card; this is quite analogous to the solder-bump attachment of a chip to a ceramic substrate, mentioned in Section 4.1 of Chapter 1.

Pin grid array (PGA) construction predominantly makes use of an "area-array" of pins, facilitating an order of magnitude more input-output circuits

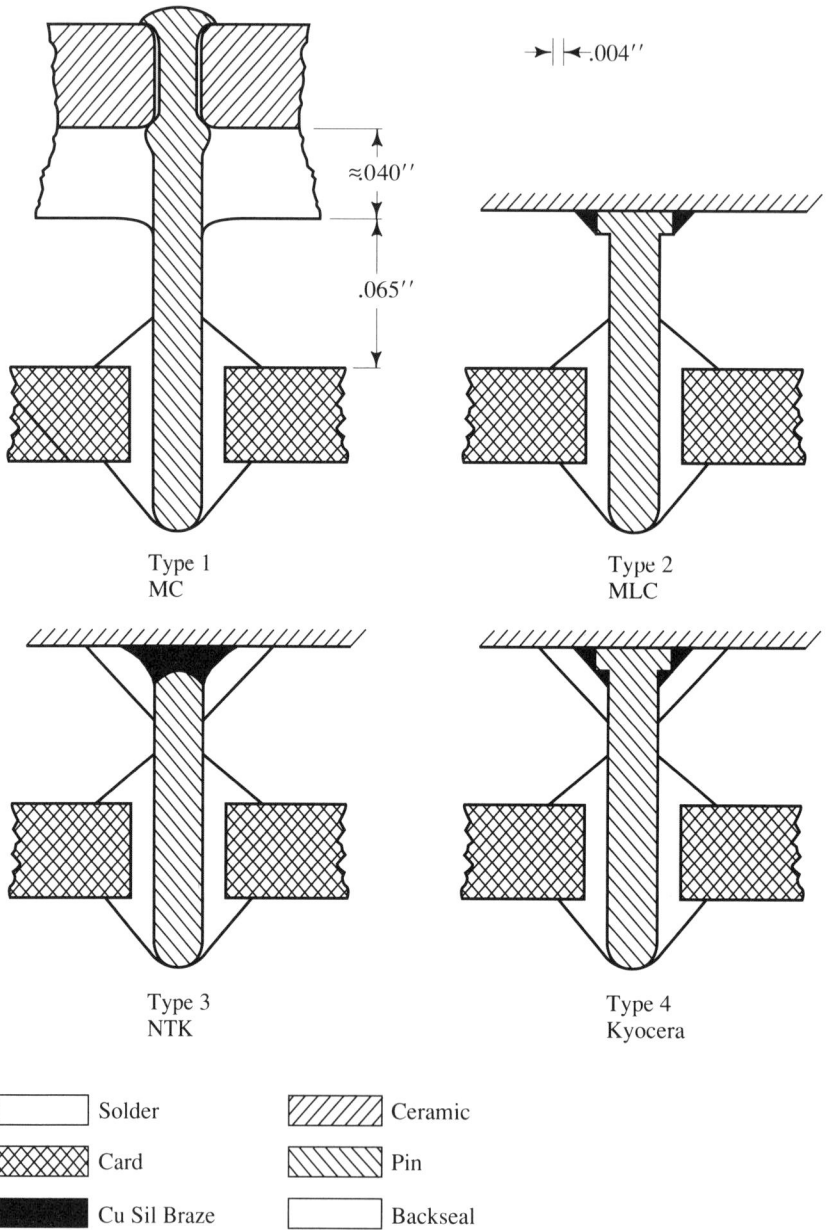

FIGURE 3.1. Cross section of four module configurations [4]. Dimensions are in inches, multiply by 25.4 to convert to millimeters.

(I/Os) than do peripheral module attachment methods. The lead, or pin, has enough length to allow some beam action. The leads, connected to metalization in the substrate, may either be swaged or impact-pinned to holes in the ceramic. Brazing of pins to the ceramic substrate is also a frequent practice [4]; See Fig. 3.1. The bottom of pins is ordinarily embedded in solder, which is usually a low-melting-point 63Sn/37Pb (tin-lead) eutectic mixture [5]. In the wave-solder (solder fountain) manufacturing method, molten solder is sucked up at 183 °C into the clearance space of the card hole by surface tension. Elastic and strength properties for tin-lead solders are quoted in Table 3.1.

Compliant leaded modules have been around for well over 25 years. Leads are extensions of the lead frame, bent down in two principal shapes (Fig. 3.2): the J-lead and the gull wing formations. Surface-mount technology (SMT) ordinarily uses eutectic Sn/Pb solder under the vapor-phase reflow (215 °C) or infrared (220 °C) methods. Ever smaller lead widths are designed (about 0.04 mm or 0.01 in. is not rare), spaced ever denser (a spacing of $s = 0.5$ mm or 0.012 in. is becoming frequent in fine-pitch technology). Design limitations of geometrical parameters are described by Jahsman [6]. The curved shaped leads assure more flexibility for taking up differential thermal expansion tendencies from module to circuit card. Figure 3.3 shows various popular modules [7]. Table 3.2 lists the dimensions of various square-outline PLCC module sizes of the Motorola semiconductor products sector; it refers to the specifications of the JEDEC (Joint Electronic Devices Engineering Council). The SOIC (small outline integrated circuit) is of rectangular plan, with leads marching along the long sides; several other "small outline" modules such as the SOJ and SOT modules are also rectangular. They are similar to DIP

TABLE 3.1. Mechanical properties of bulk Pb/Sn solder alloys at room temperature (*Metals Handbook*, ASM, 1984).

Tin content	Tensile σ_u strength	Shear τ_u strength	Modulus of elasticity	Impact strength	Stress to produce creep rate of 0.0001 in./in./day
%	psi (MPa)	psi (MPa)	kpsi (MPa)	J	psi (kPa)
0	1799(12.4)	1799(12.4)	2608(17973)	8.13	249(1716)
5	3998(27.5)	2099(14.5)	–	9.49	200(1378)
10	4398(30.3)	2399(16.5)	2758(19007)	10.85	–
20	4796(33.0)	2997(20.6)	2897(19965)	14.91	–
30	4997(34.4)	3998(27.5)	3048(21006)	16.27	14(96)
40	5396(37.2)	4597(31.7)	3338(23004)	18.98	–
50	5997(41.3)	5197(35.8)	–	20.34	125(861)
60	7595(52.3)	5596(38.6)	4346(29951)	20.34	–
63	7795(53.7)	5396(37.2)	–	20.34	335(2309)
70	7795(53.7)	5396(37.2)	5077(34989)	18.98	–

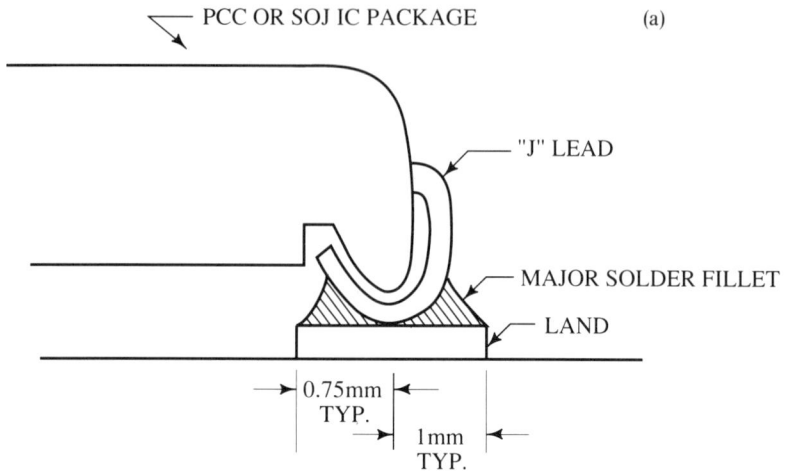

PCC OR SOJ PACKAGES "J" LEAD CONTACT POINT
AND FILLET INFORMATION

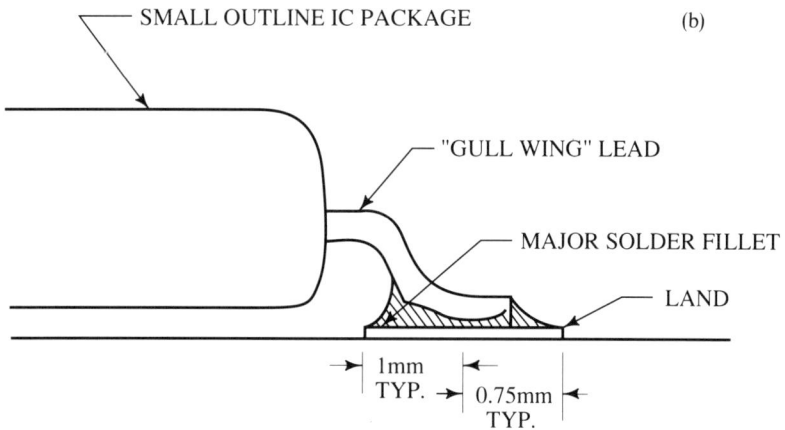

SO "GULL WING" CONTACT POINT AND FILLET INFORMATION

FIGURE 3.2. Popular surface mount leads. (a) J-lead; (b) gull wing.

(dual in-line package) modules, which have two parallel rows of leads soldered through card holes.

Modules are highly inhomogeneous and anisotropic structures, but in order to analyze their mechanical role in a circuit card system, it usually suffices to have their "equivalent" orthotropic or isotropic elastic constants. The elastic property of flexural modulus E for a PLCC can be measured by a three-point bending test: two parallel sides are supported on a knife edge, and a force P is brought down over the midsection as a line load (Fig. 3.4).

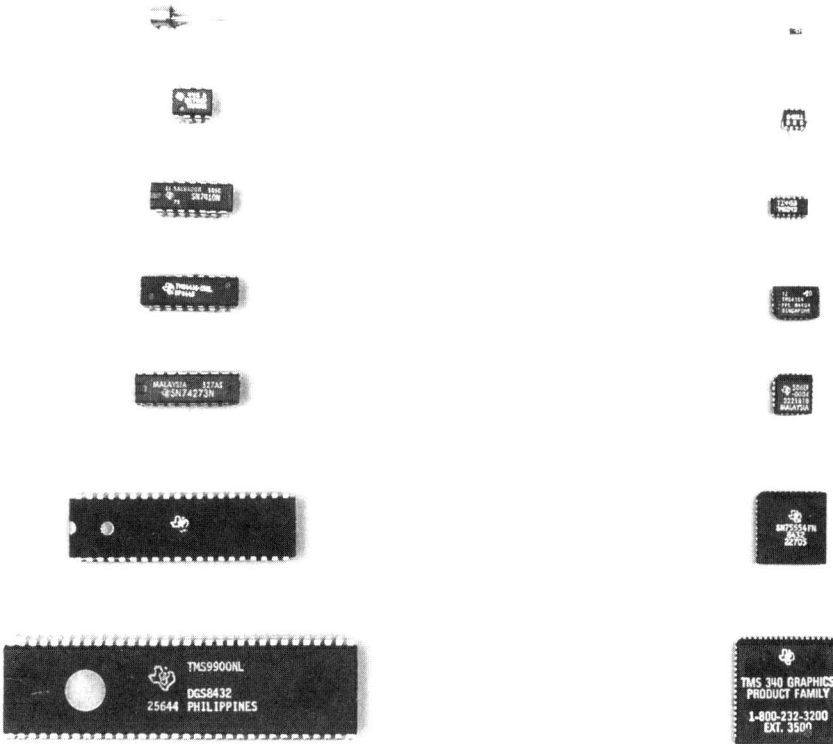

FIGURE 3.3. Some modules popular in the electronics industry (courtesy of Charles Hutchins).

From cylindrical bending, we get the spring stiffness

$$S = P/w = 48\,Db/a^3 \tag{3.1}$$

where the plate constant D is defined:

$$D = Eh^3/[12(1 - v^2)]\,, \tag{3.2}$$

and the bending rigidity for the total width is Db.

Approximating v as that of the substrate or plastic material, the equivalent E can be determined in the x and y directions from Eqs. (3.1) and (3.2). A second elastic constant, such as v, can be obtained from performing a torsion test, with point loads P applied at all module corners. The torque $T = P \cdot b$ causes a twist angle ϕ, from which, based on Fig. 3.4, the torsional stiffness by St. Venant's torsion theory [8] is

$$k_T = T/\phi = 2GJ/a = Ebh^3/[3a(1 + v)]\,. \tag{3.3}$$

The isotropic elastic equivalent constants E and v can be calculated from the preceding two measurements. (For some physical constants of materials

TABLE 3.2. PLCC sizes: Motorola semiconductor products sector.

I/O	Plastic Width mm (in.)		Total Width mm (in.)		Plastic Height mm (in.)		Total Height	
	Min.	Max.	Min.	Max.	Min.	Max.	Min.	Max.
20	8.89 (0.350)	9.04 (0.356)	9.37 (0.369)	10.19 (0.401)	–	4.57 (0.180)	–	5.08 (0.200)
28	11.43 (0.450)	11.58 (0.456)	11.91 (0.469)	12.73 (0.501)	–	4.57 (0.180)	–	5.08 (0.200)
44	16.51 (0.650)	16.66 (0.656)	16.99 (0.669)	17.81 (0.701)	–	4.57 (0.180)	–	5.08 (0.200)
52	19.05 (0.750)	19.20 (0.756)	19.53 (0.769)	20.35 (0.801)	–	4.57 (0.180)	–	5.08 (0.200)
68	24.13 (0.950)	24.33 (0.958)	24.61 (0.969)	25.48 (1.003)	–	4.57 (0.180)	–	5.08 (0.200)
84	29.21 (1.150)	29.41 (1.158)	29.69 (1.169)	30.56 (1.203)	–	4.57 (0.180)	–	5.08 (0.200)

The above are initial proposed package definitions for leaded plastic chip carriers by the JEDEC committee.

The "rolled-under" leads appear to be the preferred lead configuration.

Going to higher lead counts makes lead integrity a serious consideration.

More and more industry movement towards surface mounting in lieu of "hole-through" mounting exists.

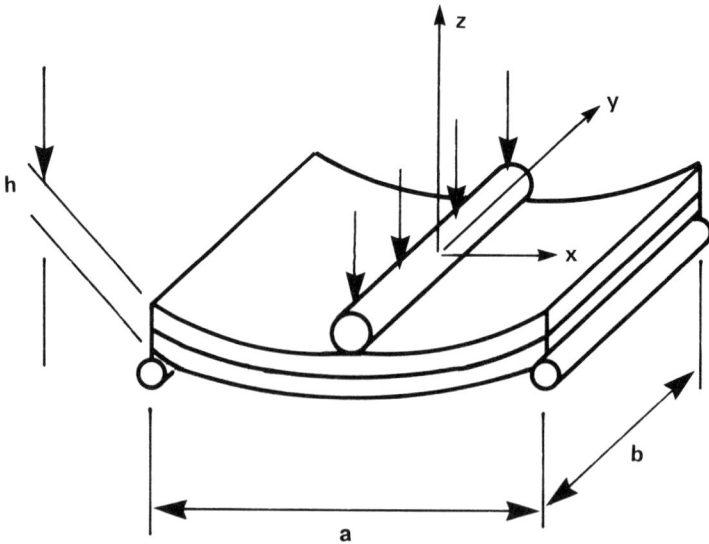

FIGURE 3.4. Three-point bending test. This test is used for modules and circuit cards to measure the equivalent modulus E or the rigidity EI.

TABLE 3.3. Engineering properties for circuit card and module materials.

Materials	Dielectric constant	CTE ppm/°C	Thermal conductivity (w/m.°C)	Elastic modulus (GPa)
Glass-ceramics	4–8	3–5	5.0	300
Cu-clad Invar (10% Cu) glass coated		3	100	200
Epoxy-Kevlar, x-y (60%)	3.6	6	0.2	
FR-4 (x-y plane)	4.7	16	0.2	14
Polyimide	3.5	50	0.2	2.7
Teflon	2.2	20	0.1	0.5
Copper		17	393	124
Gold		14	297	40
Aluminum		23	240	70
Pb-5%Sn		29	63	18
Kovar		5.3	17	138

used in electronics packaging, see Table 3.3.) The plate constant of a PLCC is calculated based on the physical height h of the module in Eq. (3.2); for example, using $h = 3.18$ mm (0.125 in.), $E = 3.29$ GPa (477 Kpsi), and $v = 0.35$:

$$D = (3290)(3.18^3)/[12(1 - 0.35^2)] = 10.05 \text{ N.mm} \quad (88.89 \text{ lb-in.}) . \quad (3.4)$$

In order to obtain equivalent orthotropic material constants (E_x, E_y, v_{xy}, v_{yx}), both the flexural and torsional tests should be performed in two perpendicular directions.

2. Circuit Cards and Boards

2.1. General Description

Printed circuit cards or simply circuit cards are, in this text, often referred to as "cards" for brevity. There is a distinction, often not in substance but in size, between "cards" and "boards." The latter name tends to be used when we mean a larger and thicker (multilayer) structure, such as the 60×70 cm (24×28-in.) IBM Clark Board [1], or mother boards ("planars") to which smaller circuit cards are connected.

Circuit cards are ordinarily made of epoxy glass (FR-4) material, which, besides its flexibility, strength, and dimensional stability, successfully simulates in its plane the coefficient of thermal expansion (CTE or α) of copper lines it carries ($\alpha = 16$–18 ppm/°C). The z-directional CTE is much higher, however: typically 50 ppm/°C. The elastic or bending moduli of FR-4 cards in ordinary use have been measured (by the same methods described in the previous section for modules) between 10 and 20 GPa (1.5–3.0 Mpsi) and the Poisson ratio between 0.23 and 0.35. When the card is drilled to receive a soldered pin, its effective modulus is reduced by the percentage of the cross-sectional area of the holes rated to the total plane cross section. The plate constant of a card segment of 1.25 mm (0.050 in.) depth is, for example,

$$D = 14,000 \times 1.25^3/[12(1 - 0.3^2)] = 2504 \text{ N.mm} \quad (22.2 \text{ lb-in.}) .$$

While the overstressing of cards is rare even under handling conditions, the circuit lines running on the surface should be stress checked. Pads where leads are attached are best grouped along with the leads when strength considerations (pull-out or adhesion) are probed; in any case, it is undesirable to make pads the weakest link in the lead structure. Severe problems are occasionally encountered due to the warping of circuit cards. This may be caused by imperfections in the relative directions or curing of the polymer prepreg layers making up the composite material [9].

For thermal stress, we note that ceramic substrates ($\alpha = 6.5$ ppm/°C) have the opposite thermal mismatch tendencies with respect to the FR-4 card as do plastic modules (>30 ppm/°C). With the individual module in mind, it is sometimes advisable to tailor the CTE to the requirements of a particular module. In order to get a very low-expansivity card such as would combine well with, e.g., a low-expansivity glass ceramic module ($\alpha = 2$–3 ppm/°C), Kevlar epoxy or copper-invar-copper cards have been used. For the latter, the combined CTE in terms of the percentage of the Cu thickness (balance Invar) is shown in Fig. 3.5. For the general case of an n-layered elastic

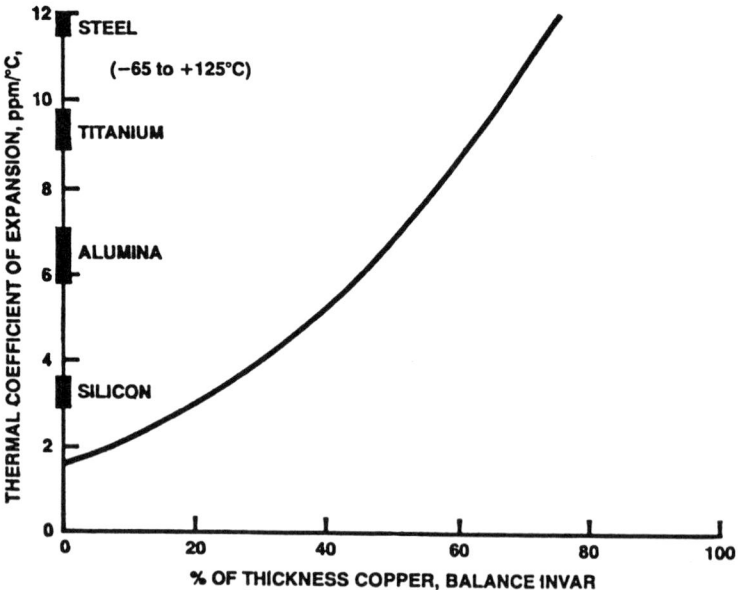

FIGURE 3.5. The coefficient of thermal expansion (CTE) versus copper content in a copper-invar-copper sandwich (courtesy of ASM).

composite, the combined in-plane modulus and CTE can be easily shown, by matching displacements on $n-1$ interfaces, to yield the expressions

$$E = \sum h_i E_i / \sum h_i \tag{3.5}$$

$$\alpha = \sum \alpha_i h_i E_i / \sum h_i E_i . \tag{3.6}$$

Example

Consider a 20-60-20 percentage-thickness combination for a Cu-Invar-Cu card; say each Cu layer is 0.5 mm thick. Solving Eqs. (3.5) and (3.6) in a tabulated form:

Mat'l	h_i (mm)	E_i (GPa)	α_i ppm/°C	$h_i E_i$	$\alpha_i h_i E_i$
Cu	1	124	17	124	2108
Invar	1.5	207	2	310	620
	2.5			434	2728

Thus $E = 434/2.5 = 174$ GPa; $\alpha = 2728/434 = 6.3$ ppm/°C (versus 5.2 ppm/°C from Fig. 3.5).

Now consider the sandwich of four 0.35-mm composites of 20-60-20 Cu-Inv-Cu, with five 0.15-mm-thick FR-4 layers:

Cu-I-Cu	1.4	174	6.3	244	1535
FR-4	0.75	14	15	10.5	158
	2.15			254.5	1693

Thus $E = 254.5/2.15 = 118$ GPa; $\alpha = 1693/254.5 = 6.65$ ppm/°C.

2.2. Properties of Laminated Construction

In the usual board manufacturing process [10] epoxy resin-impregnated and cured sheets of woven glass fabric (called "prepreg," such as FR-4) are laminated together between sheets of copper to form a core. Signal and power cores are sandwiched together under temperature and pressure. Electrical connections between different planes are formed by drilling holes through the core and copper-plating the hole walls. Circuit lines are made either by additive plating or subtractive etching of the copper sheets. A typical core dimension may be 150 μm (6 mil), with copper sheets 30 μm thick. The thinner the prepreg, the higher the electrical signal noise, but also the propagation speed becomes. To raise the latter, the dielectric constant must be kept low. Thicker copper sheets tend to generate more current intensity and produce more heat.

In modern products strength, insulation values, relative ease of fabrication, low price, and universal availability are important board design factors.

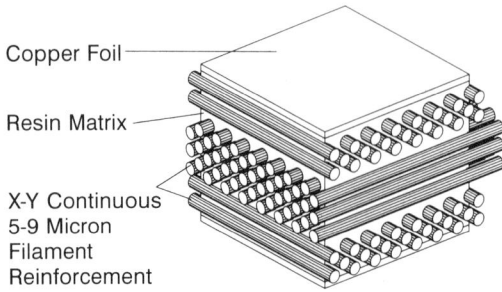

Copper Foil

Resin Matrix

X-Y Continuous
5-9 Micron
Filament
Reinforcement

FIGURE 3.6. Copper-clad *x-y* continuous fiber laminate construction (from Klimpl [11]).

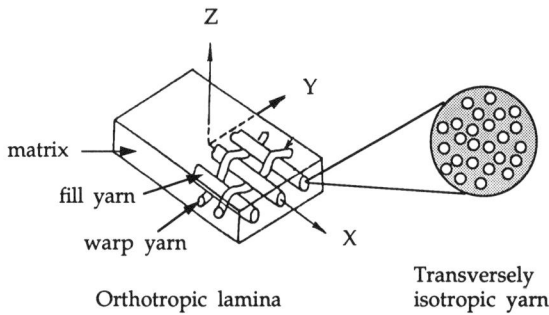

matrix

fill yarn

warp yarn

Orthotropic lamina

Transversely isotropic yarn

FIGURE 3.7. Schematic of a plain-weave laminate cross section showing woven yarn (from Dasgupta and Bhandarkar [12]).

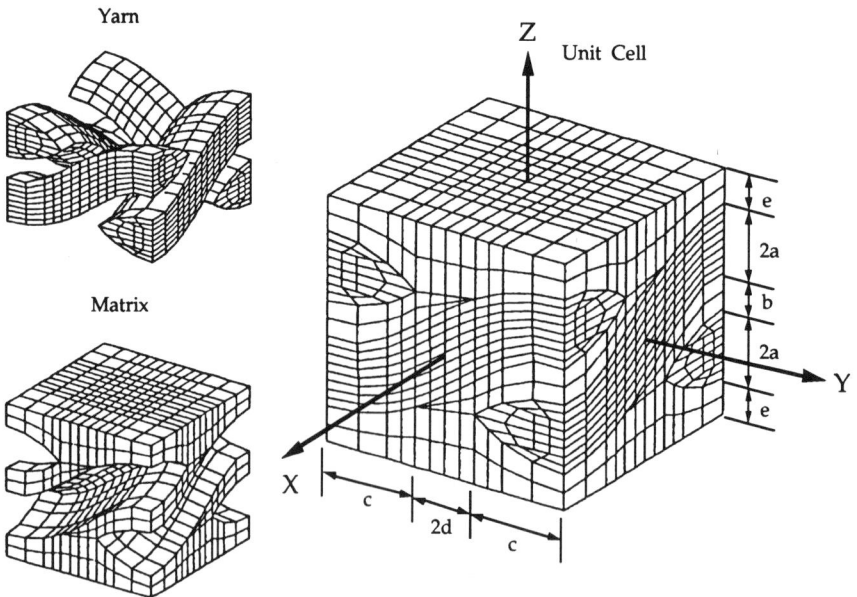

Yarn

Matrix

Unit Cell

FIGURE 3.8. Unit cell for plain-weave fabric reinforced composite (from Dasgupta and Bhandarkar [12]).

Klimpl [11] describes use of continuous precisely oriented x-y directional filament fiberglass and brominated epoxy resins, more fully realizing fiber strength (Fig. 3.6). Glass transition temperatures T_g of between 250° and 350 °C have been achieved along with a drastically reduced α_z, less than half that of FR-4.

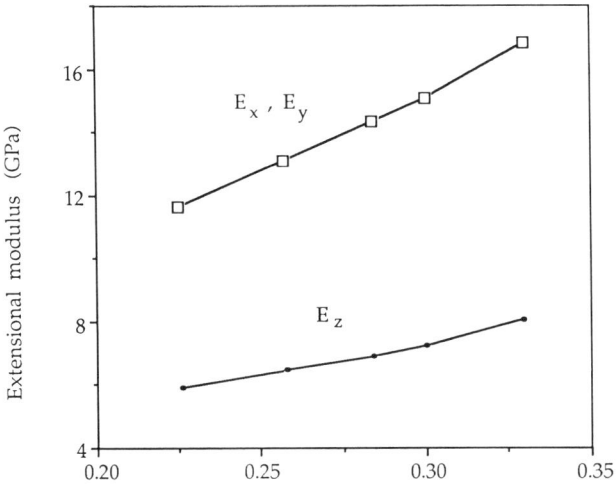

(a)

(b)

FIGURE 3.9. Elastic properties versus fiber volume fraction for epoxy glass multilayer boards (from Dasgupta and Bhandarkar [12]). (a) Extensional moduli; (b) Poisson ratios.

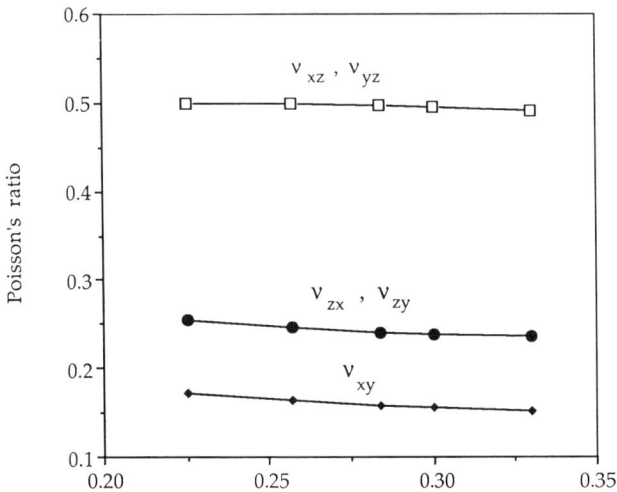

A study of composite mechanical behavior of woven composites [12] showed thermomechanical behavior varying significantly with fabric style, microstructural geometry, volume fractions of constituents, and material systems used for resin and reinforcement. Figure 3.7 shows that in a plain-weave fabric, every fill yarn is interlaced with every second warp yarn and vice versa. The orthotropic nature of the composite necessitates the use of

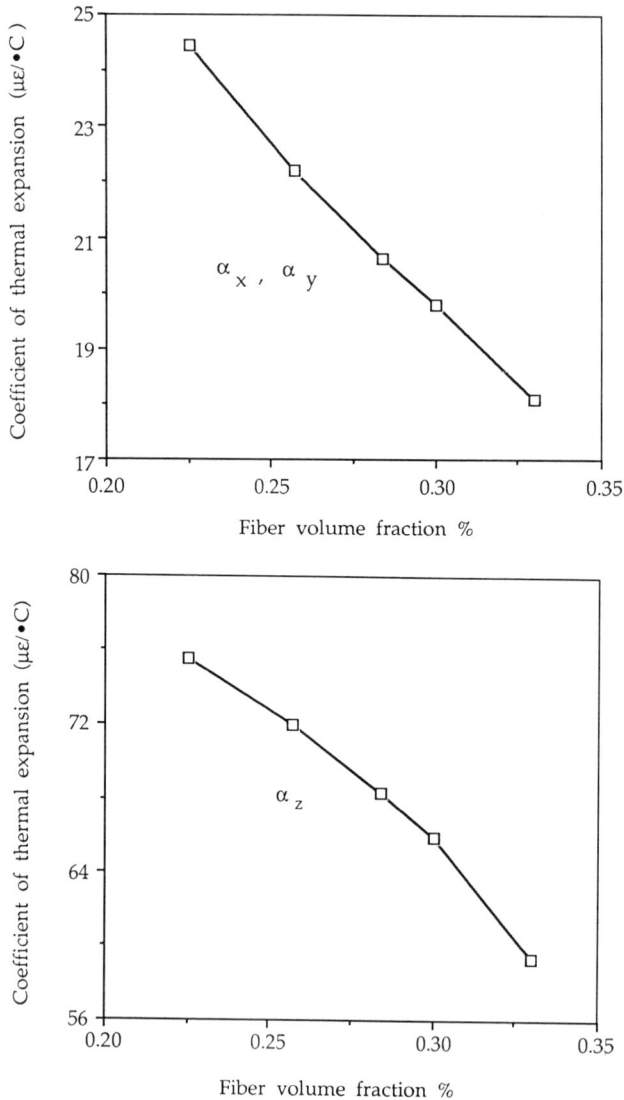

FIGURE 3.10. Coefficients of thermal expansion (CTE) for epoxy glass multilayer boards (from Dasgupta and Bhandarkar [12]).

nine independent elastic constants [say, E_x, E_y, E_z, v_{xz}, v_{yz}, v_{zx}, v_{zy}, v_{xy}, v_{yx}, and three independent CTEs (α_x, α_y, α_z)]. By a "homogenization" scheme, Dasgupta et al. [13] used an asymptotic smoothing technique to represent a characteristic volume element. A unit cell (Fig. 3.7) of the structure allows mathematical treatment of a microscale boundary value problem. This may be amenable to a finite element solution (Fig. 3.8). The extensional moduli and Poisson ratios for FR-4 versus the ratio of fiber-fill volume fraction was then plotted (Fig. 3.9). The coefficients of thermal expansion are plotted versus the ratio of fiber-fill volume in Fig. 3.10.

Thermal conductivity of multilayer boards was modeled and compared with experimental values in Ref. [14]. Electrical failure modes of printed circuit boards under temperature and humidity were described in Ref. [15].

3. Pin Leads of PGA Modules

Pins must carry axial as well shear and moment loads, as will be discussed in Chapters 5 and 6. Recognizing this need, Kelly et al. [16] suggested inclined pin pull tests at 20 degrees, with a minimum force of 40 N (9 lbs), for 0.4-mm (0.016-in.) diameter pins of 50-mm modules brazed to a ceramic substrate. Table 3.4 shows the mechanical properties of three materials, Kovar, Alloy 42, and Amzirc, suitable for pin design. Figure 3.11 shows the stress-strain curve derived from axial pull tests and moment-capacity curves for Amzirc pins of 0.4- and 0.5-mm (0.016- and 0.020-in.) diameter sizes.

Plastic deformation of the pin is likely to occur owing to intensive thermal cycling. This loading tends to cause a radial stress down the tubular solder joint, giving rise to longitudinal cracks, the type of which is seen in the cross section of Fig. 3.12. The crack extends downward below the fillet; note the copper sleeve and pad of the plated-through hole (PTH) and conductor planes of the circuit card.

TABLE 3.4. Physical properties of pin materials (σ_y, σ_u data from wire pull test).

Name	Description	E GPa (kpsi)	α (10^{-6}) °C	Annealed σ_y MPa (kpsi)	Annealed σ_u MPa (kpsi)	Annealed ε_u	Unannealed σ_y MPa (kpsi)	Unannealed σ_u MPa (kpsi)	Unannealed ε_u
Kovar	Iron-nickel-cobalt alloy	138 (20,000)	5.9	345 (50)	517 (75)	0.30	724 (105)	724 (105)	–
Alloy 42	Iron-nickel alloy	148 (21,500)	5.4	276 (40)	517 (75)	0.30	731 (106)	731 (106)	–
Amzirc	Zirconium-copper alloy	127 (18,500)	16	89.7 (13)	255 (37)		–	–	–

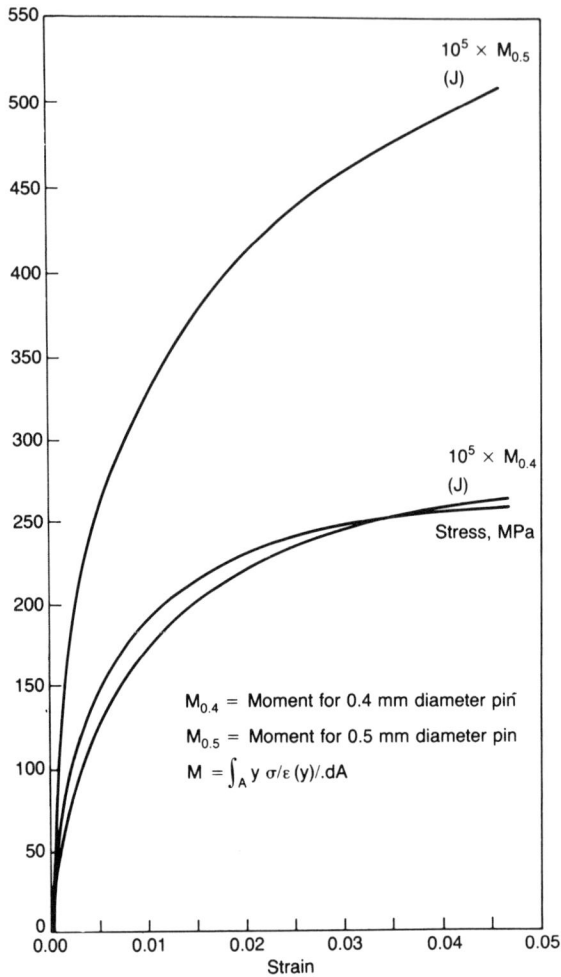

FIGURE 3.11. Stress-strain (measured) and moment capacity versus strain (computed) relations of annealed Amzirc pins.

Pin-through-hole construction is, in principle, replaceable by surface solder joining pin ends to a card. This high-I/O structure has obvious advantages in obviating holes and saving "card real estate," but difficulties in solder reflow have not yet facilitated its wider acceptance by the industry. Its viability has been shown on a laboratory basis [17]; various pin designs are possible (Fig. 3.13). Rectangular ends are preferable to rounded ends, and deeper insertion is beneficial. T-shape ends supply much more strength, although they cost "real estate."

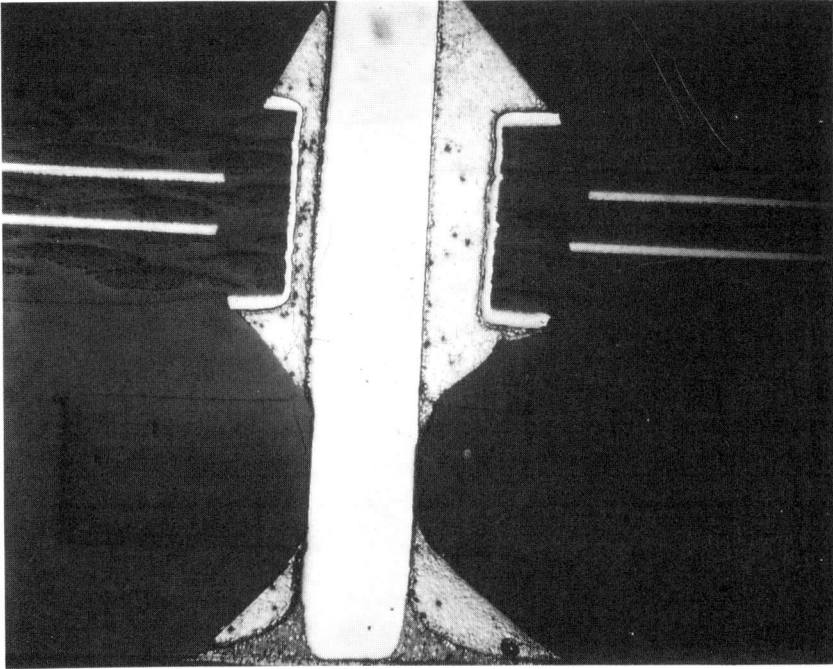

FIGURE 3.12. Cross section through soldered pin-through-hole arrangement. The braze joint to a substrate is above.

4. Strength of Compliant Leads in Surface-Mount Construction

J-leads are hooked under the package past the curved solder joint. Gull wings have flat bottoms that require uniformity in height and bend angle to achieve reliable contact along card tabs.

Both thermal and handling (flex) tests of cards with surface soldered modules attached by J-leads showed a failure mechanism (Fig. 3.14) with a crack starting at the outside of the solder joint, tending to separate it from the lead [18]. Propagation of this crack to the bottom of the lead's arc completes mechanical rupture. Such a failure can be brought about by pulling along the lead in the z-direction (see Fig. 3.15). Therefore we must maintain the axial (tensile) pull force F, resulting from module-card interaction, at safe levels. This axial force generates bending and longitudinal pull or push in the eccentrically loaded joint.

Slicing pairs of surface-soldered leads along an SOIC module (Fig. 3.16) and attaching both the module and card joints rigidly between the two faces of the MTS load tester, tensile fatigue results were generated [19] in two systems of J-leads (see Fig. 3.17). Failure always occurred as a solder joint

Round Pin **Cut Pin** **Inserted Deeper**

Step Pin

| Copper Pin......... |
| Solder................ |
| Copper Pad........ |

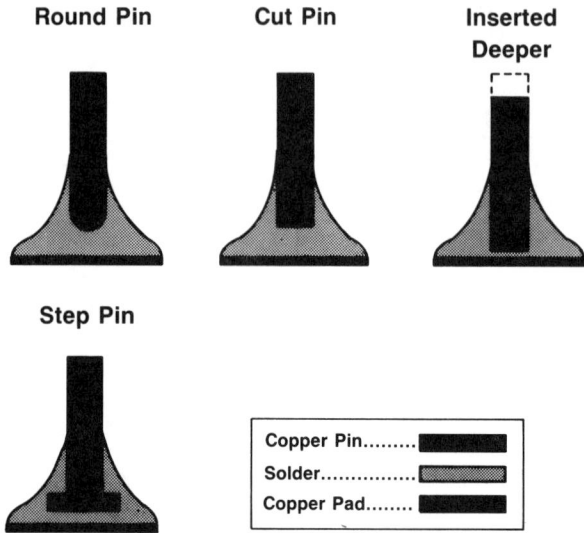

FIGURE 3.13. Surface-soldered pin ends.

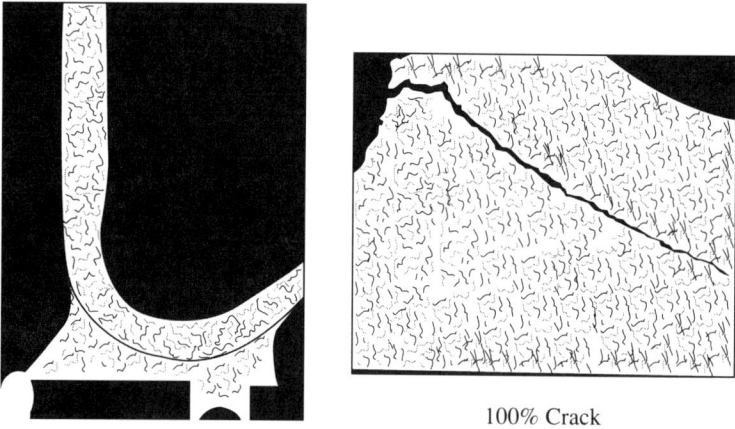

100% Crack

FIGURE 3.14. Cross-sectioned J-lead solder joint showing fatigue crack parallel to the lead-solder interface.

separation, as in Fig. 3.14. Figure 3.18 shows fatigue test results for two types of gull wings; now the failures, contrarily, were exclusively in the leads, near their exit from the module. Load cycling was done in the 10–40-Hz frequency range.

FIGURE 3.15. Pull-test scheme for a cutaway J-lead.

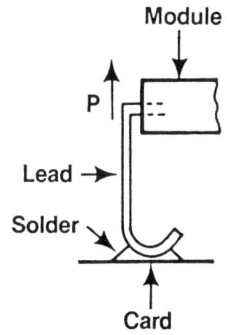

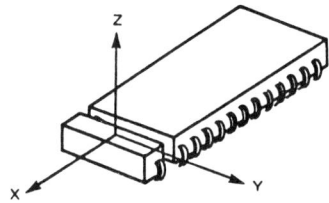

FIGURE. 3.16. Schematic of J-leaded surface-mount module. One pair of leads is cut away.

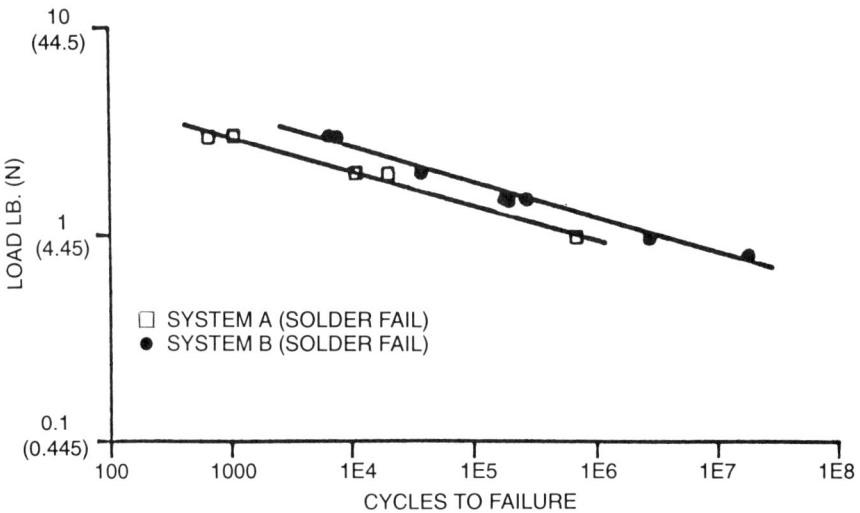

FIGURE 3.17. Fatigue curve of J-lead pairs (from Engel et al. [19]). The load was imposed by two-way displacement cycling.

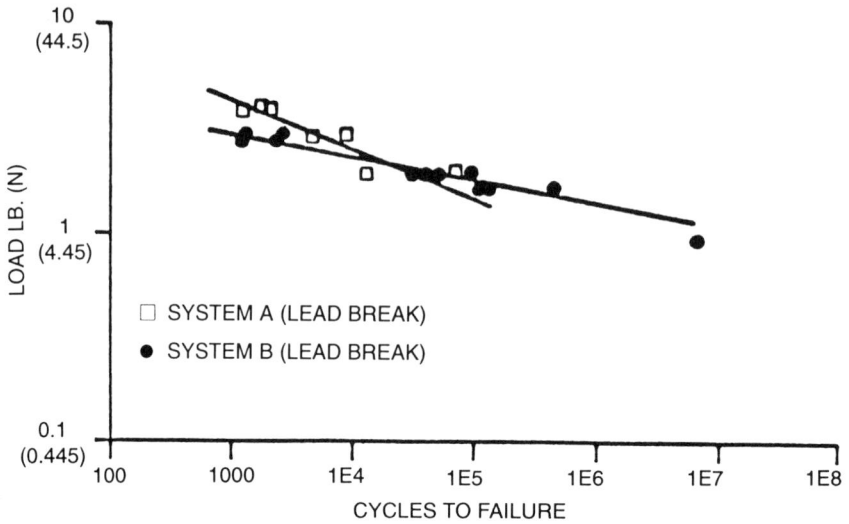

FIGURE 3.18. Fatigue curve of gull wing pairs (from Engel et al. [19]). The load was imposed by two-way displacement cycling.

TABLE 3.5. Lead stiffness test results (1 mm/min test speed).

		Per single lead			
Type of lead	Direction	K, stiffness N/mm (lb/in.)	Yield point N (lb)	Ultimate load N (lb)	Number of samples
J-lead					
2-lead assembly	z	$K_{zz} = 640$ (3650)	18.0 (4.05)	21.1 (4.75)	5
26-lead assembly		530 (3050)	15.8 (3.55)	18.0 (4.05)	5
Gull wing					
"Type 602"	z				
2-lead assembly		$K_{zz} = 540$ (3100)	20.5 (4.60)	24.5 (5.50)	6
26-lead assembly		580 (3300)	17.1 (3.85)	21.4 (4.80)	3
Gull wing					
Type "590"	z				
2-lead assembly		$K_{zz} = 630$ (3600)	17.8 (4.00)	24.5 (5.5)	3
J-Lead					
2-lead assembly	x	63 (370)			6
2-lead assembly	y	60 (340)			4
Gull wing					
2-lead assembly	x	58 (330)			6
2-lead assembly	y	33 (190)			4

5. Stiffness of Compliant Leads

For the analysis of surface-mounted circuit card systems that lies ahead in Chapters 7 and beyond, we shall need lead stiffnesses. For flexure of the board, lead stiffnesses in the transverse plane (x, z) are essential; for thermal expansion, the ones in the plane of the board are important. It must be borne in mind, of course, that a small change in lead geometry may introduce a large change in the numerical value of beam rigidities; thus lead analysis can establish limiting response parameters or be used in a comparative mode [20].

Table 3.5 shows the z-directional stiffnesses obtained from pull testing both a full module with 13 pairs of leads (Fig. 3.16) and pairs of leads sliced off [19]. Since module and card were both adhered to the test fixtures, nearly

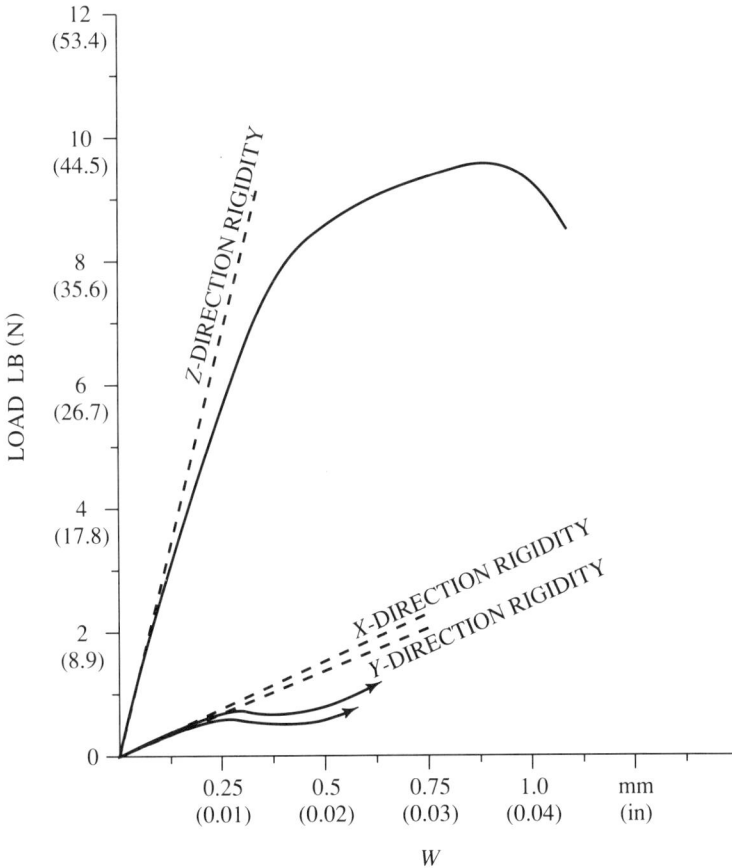

FIGURE 3.19. Load versus deflection curves for a pair of J-leads (from Engel et al. [19]).

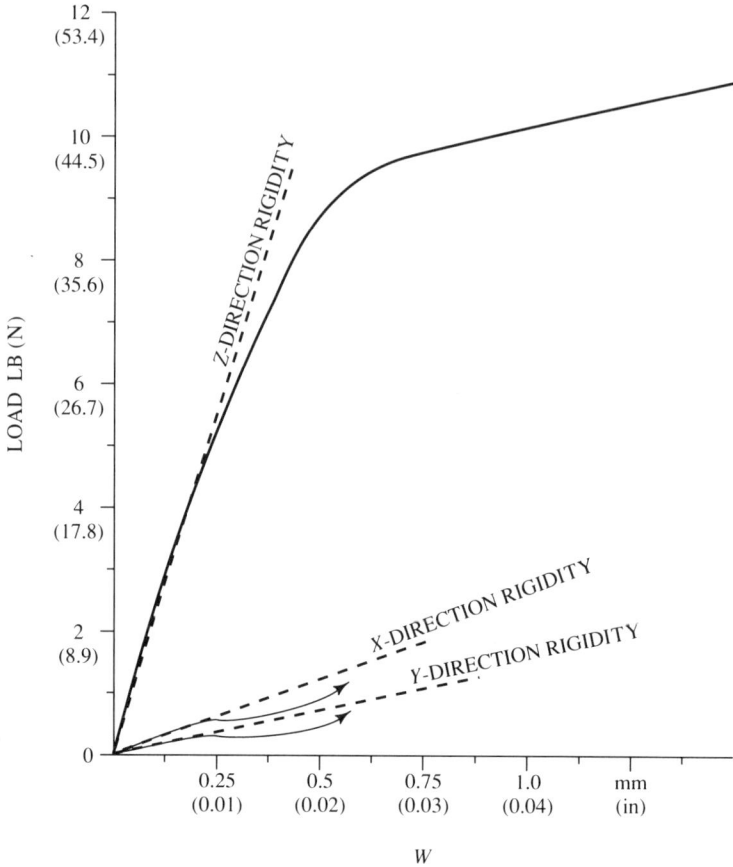

FIGURE 3.20. Load versus deflection curves for a pair of gull wing leads (from Engel et al. [19]).

fixed boundary conditions resulted on the SOJ modules above and the card below. The Instron test speed was 1 mm/min. Some x- and y-directional data are also included in Table 3.5. Graphs of load versus deflection relations are shown in Fig. 3.19 for J-leads and in Fig. 3.20 for gull wings. In both cases, the z-directional stiffness is almost an order of magnitude higher than the other two stiffnesses. Thus, in the typically displacement-controlled loading modes (involving thermal, handling, and vibration effects) causing lead stresses, the springiness of the leads in the z-direction will be extremely significant.

6. Solder Strength

A particular type of solder is often selected on the basis of a special property, such as electrical characteristics, density, thermal expansion, or fatigue resistance. Soldering methods may also be varied to achieve a desirable quality. A coarser grain structure, resulting from rapid quenching, allows less creep; fine-grained solders, on the other hand, have better fatigue resistance. The physical size of a solder joint (the "size effect") is significant: bulk specimens generally possess much less unit strength than joint-sized ones. Solders, although composed of metals, are far from being linearly elastic; at ordinary temperatures their "modulus" must be interpreted in some idealized or average sense (e.g., tangent or secant modulus).

Sn63/Pb37 eutectic solder is used in soldering leads to circuit cards more often than any other. Lau and Rice [21] collected much useful data on a close-to-eutectic configuration, Sn60/Pb40, and many other solders. The basic difference between Sn60/Pb40 and Sn63/Pb37 solders is that the former is more agreeable to plastic deformation, while the latter has more creep

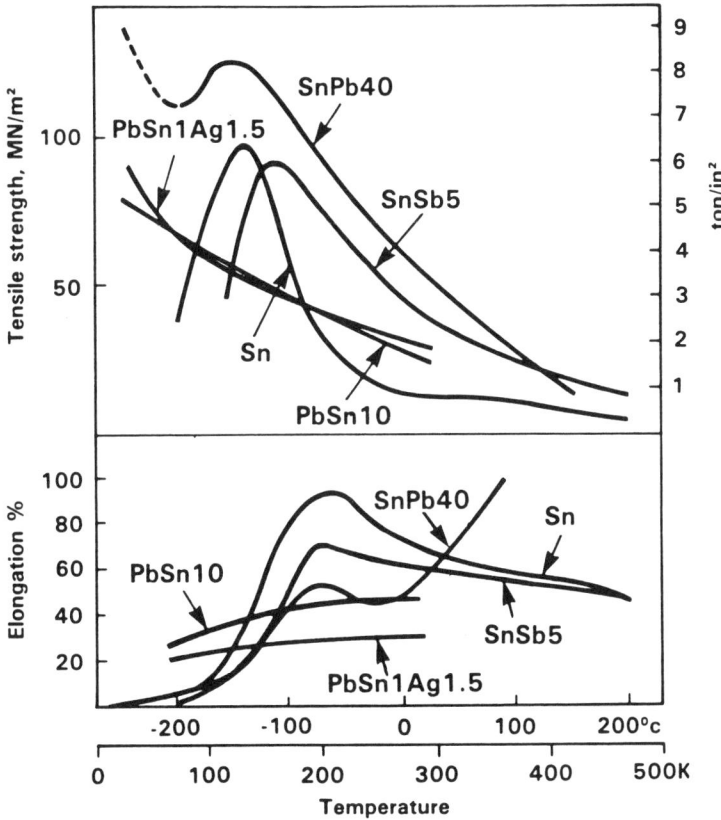

FIGURE 3.21. Influence of temperature on the tensile properties of some bulk solders (from Lau and Rice [21]).

resistance. Figure 3.21 shows the influence of temperature on the tensile properties of solders. The modulus versus temperature and frequency relationship of bulk 60Sn/40Pb samples is shown in Fig. 3.22. Tensile and shear strengths are shown in Figs. 3.23 and 3.24, respectively. Creep rupture

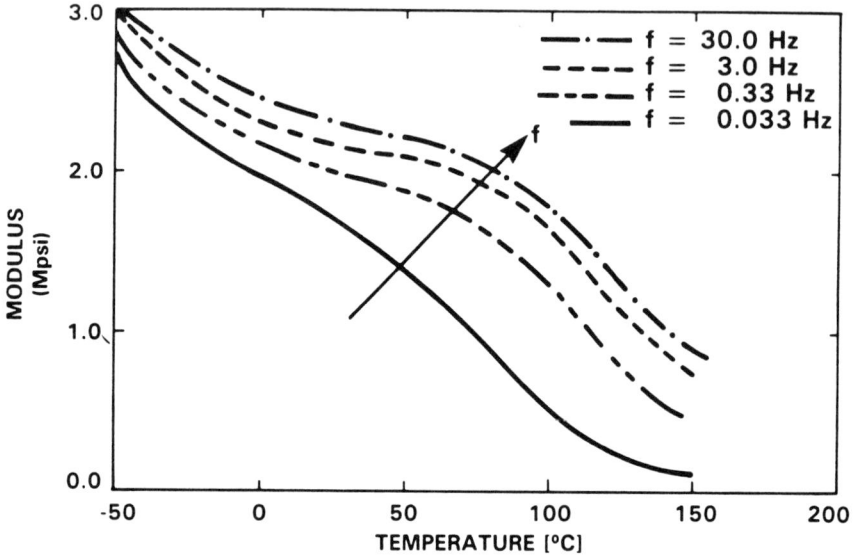

FIGURE 3.22. Modulus changes for bulk 60Sn-40Pb alloys as a function of temperature and frequency (from Lau and Rice [21]).

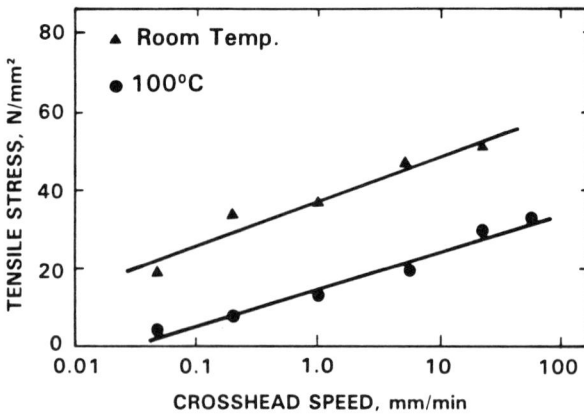

FIGURE 3.23. Tensile strength of bulk 60Sn-40Pb alloy tested at 20° and 100°C (source: International Tin Research Institute).

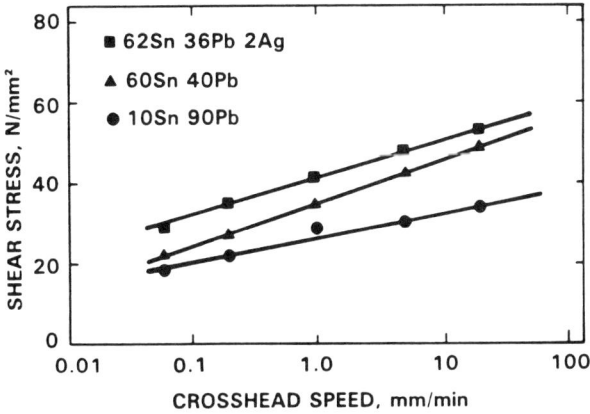

FIGURE 3.24. Shear strength of joints made with three solder alloys and tested at 20 °C (source: International Tin Research Institute).

strengths generally decline with temperature (Fig. 3.25), for both aged and unaged specimens. The same is true for fatigue strengths, as depicted in Fig. 3.26.

It is noted that a definition for the onset of fatigue of a solder joint can be made from either a mechanical or electrical point of view; a common

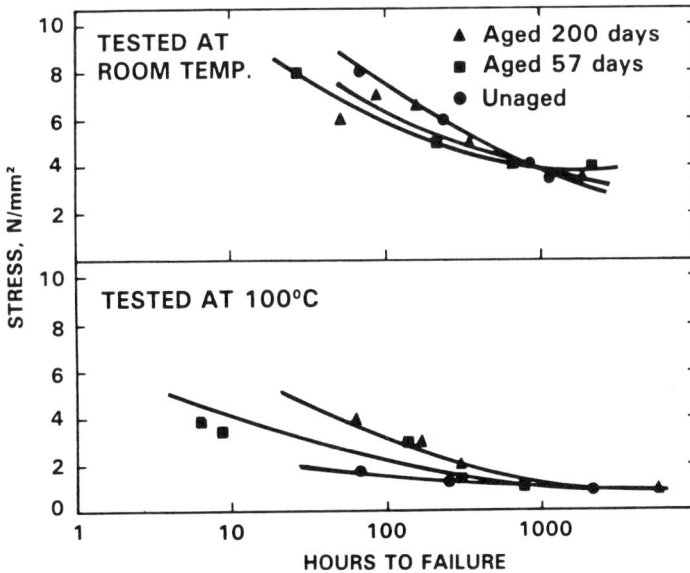

FIGURE 3.25. Creep rupture strength of joints soldered with 60Sn-40Pb alloy showing the effect of 57 and 200 days aging at 170 °C (source: International Tin Research Institute).

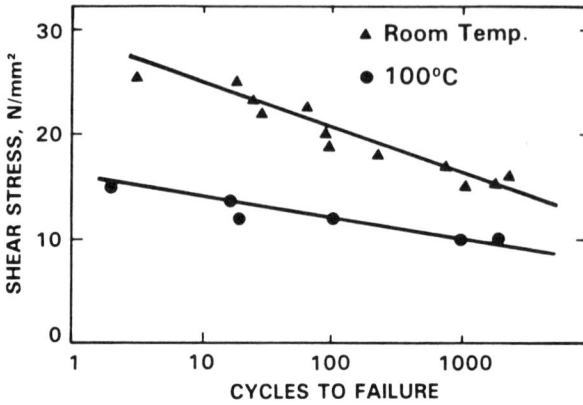

FIGURE 3.26. Fatigue strength of joints soldered with 60Sn-40Pb alloy and tested at 20° and 100 °C (source: International Tin Research Institute).

experience is apparent continued electrical functioning well beyond the full-grown crack stage. Some workers define fatigue as a specific rise of electrical resistance [22] (e.g., 0.02 Ω or 10%). Another electrical failure criterion could be the detection of a brief (e.g., 10 µs) open. A mechanical definition of solder joint fatigue may be identified with a given drop (e.g., 50%) of force amplitude in a constant displacement-amplitude test [23]; the latter is related to crack growth.

Long-term loading greatly diminishes the strength of a solder joint through creep. Thus the tensile design strength of a eutectic Sn/Pb solder under permanent load may typically be only 5% of the instantaneous strength, $\sigma_u = 55$ MPa (8000 psi), the corresponding shear strength being $\tau_u = 38$ MPa (5500 psi).

A fundamental property of solders is their great ability to stress relax under constant displacement. The master curve procedure for time-temperature shifting of Sn/Pb solders has been found valid by Baker [24].

Mechanical strength expressions for surface-mount solder joints were formulated by Chen [25]. He gave the pull strength of a butt lead joint (Fig. 3.27) as

$$F_p = (1 - M)(W_L L_L)\sigma_u + (2[1 - M]W_L + L_L)H_F\tau_u \qquad (3.7)$$

where M = percentage of component misalignment and H_F = solder fillet height.

The maximum shear strength of a similar butt joint is accordingly

$$\tau = G_s(\Delta L_T/H_s)(C_L/C_s) \qquad (3.8)$$

where G_s is the shear modulus of solder, ΔL_T the elastic displacement of the lead when subjected to shear force in the transverse direction, C_L the stiffness of the lead material, and C_s the stiffness of solder.

A thermal fatigue model of SMT solder joints is discussed in Chapter 11.

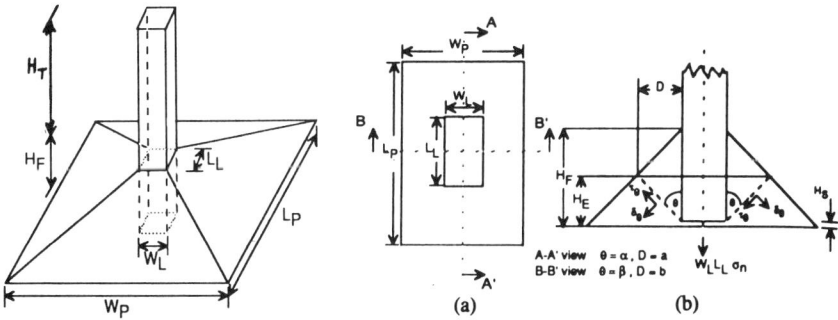

FIGURE 3.27. Pictorial illustration of a solder fillet (from Chen [25]).

References

1. Seraphim, D.P., Lasky, R.C., and Li, C-Y. (1989), *Principles of Electronic Packaging*, McGraw-Hill, New York.
2. Tummala, R.R., and Rymaszewski, E.J. (1989), *Microelectronics Packaging Handbook*, Van Nostrand, New York.
3. Chen, W.T., Raski, J.Z., Young, J.R., and Jung, D.Y. (1991), "A Fundamental Study of the Tape Automated Bonding Process," *ASME J. Elec. Packag.*, 113(3), 216–225.
4. Engel, P.A., Lim, C.K., Toda, M.D., and Gjone, R. (1984), "Thermal Stress Analysis of Soldered Pin Connectors for Complex Electronics Modules," *Computers in Mech. Eng.*, 2, 59–69.
5. Manko, H.H. (1979), *Solders and Soldering*, 2d ed., McGraw-Hill, New York.
6. Jahsman, W.E. (1989), "Leadframe and Wire Length Limitations to Bond Densification," *ASME J. Elec. Packag.*, 111(4), 289–293.
7. Motorola, Inc., Design Information Manuals.
8. Seely, F.B., and Smith, J.O. (1952), *Advanced Mechanics of Materials*, 2d ed., Wiley, New York, pp. 266–277.
9. Daniel, I.M., Wang, T.M., and Gotro, J.T. (1990), "Thermomechanical Behavior of Multilayer Structures in Microelectronics," *ASME J. Elec. Packag.*, 112(1), 11–15.
10. Johnson, E.A., and Seraphim, D.P. (1986), "Fabricating Advanced Circuit Boards," *Mech. Eng.*, 108(10), pp. 58–61.
11. Klimpl, F.E. (1992), "A New Process for Producing High-Performance Laminate," 35th Annual IPC Meeting, Bal Harbour, Fla.
12. Dasgupta, A., and Bhandarkar, S.M. (1992), "Effective Thermomechanical Behavior of Plain-Weave Fabric Reinforced Composites Using Homogenization Theory," *ASME J. Mater. and Technol.*, to be published.
13. Dasgupta, A., Bhandarkar, S.M., Pecht, M., and Barker, D.B. (1991), "Thermoelastic Properties of Woven-Fabric Composites Using Homogenization Thechniques," *Proc. 5th Tech. Conf. of the American Soc. for Composites*, pp. 1001–1010.
14. Agarwal, R.K., Dasgupta, A., Pecht, M., and Barker, D.B. (1991), "Prediction of PWB/PCB Thermal Conductivity," *Int'l J. for Hybrid Microelec.*, 14(3), 83–95.
15. Pecht, M., Wu, B-C., and Jennings, D. (1992), "Conductive Filament Formation in Printed Wiring Boards," *Proc. Int'l Electronics Manufac. Tech. Symp.*

16. Kelly, J.H., Lim, C.K., and Chen, W.T. (1984), "Optimization of Interactions Between Packaging Levels," *IBM Res. Dev.*, **28**(6), 719–726.
17. Engel, P.A., and Lee, L-C. (1986), "Surface Solder Stress Design for Thermal Loading of Modules," *Proc. NEPCON East*, Boston, Mass., pp. 263–270.
18. Engel, P.A. (1986), "Torque Stress Analysis for Printed Circuit Boards Carrying Peripherally Leaded Modules," ASME WAM, Anaheim, Calif. Paper No. 86-WA-EEP-2.
19. Engel, P.A., Caletka, D.V., and Palmer, M.R. (1991), "Stiffness and Fatigue Study for Surface Mounted Module/Lead/Card Systems," *ASME J. Elec. Packag.*, **113**(2), 129–137.
20. Barker, D.B., Sharif, I., Dasgupta, A., and Pecht, M.G. (1991), "Effect of SMC Lead Dimensional Variabilities on Lead Compliance and Solder Joint Fatigue Life," *ASME J. Elec. Packag.*, **114**(2), 177–184.
21. Lau, J.H., and Rice, D.W. (1985), "Solder Joint Fatigue in Surface Mount Technology: State of the Art," *Solid State Technol.*, **28**, 91–104.
22. Wild, R.N. (1988), "Some Factors Affecting Leadless Chip Carrier Solder Joint Fatigue Life II," *Circuit World*, **14**(4), 29–42.
23. Solomon, H.D. (1986), "Fatigue of 60/40 Solder," *IEEE Trans.*, **CHMT-9**(4), 423–432.
24. Baker, E. (1979), "Stress Relaxation of Tin-Lead Solders," *Mater. Sci. Eng.*, **38**, 241–247.
25. Chen, W. (1990), "Analyzing the Mechanical Strength of SMT Attached Solder Joints," *IEEE Trans.*, **CHMT-13**(3), 553–558.

Chapter 4

Leadless Chip Carriers

1. Loads and Materials

Leadless attachment of modules (chip carriers and various components) to printed circuit cards can be achieved with a surface solder joint. This is the simplest of the three customary methods of attachment: leadless, pin-in-hole, and leaded. While leadless peripheral mounts are efficient in "real estate" and fabrication, they have little flexibility; those thermal and mechanical stress problems we shall survey throughout this book are magnified here.

A good sign of a rugged module-to-card attachment is the survival of 1000 thermal test cycles between $-55°$ and $+125°$C, as per specification MIL STD 883; Wild [1] and Lake and Wild [2] discuss two more cycling types (Fig. 4.1), which could be even more demanding to solder for their longer time-duration abetting creep. Given the coefficients of thermal expansion (CTE) of a ceramic module and an epoxy glass (FR-4) card, Sn/Pb solders cannot carry the shear strain calculated under ordinary parameters. Thus for a 0.25-mm (0.01-in.) high solder joint, at a 10-mm (0.4-in.) distance from the module center, Eq. (1.53) yields a reflow shear strain unsupportable by the strength values listed in Table 3.1:

$$\gamma = (\Delta\alpha)(\Delta T)r/h = (10 \times 10^{-6})(180)(10)/0.25 = 0.072 \ .$$

Leadless mounts must, in the meantime, support much more severe handling (flexing) stresses [3] than comparable leaded ones (Fig. 4.2). The radii of curvature of the bent card and the stiffness of the module and card determine the z-directional stress; corner leads get the highest stress. Flexural analysis will be pursued in Chapters 7 to 10.

Solutions of the general thermal-mismatch problem between circuit card and module can be categorized as follows (mostly ceramic modules are considered):

1. **Restrained expansion.** Cards are attached to low-CTE (e.g., Cu-Invar) cores, which restrain their expansion. Hammond [4] describes an avionics application with polyimide multilayer boards stabilized by copper-clad

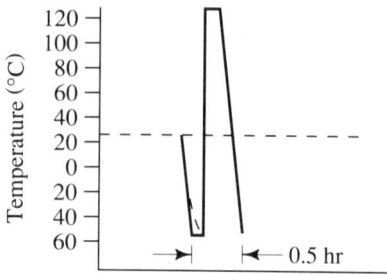

FIGURE 4.1. Thermal cycling profiles for leadless chip carriers (from Lake and Wild [2]).

a) Thermal Shock Cycle

 ○ Per MIL-STD 883
 ○ −55°C to 125°C (ΔT = 180°C)
 ○ 1/2 hr cycle, 15 min dwell

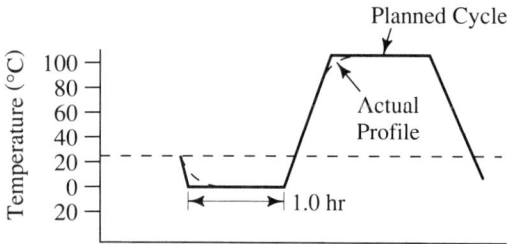

b) Shipboard Type Cycle

 ○ 0° to 105°C (ΔT = 105°C)
 ○ 3 hr cycle, 1.0 hr
  dwell, ~5° to 7°C ΔT/min

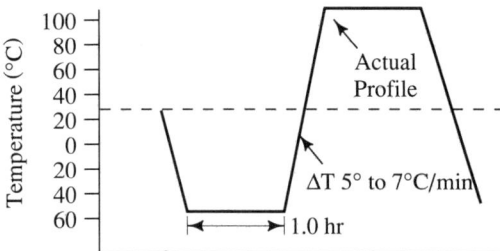

c) Avionics Type Cycle

 ○ −55°C to 105°C (ΔT = 160°C)
 ○ 3 hr cycle, 1.0 hr
  dwell, ~5°C to 7°C ΔT/min

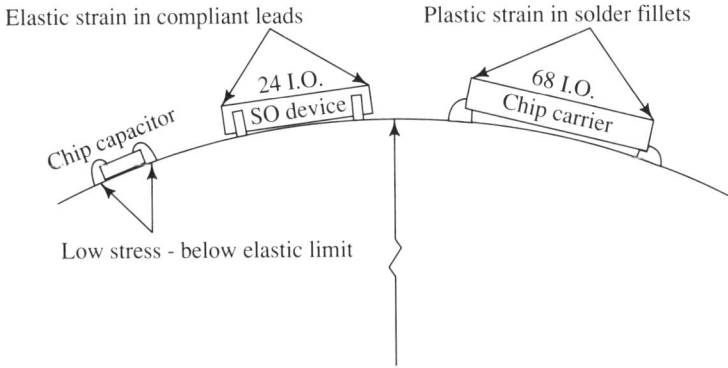

FIGURE 4.2. Solder joints of various size modules stressed by flexing the supporting circuit card (from Brierley and McCarthy [3]).

invar. He found, however, that toward the edges of the board the expansion restraint was not effective. Wild [1] observes that restrained xy-expansion tends to exacerbate barrel stresses for plated-through holes.

2. **Low-expansion cards.** Cu-Inv-Cu applications have been described in Section 3.2.
3. **Conformal coatings.** Wild [1] has used 25-μm (1-mil) epoxy Kevlar coatings, improving fatigue life.
4. **Leaded joints.** Compliant leads may reduce solder stress into the predominantly elastic range, where fatigue life is greatly increased (see Chapter 11).

While the preceding remedies can reduce differential expansion in the steady state, power cycles will always induce some transients by uneven heating. In addition, thermal expansion stresses arise between solder (with a CTE around 24 ppm/°C) and the joined members, especially ceramic modules, resulting in solder cracking. Several mathematical solutions of this mismatch problem are known [5, 6].

Faster loading rates produce less solder fatigue damage (see Chapter 3); for this reason, the thermal shock cycle of Fig. 4.1a was not found as damaging as those of Figs. 4.1b and 4.1c. The fatigue of a solder joint is also strongly connected with its metallurgy and shape.

2. Thermal Stress Analysis

Hall and co-workers [7] used two kinds of experimental measurement techniques for thermal strains: strain gauges attached to module and card, and holographic interferometry. They then proved that not only was in-plane shear γ of a solder post significant, but so was its bending (out-of-plane rotation) ϕ and its stretch (out-of-plane displacement), ε_z (Fig. 4.3). With

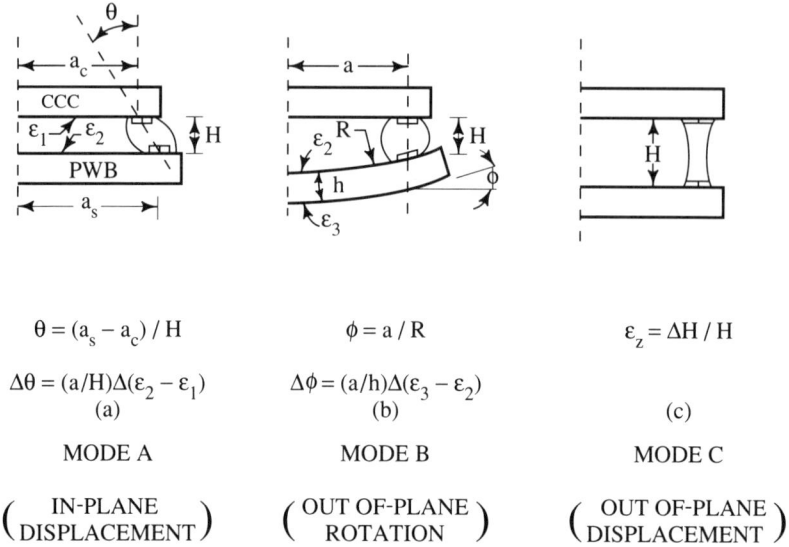

$$\theta = (a_s - a_c) / H \qquad \phi = a / R \qquad \varepsilon_z = \Delta H / H$$

$$\Delta\theta = (a/H)\Delta(\varepsilon_2 - \varepsilon_1) \qquad \Delta\phi = (a/h)\Delta(\varepsilon_3 - \varepsilon_2)$$

$$(a) \qquad\qquad (b) \qquad\qquad (c)$$

MODE A   MODE B   MODE C

$$\begin{pmatrix} \text{IN-PLANE} \\ \text{DISPLACEMENT} \end{pmatrix} \begin{pmatrix} \text{OUT OF-PLANE} \\ \text{ROTATION} \end{pmatrix} \begin{pmatrix} \text{OUT OF-PLANE} \\ \text{DISPLACEMENT} \end{pmatrix}$$

FIGURE 4.3. Three deformation modes for a solder post (from Hall et al. [7], © 1983 IEEE).

strain gauges attached to the top and bottom of a ceramic module (ε_1 and ε_2) and to the top and bottom of a card (ε_3 and ε_4) (Fig. 4.4), γ and ϕ could be measured, and related to the moments M and shears V arising in the solder joints. Having done this, Hall could experimentally show the role of the solder joint as a function of temperature [8].

In Ref. [8] a square module of size a is idealized as a polar symmetrical one (Figs. 4.5a and 4.5b). The advantage is that elastic plate displacements due to

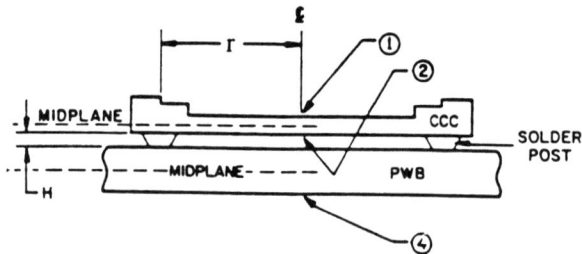

FIGURE 4.4. Strain gauge arrangement for leàdless chip carrier and circuit card (from Hall [8], © 1984 IEEE).

plate moments are readily available. A circular equivalent module of the same area has a radius $r = a/\pi^{1/2}$ and solder joint spacing s. Its curvature is calculated from the radial plate moment acting on the module, $m_m = (M_m + V_m h_m/2)/s$:

$$\frac{1}{R_m} = \frac{12(1 - v_m)}{E_m h_m^3} m_m = \frac{12(1 - v_m)}{E_m h_m^3 s}\left(M_m + \frac{V_m h_m}{2}\right) \tag{4.1}$$

whereas for the card, an infinite plate, subjected to plate moment $m_c = (M_c + V_c h_c/2)/s$ at the radius r:

$$\frac{1}{R_c} = \frac{6(1 - v_c^2)}{E_c h_c^3} m_c = \frac{6(1 - v_c^2)}{E_c h_c^3 s}\left(M_c + \frac{V_c h_c}{2}\right). \tag{4.2}$$

The latter is easily derived from adding two moments: m_1, due to a circular plate of radius r [Eq. (4.1)], and the other, m_2, due to the bending of an infinite card with a hole r cut out of the middle:

$$m_1 = \frac{E_c h_c^3}{12(1 - v_c)}\frac{1}{R_c}; \qquad m_2 = \frac{E_c h_c^3}{12(1 + v_c)}\frac{1}{R_c}. \tag{4.3}$$

By strain gauge measurements, one gets in the meantime (Fig. 4.4):

$$1/R_m = (\varepsilon_2 - \varepsilon_1)/r \tag{4.4}$$

and

$$1/R_c = (\varepsilon_4 - \varepsilon_3)/r. \tag{4.5}$$

In order to solve for the shear force on an individual solder post, we note that, from equilibrium (Fig. 4.6):

$$V_m = V_c = V \tag{4.6}$$

and

$$M_m + M_c = V \cdot H. \tag{4.7}$$

We then obtain the solder shear force by equating the curvatures [Eqs. (4.1) and (4.2)] with the measured values in Eqs. (4.4) and (4.5):

$$V = \frac{s}{12\left[H + \dfrac{h_m}{2} + \dfrac{h_c}{2}\right]}\left\{\frac{E_m h_m^2(\varepsilon_2 - \varepsilon_1)}{1 - v_m} + \frac{2E_c h_c^2(\varepsilon_4 - \varepsilon_3)}{1 - v_c^2}\right\}. \tag{4.8}$$

With this apparatus, Hall was able to plot the total hysteresis loop of a temperature cycle, beginning from $-25°$ to $+125°$C (Figs. 4.7a and 4.7b). He actually repeated the cycle, without getting significant distortion of the hysteresis loop. The adoption of a "zero" bend angle ϕ (Fig. 4.7a) corresponding to high temperatures (over $120°$C) was made since here solder is very pliable and exerts negligible force on the module and card.

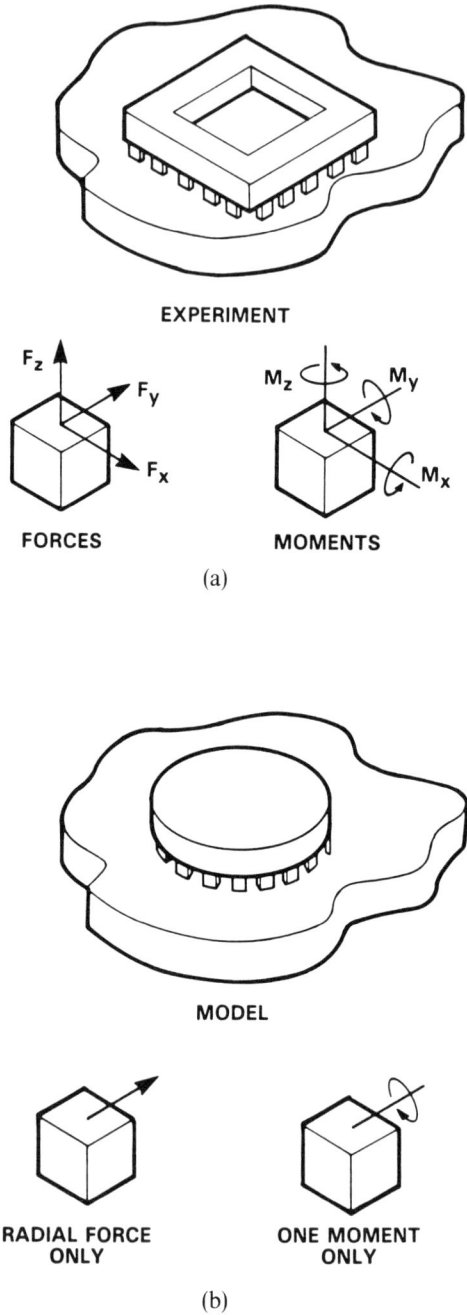

FIGURE 4.5. Forces and moments arising from thermal loading acting on solder posts (from Hall [8], © 1984 IEEE). (a) Real (square) chip carrier; (b) idealized (equivalent circular) chip carrier.

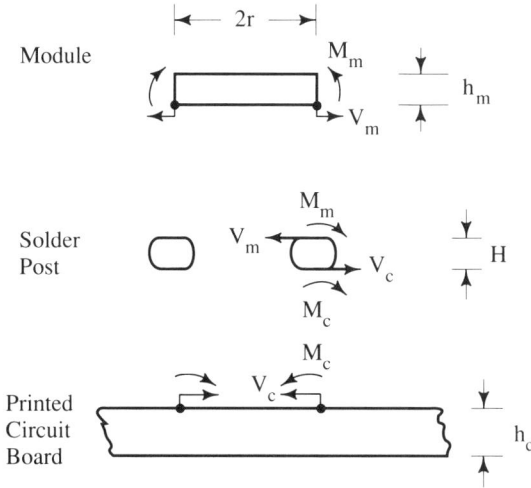

FIGURE 4.6. Free-body diagrams of module, solder post, and circuit card.

Coming up from low temperature, the ϕ versus T curve of Fig. 4.7a starts straight. As it rises, it corresponds quite well to bimetal thermostat behavior (Chapter 1, Section 4.2), averaging the slope of the module and card. In this treatment, the standoff H may be simply added to the thermostat thickness, which becomes $h_m + h_c + H$. As a matter of fact, Eqs. (4.1) and (4.2) plus the bimetal thermostat equation (1.63) suffice to derive V entirely without experiment. We now get, by stipulating $1/R = 1/R_m = 1/R_c$:

$$V = \frac{s/h\left[\dfrac{E_m h_m^3}{2(1 - v_m)} + \dfrac{E_c h_c^3}{1 - v_c^2}\right]}{H + (h_m + h_c)/2} \frac{(1 + m)^2 (\alpha_c - \alpha_m)\,\varDelta T}{3(1 + m)^2 + (1 + my)(m^2 + 1/my)} = C\varDelta\alpha\varDelta T$$

(4.9)

which, using the experimental geometry of Ref. [8]:

$$\alpha_m = 4.6 \times 10^{-6}/°\text{C}, \qquad H = 0.28 \text{ mm}, \qquad E_m = 255 \text{ GPa}, \qquad h = 2.26 \text{ mm}$$

$$\alpha_c = 18.4\,°\text{C}, \qquad h_m = 0.51 \text{ mm}, \qquad v_m = 0.3, \qquad m = h_c/h_m = \frac{1.47}{0.51} = 2.88$$

$$s = 0.635 \text{ mm}, \qquad h_c = 1.47 \text{ mm}, \qquad E_c = 10.9 \text{ GPa}, \qquad y = \frac{E_c}{E_m} = 0.0427$$

$$v_c = 0.28$$

results in good agreement with measurements of $V \approx 0.01\ \varDelta T$ (lbs).

The concave-from-below curvatures (Fig. 4.7a) of cold temperatures give way to straightness, a lack of bending, at about 40 °C. Beyond this, the curvature changes to convex, and convexity increases for a while until

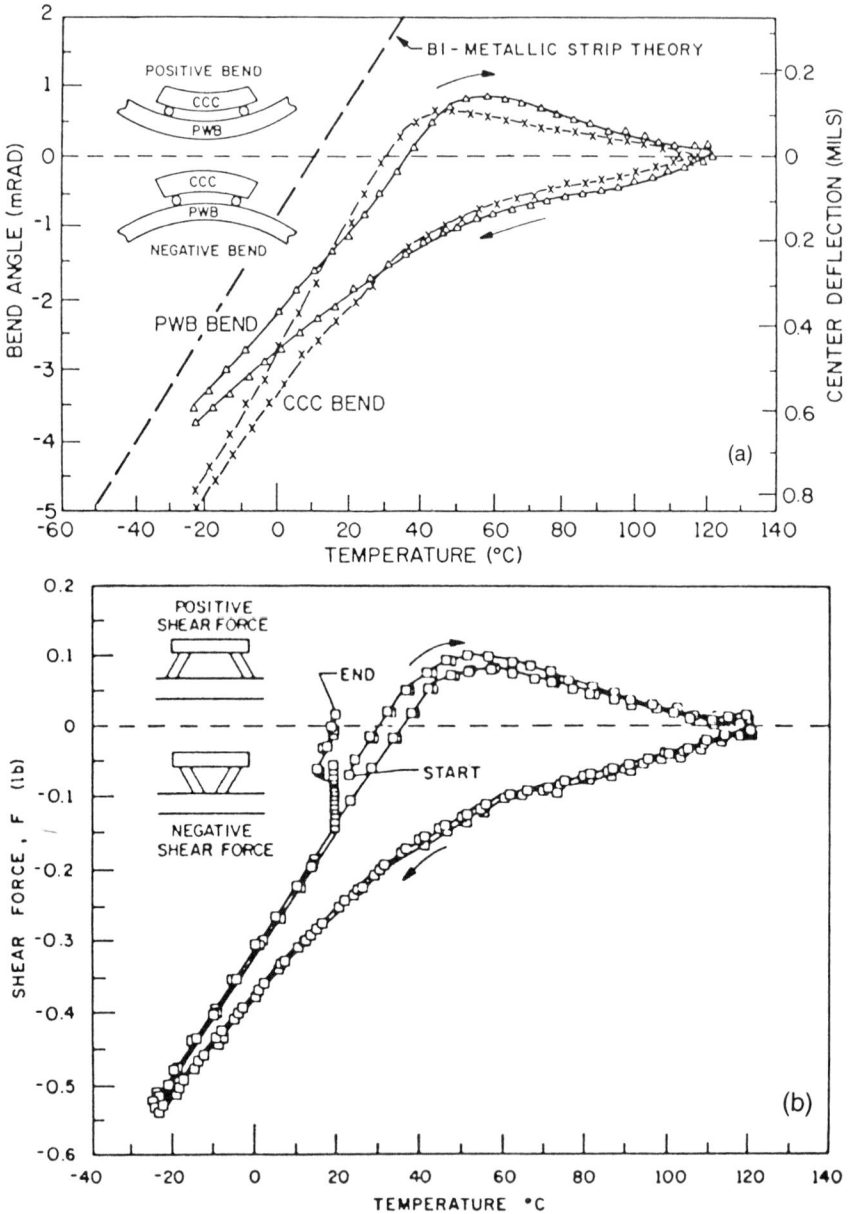

FIGURE 4.7. Hall's measurement of leadless chip carrier deformation and solder forces from thermal cycling (from Hall [8], © 1984 IEEE). (a) Bend angle and central deflection versus temperature; (b) solder post shear force versus temperature.

the solder softens enough; at that point, it transmits no more force and the curvature disappears. Reversing the temperature to cooler now, the assembly has a concave shape throughout the region. The V versus ϕ curve stiffens at 40 °C, and an essentially elastic solder behavior is seen.

This investigation showed that there was appreciable flexibility in the module; the module bent with the card throughout most of the temperature range, except at the coldest temperatures where there was bimetallic behavior. At higher temperatures, solder shear displacement was the dominant response. The midplanes of the module and card moved freely, as if subject to thermal expansion alone. For the z-directional action, the holographic technique showed large corner post forces (compressive at $T > 40\,°C$) tending to balance out the rest of the (tensile) solder forces.

An equivalent but more detailed solution for two-dimensional beam geometries with interconnecting solder posts was derived by Goldmann [10].

3. The Influence of Solder Joint Shape

The preceding analysis considered rectangular solder posts with averaged stresses acting upon it at its faces. In reality, leadless solder mounts have variegated geometric and shape parameters; some of these have great influence upon the failure mechanism, crack formation, and fatigue life of the joint.

Investigating the propagation of a crack underneath a ceramic module, Wild [1] recommends a bulbous solder joint; a crack, propagating from the inside past the module corner, would then take longer to reach the fillet surface. Solomon [11] describes the crack propagation process with both load drop and electrical resistance change ΔR (Fig. 4.8).

Sherry et al. [12] studied three specific solder joint shapes by finite element techniques; in Figs. 4.9a–4.9c they are called D-, G-, and B-joints, respectively. In the plane strain two-dimensional analysis, they allowed solder to deform plastically, but the ceramic on top and a copper layer below would be considered in the elastic range; this was later justified by the computational results. They performed solder shearing load tests at strain rates of 10^{-2}, 10^{-3}, 10^{-4}, and $10^{-5}\,s^{-1}$ to provide solder data for the finite element analysis. Figures 4.10a–4.10c show a sequence of strain distribution in a D-joint, upon increasing the applied shear displacement. The D-joint was found least satisfactory, developing highest strains at the solder/ceramic

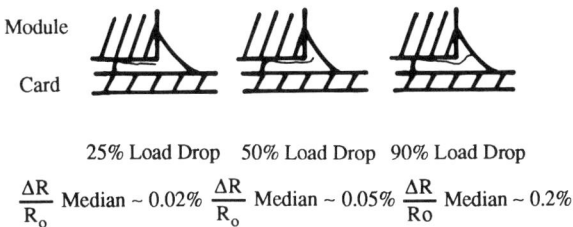

Module

Card

25% Load Drop   50% Load Drop   90% Load Drop

$\dfrac{\Delta R}{R_o}$ Median ~ 0.02% $\dfrac{\Delta R}{R_o}$ Median ~ 0.05% $\dfrac{\Delta R}{Ro}$ Median ~ 0.2%

FIGURE 4.8. Sequence of solder joint failure (after Solomon [11], © 1989 IEEE).

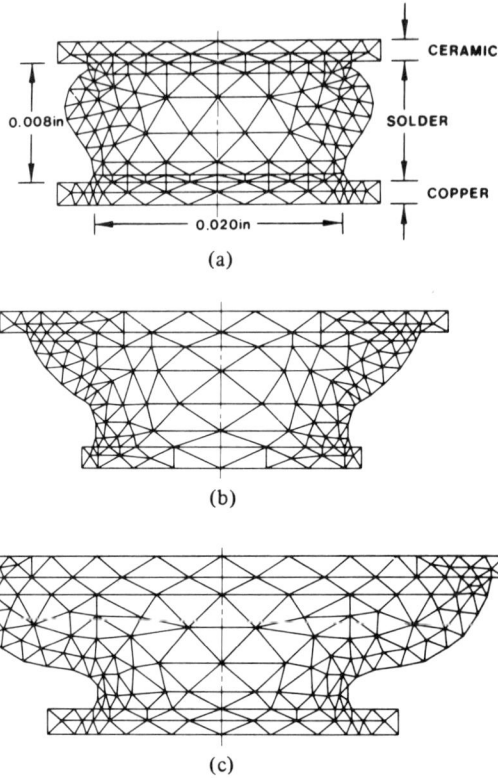

FIGURE 4.9. Finite element maps derived from longitudinal metallographic cross sections of solder joints. (a) D-joint; (b) G-joint; (c) B-joint. Maps include ceramic, solder, and copper elements of joint (from Sherry et al. [12], © 1985 IEEE).

interface; both other solder joint models tended to fail at the solder/copper interface.

Shah [13] used two-dimensional elastic finite element analysis to investigate the effect of several geometric parameters upon the stress in a solder joint connecting a ceramic chip capacitor and a multilayer board; for the latter, glass epoxy (FR-4 or G-10), glass-reinforced polyimide (G-30), and alumina were considered, and general laminate theory [14] was included in the FE analysis. Figure 4.11 shows the geometric parameters varied, among them: standoff height of solder; height of the solder fillet; ceramic rotation; and chip translation. It was seen that the standoff height (T1 in Fig. 4.11) of solder between the ceramic termination cap and the board mounting pad had the most influence on decreasing solder stress values.

Among many studies on the influence of solder shape on thermal and stress response of leadless solder mounts, we note References [15–17].

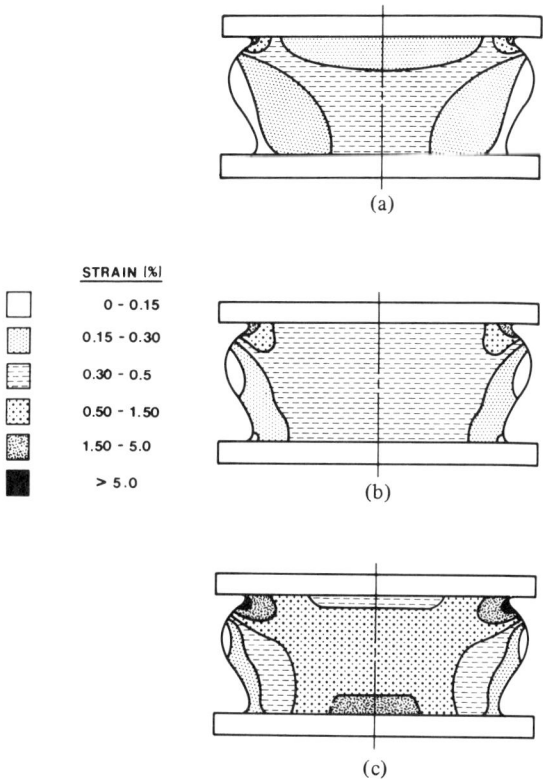

FIGURE 4.10. Finite element strain distributions in a D-joint as a function of the displacement. (a) 1.2 μm; (b) 1.6 μm; (c) 2.1 μm (from Sherry et al. [12], © 1985 IEEE).

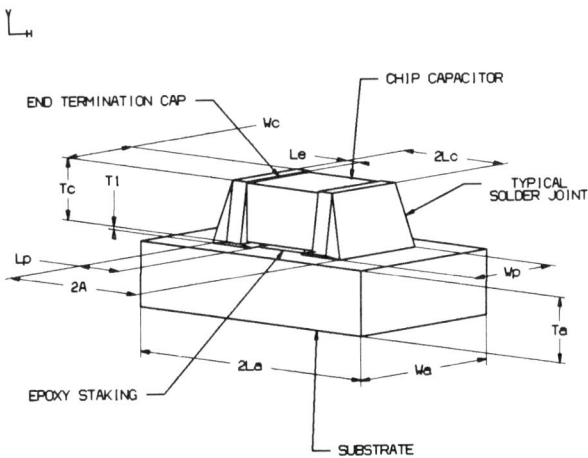

FIGURE 4.11. Leadless chip capacitor mounting configuration (from Shah [13]).

4. Constitutive Equation for Solder Mount

Hall [9] modeled the behavior of a thermomechanically loaded compliant-leaded solder joint. At any temperature, the solder undergoes simultaneous stress relaxation and creep, a process he called "stress reduction." Figure 4.12 shows the schematic of a solder joint creeping under constant weight (a) and relaxing under constant displacement (b). In real compliant leaded solder joints, the interaction of the card and the module supplies the elastic spring element k (Fig. 4.12c) and the total shear strain of the solder is the angle γ_t. The elastic shearing spring force V is written

$$V = k(x - x_0) . \tag{4.10}$$

The spring length x under load is written in terms of the solid element L, as yet unidentified in terms of our structure:

$$x = L - H\gamma_t \tag{4.11}$$

and we choose $x_0 = L_0, \gamma_t = 0$, corresponding to the spring position and shear angle, respectively, at $\Delta T = 0$. Since the "solid frame" linearly expands with the temperature,

$$L = L_0(1 + \Delta\alpha \cdot \Delta T) , \tag{4.12}$$

and we get the fundamental equation for the shear force:

$$V = k \cdot \Delta\alpha \cdot L_0 \cdot \Delta T - kH\gamma_t . \tag{4.13}$$

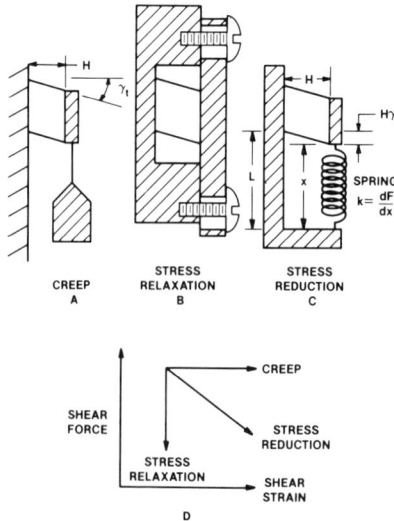

FIGURE 4.12. Idealized conditions of creep, stress relaxation, and stress reduction in shear (from Hall [9], © 1987 IEEE).

The time derivative of this system gives force and displacement rates:

$$\dot{V} = k \cdot \Delta\alpha \cdot L_0 \Delta\dot{T} - kH\dot{\gamma}_t . \tag{4.14}$$

For the special case of isothermal stress reduction, Fig. 4.13 illustrates the relationships graphically. It helps in identifying two constants: k and L_0, of the module/solder-joint/card system. For example, at cold temperatures $\Delta\gamma_t = 0$, and so the bimetallic equation (4.9) governs. Then the shear force increase for raising the temperature by ΔT is

$$\Delta V = C\Delta\alpha \cdot \Delta T , \tag{4.15}$$

and equating ΔV with the first term of Eq. (4.13), we get

$$k = C/L_0 . \tag{4.16}$$

Note that when we relax the solder at the new temperature $T_1 = T + \Delta T$ the isotherm of Fig. 4.13a takes us to point B, and so

$$\gamma_t = \Delta V/kH = \Delta\alpha \cdot \Delta T \cdot L_0/H . \tag{4.17}$$

Thus L_0 turns out to be the distance r between the neutral point of the module assembly and the solder joint.

To determine the plastic shear rate $\dot{\gamma}_p$ of the solder, we start from $\gamma_t = \gamma_e + \gamma_p$. Taking an effective shear stress as $\tau = V/A$, where A is the area of the joint, the elastic strain is

$$\gamma_e = V/GA \tag{4.18}$$

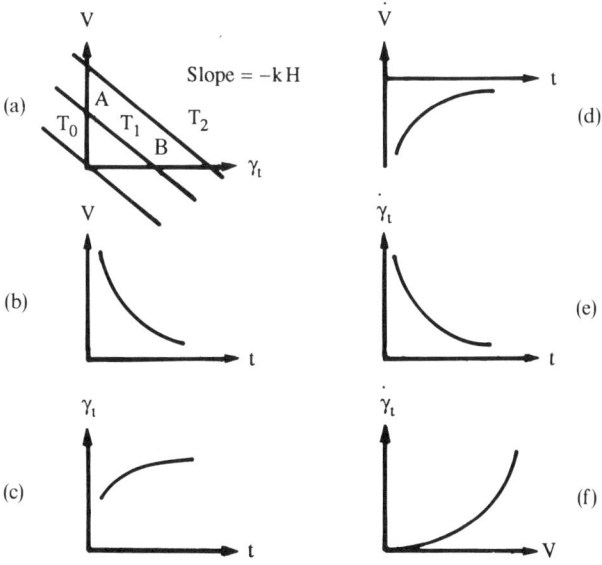

FIGURE 4.13. Isothermal stress reduction (from Hall [9], © 1987 IEEE).

so that Eq. (4.13) yields

$$V = kL_0 \cdot \Delta\alpha \cdot \Delta T - kH(\gamma_e + \gamma_p) = kL_0 \Delta\alpha \Delta T - kH\left(\frac{V}{GA} + \gamma_p\right)$$

and

$$V\left(1 + \frac{kH}{GA}\right) = kL_0 \Delta\alpha \cdot \Delta T - kH\gamma_p.$$

Taking time derivatives:

$$kH\dot{\gamma}_p = kL_0 \Delta\alpha \cdot \dot{T} - \dot{V}\left(1 + \frac{kH}{GA}\right),$$

and thus

$$\dot{V} = \frac{kL_0 \Delta\alpha \cdot \dot{T} - kH\dot{\gamma}_p}{1 + kH/GA}. \tag{4.19}$$

Also,

$$\dot{V}/GA = \dot{\gamma}_t - \dot{\gamma}_p,$$

and by further manipulation we get

$$\dot{\gamma}_p = \dot{\gamma}_t\left(1 + \frac{kH}{GA}\right) - \frac{kL_0 \cdot \Delta\alpha \cdot \dot{T}}{GA}. \tag{4.20}$$

Thus measuring $\dot{\gamma}_t$ and $\dot{T}$ in a particular experiment, we can obtain $\dot{\gamma}_p$, the sought plastic strain rate.

In isothermal stress reduction, $\dot{T} = 0$, Eq. (4.20) reduces to

$$\dot{V} = -C_1\dot{\gamma}_p \tag{4.21}$$

where

$$C_1 = kH/(1 + kH/GA). \tag{4.22}$$

Two extreme conditions may exist: for $k \ll GA/H$ ("creep condition"), $C_1 = kH$. On the other hand, for $k \gg GA/H$ ("stress relaxation condition"), $C_1 = GA$.

It is noted that in a purely isothermal test, under stress relaxation conditions, by Eq. (4.20) we get:

$$\dot{\gamma}_e = \dot{\gamma}_t - \left(\dot{\gamma}_t\left[1 + \frac{kH}{GA}\right]\right) \approx -\dot{\gamma}_t\frac{kH}{GA} \approx -\dot{\gamma}_p, \tag{4.23}$$

meaning that the elastic strain rate is opposite in sign and approximately equal in magnitude to the plastic strain rate.

5. Conclusions

Leadless attachments tend to be highly stressed when used in a thermally mismatched module/card assembly; various measures, such as "CTE tailoring" have been used to adapt them. Hall's experimental-analytical model, under some idealized (polar symmetrical) conditions, showed the influence of temperature on solder behavior, going from the bimetallic behavior of cold temperatures to unrestrained shear at high temperatures. Solder mount shape has great influence on failure modes. Hall further developed his model to determine constitutive laws for solder.

6. Exercises and Questions

1. Discuss the relative merits of short-duration (thermal "shock") versus long-duration thermal stress tests (Fig. 4.1) on the solder joints of a lead-less chip carrier.
2. Discuss the merits of an "equivalent" circular module idealization for the thermal stress analysis of a square chip carrier.
3. Using the data given under Eq. (4.9), calculate the shears, moments, and displacements of a leadless chip carrier at $\Delta T = 20\,^{\circ}\mathrm{C}$.
4. How would you calculate the constants k, L_0, H, G, and A of Eqs. (4.10) and (4.11) from the data of Exercise 3?

References

1. Wild, R.N. (1988), "Some Factors Affecting Leadless Chip Carrier Solder Joint Fatigue Life II," *Circuit World*, **14**(4), 29–41.
2. Lake, J.K., and Wild, R.N. (1983), "Some Factors Affecting Leadless Chip Carrier Solder Joint Fatigue Life," *Proc. 28th National SAMPE Symposium*, Anaheim, Calif.
3. Brierley, C.J., and McCarthy, J.P. (1986), "The Reliability of Surface Mounted Solder Joints Under PWB Cyclic Mechanical Stresses," *Circuit World*, **12**(3), 16–19.
4. Hammond, R.J. (1986), "Digital Electronic Engine Controls and Leadless Chip Carriers," *Proc. IEEE/AIAA 7th Digital Avionics Systems Conf.*, Fort Worth, Tex., pp. 625–632.
5. Robert, M., and Keer, L.M. (1987), "The Elastic Circular Cylinder with Displacement Prescribed at the Ends – Axially Symmetric Case," *J. Mech. Appl. Math.*, **40**, Pt. 3, 339–363.
6. Suhir, E. (1992), "Mechanical Behavior and Reliability of Solder Joint Interconnections in Thermally Matched Assemblies," *Proc. 42nd ECTC*, San Diego, Calif., pp. 563–572.
7. Hall, P.M., Dudderar, T.D., and Argyle, J.F. (1983), "Thermal Deformations Observed in Leadless Ceramic Chip Carriers Surface Mounted to Printed Wiring Boards," *IEEE Trans.*, **CHMT-6**(4), 544–552.
8. Hall, P.M. (1984), "Forces, Moments, and Displacements During Thermal Chamber Cycling of Leadless Ceramic Chip Carriers Soldered to Printed Wiring Boards," *IEEE Trans.*, **CHMT-7**(4), 314–326.
9. Hall, P.M. (1987), "Creep and Stress Relaxation in Solder Joints in Surface-Mounted Chip Carriers," *Proc. IEEE ECC Conf.*, pp. 579–588.

10. Jeannotte, D.A., Goldmann, L.S., and Howard, R.T. (1989), "Package Reliability," Ch. 5 in Tummala, R.R., and Rymaszewski, E.J., eds., *Microelectronics Packaging Handbook*, Van Nostrand, New York.

11. Solomon, H.D. (1989), "Low Cycle Fatigue of Surface-Mounted Chip Carrier/ Printed Wiring Board Joints," *IEEE Trans.*, **CHMT-12**(4), 473–479.

12. Sherry, W.M., Erich, J.S., Bartschat, M.K., and Prinz, F.B. (1985), "The Effect of Joint Design on the Thermal Fatigue Life of Leadless Chip Carrier Solder Joints," *IEEE Trans.*, **CHMT-8**(4), 417–424.

13. Shah, M.K. (1990), "Analysis of Parameters Influencing Stresses in the Solder Joints of Leadless Chip Capacitors," *ASME Trans. J. Elec. Packag.*, **112**(2), 147–153.

14. Tsai, S.W., and Hahn, H.T. (1980), *Introduction to Composite Materials*, Technomic Publishing Co., Westport, Conn.

15. Waller, D.L., Fox, L.R., and Hannemann, R.J. (1983), "Analysis of Surface Mount Thermal and Thermal Stress Performance," *IEEE Trans.*, **CHMT-6**(3), 257–266.

16. Charles, H.K., Jr., and Clatterbaugh, G.V. (1990), "Solder Joint Reliability – Design Implications from Finite Element Modeling and Experimental Testing," *ASME Trans. J. Elec. Packag.*, **112**(2), 135–146.

17. Cooley, W.T., and Razani, A. (1991), "Thermal Analysis of Surface Mounted Leadless Chip Carriers," *ASME Trans. J. Elec. Packag.*, **113**(2), 156–163.

Chapter 5

Thermal Stress in Pin-Grid Arrays: Primary Analysis of Pins

1. Introduction

A differential in-plane thermal expansion between a module and circuit card tends to induce flexural stress in the pins bridging them in the pin-grid array (PGA) structure. We shall treat the uniform temperature rise (or fall) ΔT of the assembly. It is understood from Eqs. (1.51) and (1.52) that the thermal strain terms $(\alpha_1 \Delta T_1 - \alpha_2 \Delta T_2)$ for nonuniform and $(\alpha_1 - \alpha_2)\Delta T$ for uniform temperature pose no difference to mathematical handling. Pins may be brazed or swaged to a ceramic module constituting a fixed joint on top (Fig. 3.1) [1]; on the bottom they are assumed to be soldered into a card hole [pin-in-hole (PIH) construction], making for an "elastic support." Figure 5.1

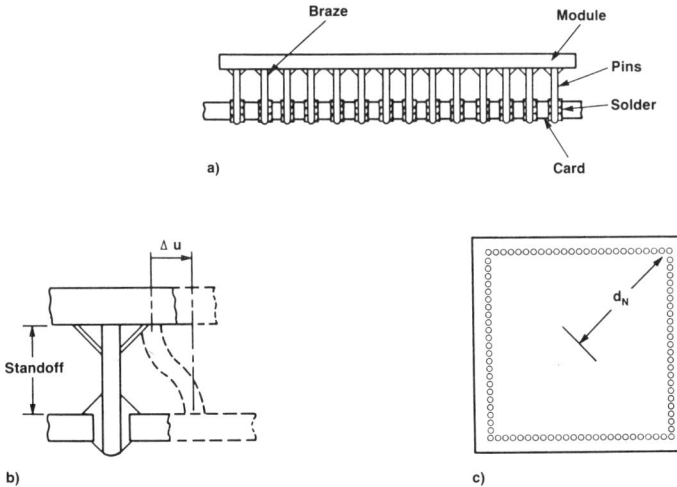

FIGURE 5.1. General arrangement of a pin grid array (PGA). (a) Cross section; (b) enlargement, showing deformation due to thermal mismatch Δu between module and card; (c) plan view of a pin arrangement (shown as peripheral, for simplicity).

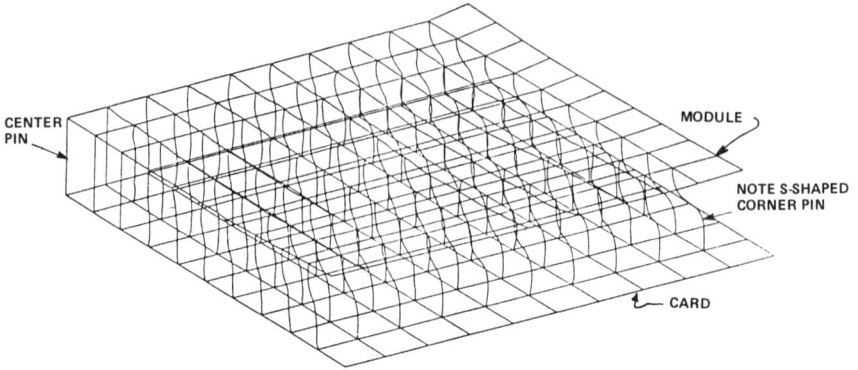

FIGURE 5.2. NASTRAN quarter model of a PGA deformed by thermal loading.

shows the physical arrangement with deformations typical of thermal loading. (For simplicity, only a peripheral row of pins is shown in Fig. 5.1c.)
Figure 5.2 is the doubly symmetrical thermal deformation pattern plotted by
a finite element (NASTRAN) program for a square module.

The displacements of the system may be modeled in polar symmetry
around a neutral point, as was done for leadless chip carriers in Chapter 4,
Section 2. Here, however, the pins enter as flexural elements bent in a radial
plane. If both ends of the pin were fixed and the module and card were
undeformable, then the bending moments M_1 and M_2 acting on a pin located
at a radial distance r from the neutral point would be the fixed-end moments
determined by

$$M_1 = M_2 = 6EIu/L^2 \tag{5.1}$$

where u is the thermal mismatch; for uniform temperature rise:

$$u = \Delta\alpha \cdot \Delta T \cdot r . \tag{5.2}$$

Equation (5.1) gives an overly simplified and conservative analysis
of the pin. Most importantly, solder is able to adjust to load by readily
deforming and acting as an elastic foundation, reducing stress. A substantially bent pin, held to a constant span length ("standoff") L, will develop
a tensile axial force F responding to an effective increment ΔL in length. In
addition, to render a more realistic description, we can include the plastic
deformation induced in the pin by bending. On the other hand, the module
and card are flexible plates that permit relieving some displacement
constraints imposed by an assumed fixity; this effect will be discussed in
Chapter 6.

2. Elastic Foundation Modulus of a Soldered Pin

The pin is embedded in solder, usually eutectic 63–37 SnPb, below the
standoff region (Fig. 5.3). The clearance space in the card hole is filled out

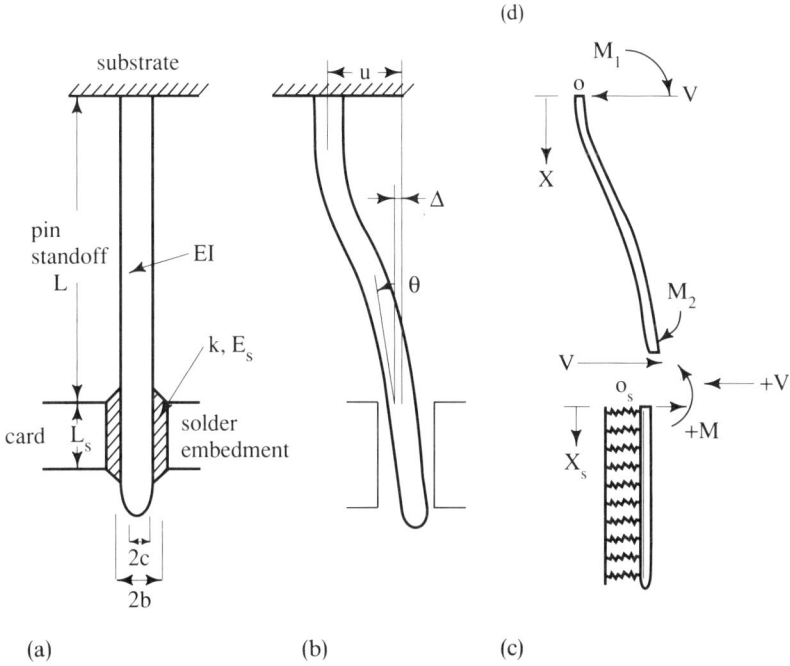

FIGURE 5.3. Model of a single pin. (a) Undeformed pin; (b) thermal displacement; (c) solder embedment modeled as a beam on elastic foundation; (d) pin standoff under end forces.

with solder by the "wave solder" manufacturing process, which relies on sucking up the molten metal by surface tension around the pins. This solder embedment, acting as an elastic foundation, allows the pin much greater flexibility than would a fixed joint.

We shall consider the foundation constant k for a circular beam of radius c and rigidity EI inserted concentrically in an elastic material filling out the clearance space between $r = c$ and the hole radius $r = b$. Assuming rigid card, rigid pin, and elastic solder with modulus E_s, Engel et al. [1] derived an approximate foundation modulus (Fig. 5.4):

$$k = 4E_s c/(b - c) . \tag{5.3}$$

Pursuing an application to buckling of embedded optical fibers, Vangheluwe [2] derived an expression valid for rigid hole walls using Muskhelishvili's complex variable approach [3]:

$$k = \frac{4\pi E_s (1 - v_s)(3 - 4v_s)}{(1 + v_s)\left[(3 - 4v_s)^2 \ln b/c - \dfrac{(b/c)^2 - 1}{(b/c)^2 + 1} \right]} . \tag{5.4}$$

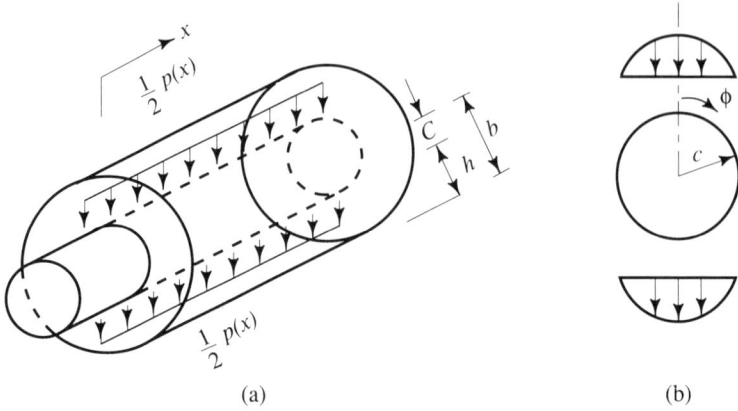

(a)   (b)

FIGURE 5.4. Assumptions made in deriving an approximate foundation constant and solder pressure analysis (from Engel et al. [1]). (a) Pressure on both sides of pin causing a deflection $\approx ph/2cE_s$; (b) assumed pressure distribution $\sigma_z \approx (p/2c)\cos\phi$ around the pin ($\sigma_{av} \approx 2\sigma_{max}/3$).

Suhir [4] refined Vangheluwe's optical fiber treatment, including nonrigid elastic behavior for the housing (E_c, v_c) as the hole wall material of external radius a, obtaining

$$k = \frac{4\pi E_s E_c (1 - v_s)(1 - v_c)}{C_{11}\beta_1 - C_{21}\beta_2 - C_{31}} \tag{5.5}$$

where

$$C_{11} = a^2 - c^2 - v_s(a^2 - b^2) - v_c(b^2 - c^2)$$

$$C_{12} = 8(1 - v_s)(1 - v_c)(b^4 + (3 - 4v_s)c^4$$

$$C_{13} = 8(1 - v_s)(1 - v_c)[b^4 + (3 - 4v_c)a^4]$$

$$C_{21} = E_s(1 - v_s)(1 + v_c)(a^2 - b^2)$$
$$\qquad + E_c(1 + v_s)(1 - v_c)(b^2 - c^2)$$

$$C_{22} = 8E_c(1 - v_s^2)(1 - v_c)(3 - 4v_s)(b^4 - c^4)$$

$$C_{23} = 8E_s(1 - v_s)(1 - v_c^2)(3 - 4v_c)(a^4 - b^4)$$

$$C_{31} = E_s(1 - v_s)(1 + v_c)(3 - 4v_c)\ln\frac{b}{a}$$

$$\qquad + E_c(1 + v_s)(1 - v_c)(3 - 4v_s)\ln\frac{c}{b}$$

$$C_{32} = 8E_c(1 - v_s^2)(1 - v_c)(b^2 - c^2)$$

$$C_{33} = 8E_s(1 - v_s)(1 - v_c^2)(a^2 - b^2)$$

$$\beta_1 = \frac{C_{23}C_{32} - C_{22}C_{33}}{C_{12}C_{23} + C_{22}C_{13}}, \qquad \beta_2 = \frac{C_{12}C_{33} + C_{13}C_{32}}{C_{12}C_{23} + C_{22}C_{13}} .$$

Figure 5.5 features a comparison between the three elastic foundation formulations [Eqs. (5.3)–(5.5)] for a $2c = 0.4$-mm diameter. (0.016-in.) Kovar pin ($E = 138$ GPa or 20 Mpsi, $v = 0.3$) inserted into a eutectic Sn/Pb solder filled hole of an FR-4 card. The practical region of solder clearance is $1 < b/c < 2$. Note that the value of k, by definition of the foundation constant λ in Eqs. (1.9) and (1.10), occurs under the fourth root in elastic-foundation analysis, so it does not usually require great accuracy. Figure 5.5 shows Vangheluwe's foundation modulus k consistently above Suhir's. Engel's k values start in between, but at $b/c = 1.65$ they dip under Suhir's curve. Figure 5.6 is for a constant soldered hole diameter $2b = 1$ mm (40 mil), and the effect on λ of increasing c is demonstrated. The formulations yield smoothly declining values with an increasing pin diameter up to 0.72 mm. Beyond this pin diameter, however, Vangheluwe's rigid foundation assumption makes for an increasing λ value. Vangheluwe's formula is well applicable to compliant solder joints (meaning smaller E_s), while for stiffer solder joints the more complicated Suhir formula should have primacy.

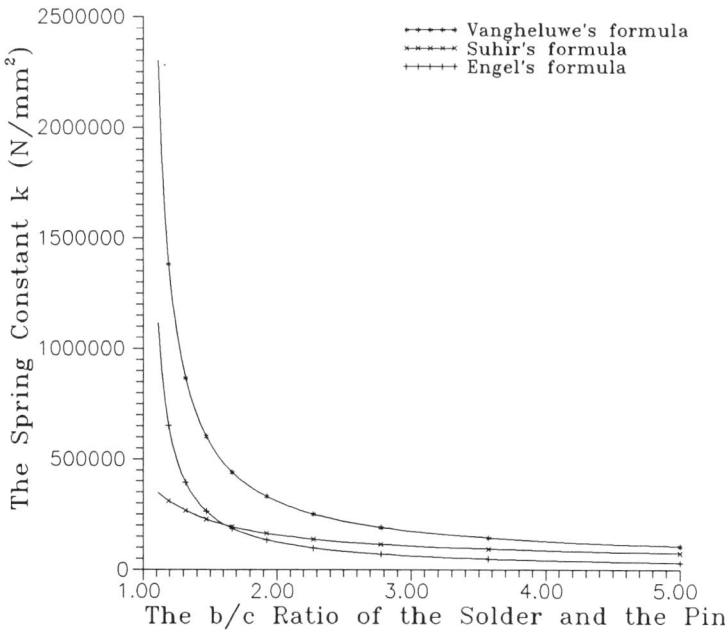

FIGURE 5.5. Plot of the foundation modulus k versus the b/c ratio of the solder and pin.

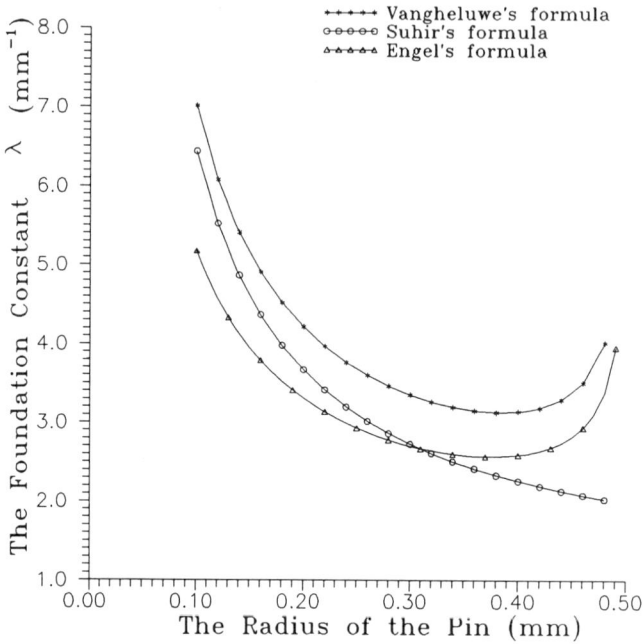

FIGURE 5.6. Plot of the foundation constant λ versus the radius of the pin.

For square pins of side c embedded in a circular soldered hole, no exact formulation is possible, and a finite element approach (Fig. 5.7) is fruitful. The "equivalent radius" c_{eq} corresponding to a square pin of side c was defined [5] as

$$c_{eq} = (c^2/\pi)^{1/2} . \tag{5.6}$$

The k-factors for the Kovar material and eutectic solder considered before, in conjunction with a rigid card, were evaluated by the ANSYS finite element program. Good correspondence to the Vangheluwe formula (5.4) was found when the latter formula was used combined with a factor $f = \sqrt{2}/2$:

$$k_{eq} = f \cdot k . \tag{5.7}$$

FIGURE 5.7. ANSYS modeling of a square pin. (a) Plan view; (b) Front view.

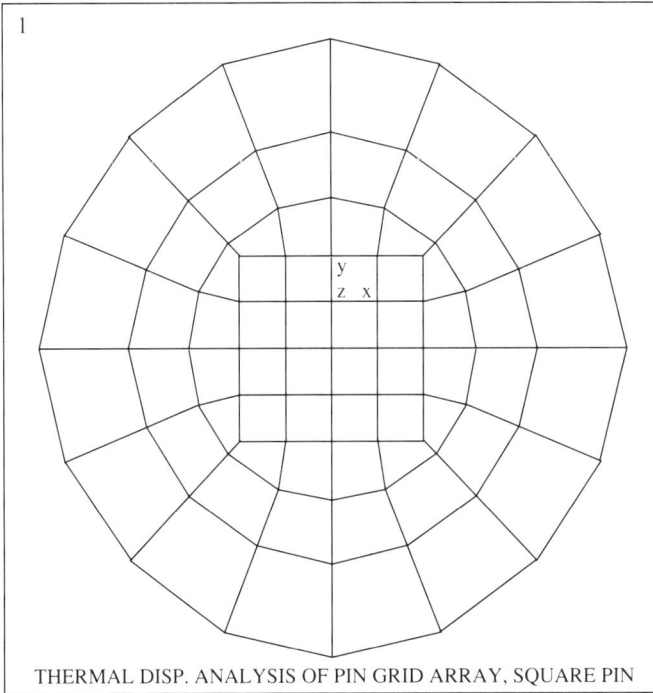

ANSYS 4.4
UNIV VERSION
APR 1 1991
16: 22: 38
POST1
ELEMENTS
TYPE NUM

ZY = 1
DIST = 0.55
ZF = 1.44

THERMAL DISP. ANALYSIS OF PIN GRID ARRAY, SQUARE PIN

(a)

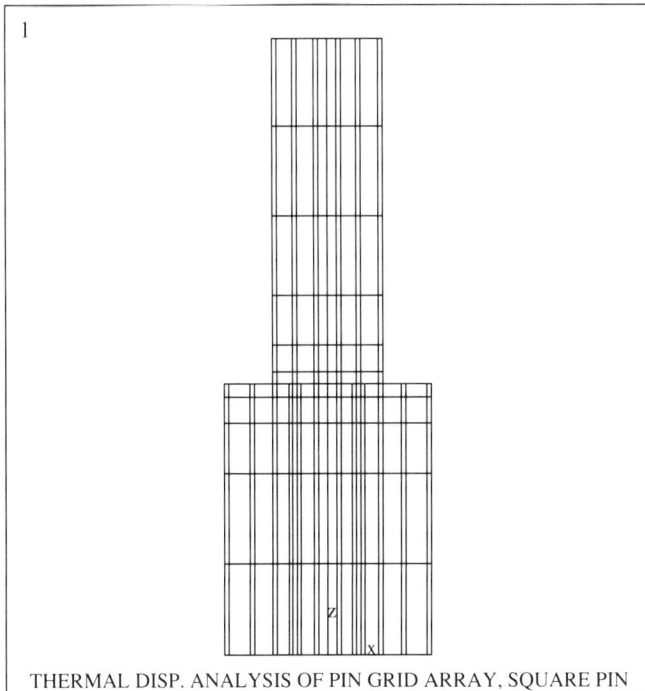

ANSYS 4.4
UNIV VERSION
APR 1 1991
16: 32: 32
POST1
ELEMENTS
TYPE NUM

YV = −1
DIST = 1.584
ZF = 1.44

THERMAL DISP. ANALYSIS OF PIN GRID ARRAY, SQUARE PIN

(b)

3. Elastic Foundation Treatment for the Embedded Pin

The embedded part of the pin is considered obeying the equation of a beam on an elastic foundation as derived in Section 2 of Chapter 1. This beam segment is acted upon by a shear V and moment M_2 from the standoff portion. For a total pin solution it will be useful to relate these internal forces to the transverse displacement Δ and rotation θ at the top of embedment, $x_s = 0$ (Fig. 5.3). While the derivation of the embedment "stiffness" matrix in Section 2 of Chapter 1 was done for general parameters, in the following treatment we shall assume a semi-infinite pin, as in Eq. (1.19); only the sign convention is changed to reflect positive V, M, Δ, and θ as expected by Fig. 5.3 on the embedded beam:

$$\left\{ \begin{array}{c} V \\ M \end{array} \right\} = [K] \left\{ \begin{array}{c} \Delta \\ \theta \end{array} \right\} \tag{5.8}$$

$$[K] = 2EI\lambda \begin{bmatrix} 2\lambda^2 & \lambda \\ \lambda & 1 \end{bmatrix}. \tag{5.9}$$

To evaluate the moments M_1, and M_2 and shear V on the upper standoff-portion L of the pin, we write the elastic equations for that portion (Fig. 5.3c). Note that at the embedment $x = L$, the beam displacement will be $u - \Delta$, instead of just u as for a fixity.

$$V = \frac{12EI(u - \Delta)}{L^3} - \frac{6EI\theta}{L^2} \tag{5.10}$$

$$M_2 = \frac{6EI(u - \Delta)}{L^2} - \frac{4EI\theta}{L}. \tag{5.11}$$

Equating the internal forces V and M_2 from Eqs. (5.10) and (5.11) with the corresponding elastic foundation forces of Eq. (5.8), Δ and θ can be calculated from the algebraic equations

$$([K] + [K']) \left\{ \begin{array}{c} \Delta \\ \theta \end{array} \right\} = 6EIu/L^3 \left\{ \begin{array}{c} 2 \\ L \end{array} \right\} \tag{5.12}$$

where

$$[K'] = 2EI/L^3 \begin{bmatrix} 6 & 3L \\ 3L & 2L^2 \end{bmatrix}. \tag{5.13}$$

From the preceding, assuming $\lambda L_s > 3$, the elastic pin analysis can be neatly summarized in terms of five nondimensional functions of $\gamma = \lambda L_s$. These five functions are $U_1(\gamma)$, $U_2(\gamma)$, $U_v(\gamma)$, $U_\Delta(\gamma)$, and $U_\theta(\gamma)$, nondimensionalizing M_1, M_2, V, Δ, and θ as follows:

$$M_1 = \frac{6EIu}{L^2} U_1(\gamma); \qquad U_1(\gamma) = \frac{\gamma^2(\gamma + 1)}{\mathscr{D}} \tag{5.14}$$

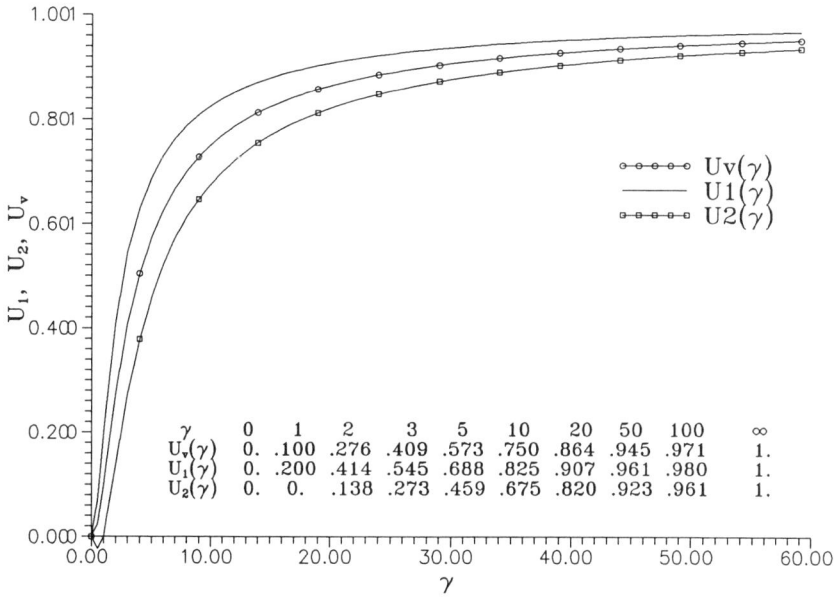

γ	0	1	2	3	5	10	20	50	100	∞
$U_v(\gamma)$	0.	.100	.276	.409	.573	.750	.864	.945	.971	1.
$U_1(\gamma)$	0.	.200	.414	.545	.688	.825	.907	.961	.980	1.
$U_2(\gamma)$	0.	0.	.138	.273	.459	.675	.820	.923	.961	1.

FIGURE 5.8. Plot of the nondimensional functions of M_1, M_2, and V versus $\gamma = \lambda L_s$.

γ	0	1	2	3	5	10	20	50	100	∞
$U_\Delta(\gamma)$	.333	.200	.103	.061	.028	.008	.002	.0004	.0001	0.
$U_\Theta(\gamma)$	0.	.100	.138	.126	.115	.080	.040	.019	.001	0.

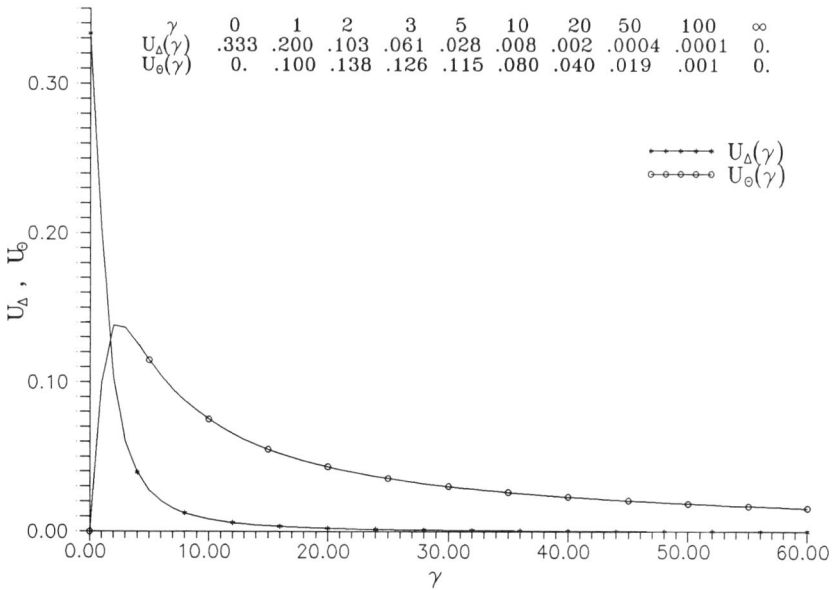

FIGURE 5.9. Plot of the nondimensional Δ and θ functions versus γ.

$$M_2 = \frac{6EIu}{L^2} U_2(\gamma); \qquad U_2(\gamma) = \frac{\gamma^2(\gamma - 1)}{\mathscr{D}} \tag{5.15}$$

$$V = \frac{12EIu}{L^3} U_v(\gamma); \qquad U_v(\gamma) = \frac{\gamma^3}{\mathscr{D}} \tag{5.16}$$

$$\varDelta = 3u U_\varDelta(\gamma); \qquad U_\varDelta(\gamma) = (\gamma + 1)/\mathscr{D} \tag{5.17}$$

$$\theta = \frac{6u}{L} U_\theta(\gamma); \qquad U_\theta(\gamma) = \gamma/\mathscr{D} \tag{5.18}$$

where

$$\mathscr{D} = (\gamma + 1)^3 + 2; \qquad \gamma = \lambda L_s . \tag{5.19}$$

The pin stresses follow from the expressions of moments and shears. For example, the maximum bending stress, which is obtained at the top of the standoff, is

$$\sigma = (6Ec/L^2)u U_1(\gamma) \tag{5.20}$$

and the allowable thermal displacement follows as a function of the allowable pin stress,

$$u_{all} = L^2 \sigma_{all}/6Ec U_1(\gamma) . \tag{5.21}$$

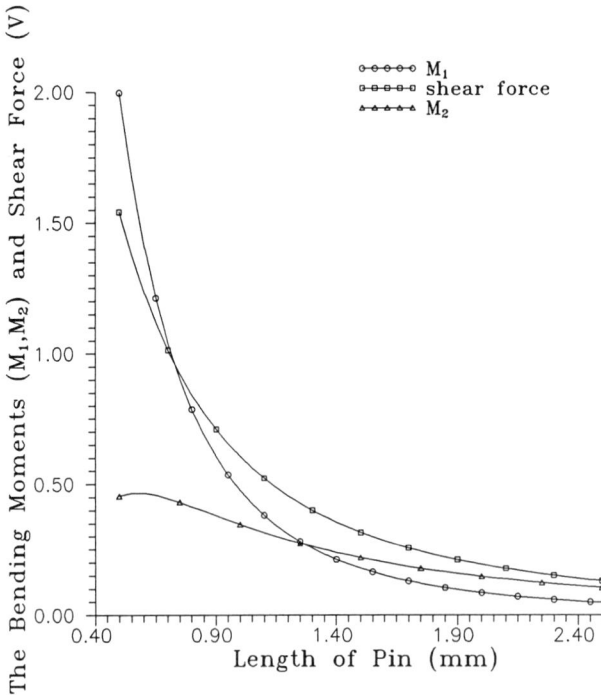

FIGURE 5.10. The bending moments M_1, M_2, and shear force V versus the length L of a circular pin.

The nondimensional functions $U_1(\gamma)$, $U_2(\gamma)$, and $U_v(\gamma)$ converge to unity while the $U_\Delta(\gamma)$ and $U_\theta(\gamma)$ functions tend to zero when the spring foundation modulus k approaches infinity (the fixed-fixed beam situation); see Figs. 5.8 and 5.9.

For the Kovar pin embedded in eutectic solder, serving as an example in Section 2, variations of M_1, M_2, and V corresponding to varying pin lengths are shown in Fig. 5.10. The stress reduction gained from lengthening the pin is apparent. Another obvious source of stress reduction lies in reducing the pin diameter.

4. Solder Pressure Calculation

For a circular pin, the maximum solder pressure σ may be estimated by assuming a monotonically increasing cosine-type variation of the radial pressure on one side and a similar distribution of radial tension on the other

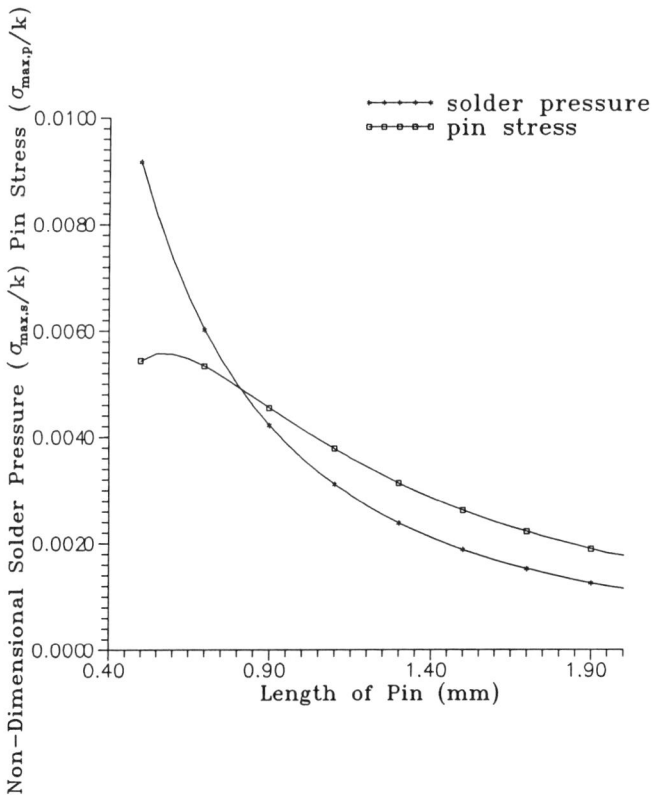

FIGURE 5.11. Nondimensional solder pressure and flexural pin stress versus pin length.

side of the pin (Fig. 5.4). For $\phi = 0$ (in the plane of symmetry), σ is maximum; and at $\phi = \pi/2$ (at the edge of the pin), σ is zero.

Alternatively, a Hertz contact pressure distribution [6] may be assumed, which has a 50 percent increase from the average to the maximum pressure in the middle. Then

$$\sigma_{max} = (3/2)\sigma_{av} \ . \tag{5.22}$$

From the distributed load $p(x) = kw(x)$ on the embedded pin supported by the elastic foundation, we have

$$p_{max} = kw(0) = k \cdot \Delta \ . \tag{5.23}$$

Then the average pressure for each pin surface, forward and backward, is

$$\sigma_{av} = p_{max}/4c \tag{5.24}$$

so that by Eq. (5.22) the maximum solder pressure (and, on the opposite side, the maximum solder tension) becomes

$$\sigma_s = \sigma_{max} = (3/8)k \cdot \Delta/c \ . \tag{5.25}$$

Figure 5.11 shows plots of solder pressure and flexural pin strain versus the pin length L, for other dimensions as given in the example of Section 2.

5. Plastic Analysis of the Pin

The thermal mismatch u as computed from Eq. (5.2) may turn out to be excessive, generating plastic deformation in the pin. At the top (braze or swage joint with the substrate), the moment M_1 will be bigger than M_2 arising at the elastic solder embedment. Thus at the lower end of the standoff, elastic conditions will be likely to prevail while the top region has already deformed plastically.

This complicated problem has been solved iteratively [5]. The end moments M_1 and M_2 are guessed first. The displacements of the standoff part, resulting from the guess solution, are matched at the bottom of the standoff, with those from the embedment part. This way, improved displacement parameters for the next iteration cycle are obtained.

We start by assuming the bending moments at the two ends of the standoff in terms of two as-yet-undetermined numerical coefficients α, β defined as follows:

$$M_1 = \alpha M_p, \qquad M_y/M_p < \alpha < 1 \tag{5.26}$$

$$M_2 = \beta M_y, \qquad 0 < \beta < 1 \tag{5.27}$$

where M_y is the yield moment and M_p the fully plastic moment (see Chapter 1, Section 5).

Choosing ξ as the axial coordinate from the top of the standoff (Fig. 5.12), we get for the length of the elasto-plastic region:

$$\xi_e = L(M_1 - M_y)/(M_1 + M_2) . \tag{5.28}$$

Then the bending moment variation is

$$M(\xi) = M_1 - (M_1 - M_y)(\xi/\xi_e) . \tag{5.29}$$

Introducing the ratio of fully plastic to elastic limit moments, which were defined in Chapter 1, Section 5:

$$m = M_p/M_y , \tag{5.30}$$

the nondimensional moment $\mu(\xi) = M(\xi)/M_y$ is written

$$\mu(\xi) = m\alpha - (m\alpha - 1)\cdot(\xi/\xi_e) . \tag{5.31}$$

The plasticity distance from Eq. (1.70) is

$$\eta(\xi) = \sqrt{A - B\mu(\xi)} = \sqrt{A - Bm\alpha + \frac{m\alpha - 1}{\xi_e}B\xi} . \tag{5.32}$$

Integrating Eq. (1.71), the slope $dw/d\xi$ becomes, in the elasto-plastic portion of the standoff:

$$\frac{dw}{d\xi} = \frac{\sigma_y}{Ec}\int_0^{\xi}\frac{d\xi'}{\sqrt{A - Bm\alpha + (m\alpha - 1)B\xi'/\xi_e}} . \tag{5.33}$$

By yet another integration, we can evaluate the displacement variation; thus the displacement $w(\xi_e)$ and slope $w'(\xi_e)$ become available. These supply the initial conditions for integrating the beam equation over the elastic

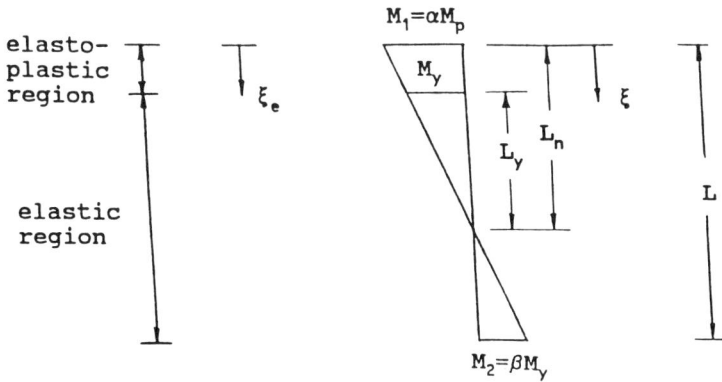

FIGURE 5.12. Moment diagram assumed along standoff-length of pin, for iterative elasto-plastic analysis of pin.

portion of the standoff, where,

$$M(\xi) = \frac{M_y}{L - \xi_e}[-(1 + \beta)\xi + L + \beta\xi_e].$$ (5.34)

Integration of the elastic beam equation finally yields the displacement and slope at the embedment, $\xi = L$:

$$\theta = \frac{M_y}{EI(L - \xi_e)}\left[\frac{L^2}{2} - L\xi_e + \frac{\xi_e^2}{2} + \left(L\xi_e - \frac{L^2}{2} - \frac{\xi_e^2}{2}\right)\beta\right]$$

$$+ \frac{2M_y\xi_e}{EI(Bm\alpha - B)}(\sqrt{A - B} - \sqrt{A - Bm\alpha})$$ (5.35)

$$w_0 = \frac{M_y}{EI(L - \xi_e)}\left[\frac{L^3}{3} - L^2\xi_e + L\xi_e^2 - \frac{\xi_e^3}{3} + \left(\frac{\xi_e^3}{6} - \frac{L\xi_e^2}{2} + \frac{L^2\xi_e}{2} - \frac{L^3}{6}\right)\beta\right]$$

$$+ \frac{2M_y\xi_e}{EI(Bm\alpha - B)} \cdot [(\sqrt{A - B} + \sqrt{A - Bm\alpha})L$$

$$+ (\sqrt{A - B} - \sqrt{A - Bm\alpha})\xi_e]$$

$$+ \frac{2M_y\xi_e^2}{EI(Bm\alpha - B)}\left[-\sqrt{A - Bm\alpha} + \frac{2(A - B)^{3/2}}{3(Bm\alpha - B)} - \frac{2(A - Bm\alpha)^{3/2}}{3(Bm\alpha - B)}\right]$$

(5.36)

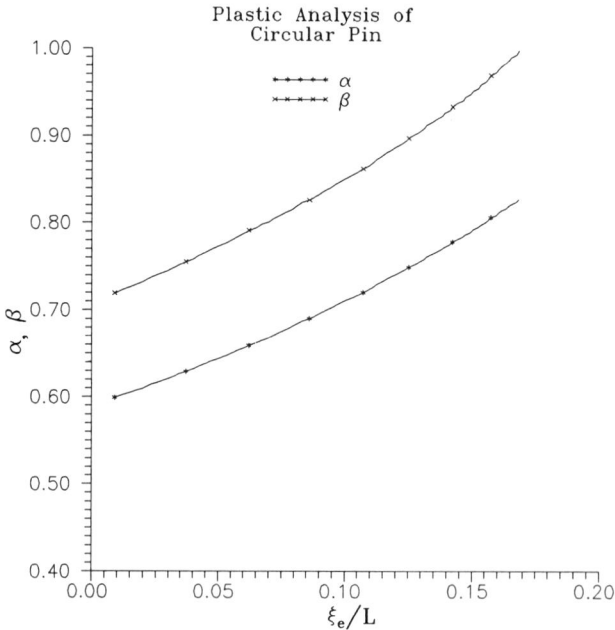

FIGURE 5.13. Plastic parameters (α, β) of the pin versus the ratio of the plastic region (ξ_e) and the length L of the pin.

Over the embedment region, the elastic equation (5.9) is valid. Substituting, from Eqs. (5.26) and (5.27):

$$\left\{ \begin{array}{c} V \\ M_2 \end{array} \right\} = \left[\begin{array}{cc} M_p/L & M_y/L \\ 0 & M_y \end{array} \right] \left\{ \begin{array}{c} \alpha \\ \beta \end{array} \right\}. \tag{5.37}$$

Equation (5.37) can now be inverted to yield a new, improved set of α and β:

$$\left\{ \begin{array}{c} \alpha \\ \beta \end{array} \right\} = \left[\begin{array}{cc} L/M_p & -1/M_p \\ 0 & 1/M_y \end{array} \right] \left\{ \begin{array}{c} V \\ M_2 \end{array} \right\} \tag{5.38}$$

or, by use of Eq. (5.9):

$$\left\{ \begin{array}{c} \alpha \\ \beta \end{array} \right\} = \left[\begin{array}{cc} L/M_p & -1/M_p \\ 0 & 1/M_y \end{array} \right] [K] \left\{ \begin{array}{c} \Delta \\ \theta \end{array} \right\} \tag{5.39}$$

These new values of (α, β) are then used for the next iteration cycle, with rapid convergence in cases of practical pin and solder geometries.

In Fig. 5.13 the converged parameters α, β of a circular pin are plotted against ξ_e/L ratios ranging from 0 to 0.15. The parameter β grows at

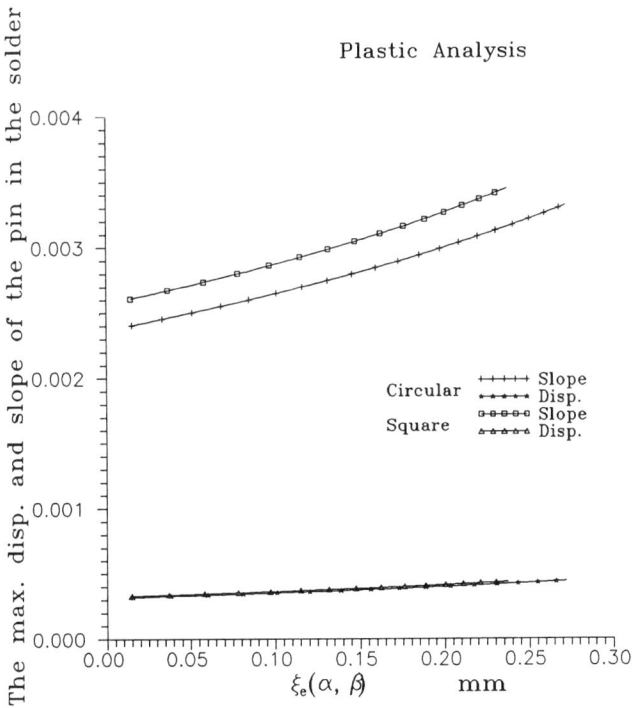

FIGURE 5.14. The maximum displacement Δ and slope θ versus the length ξ of the plastic region.

approximately the same rate as α, showing that the end moments increase equally as u increases. Figure 5.14 shows that the pin slope θ grows more rapidly than the displacement Δ at the embedment.

6. Axial Pin Force Due to Flexure

The pin must stretch in order to bend into an S-like shape under thermal mismatch. If the transverse displacement is $w(x)$ along the pin, then the stretch ΔL is expressed (Fig. 5.15) as

$$\Delta L = \int_0^L ds - L = \int_0^L dx \sqrt{1 + w'^2} - L .$$ (5.40)

For small slopes

$$(1 + w'^2)^{1/2} \approx 1 + \frac{w'^2}{2} ,$$ (5.41)

resulting in

$$\Delta L = \frac{1}{2} \int_0^L \left(\frac{dw}{dx} \right)^2 \cdot dx .$$ (5.42)

The stretch is then equated to the elastic axial elongation FL/AE from which the axial force F can be evaluated.

As a simple example, a fixed-fixed pin with boundary conditions

$$
\begin{aligned}
x = 0: \quad & w = 0, \quad w' = 0 \\
x = L: \quad & w = 0, \quad w' = 0
\end{aligned}
$$ (5.43)

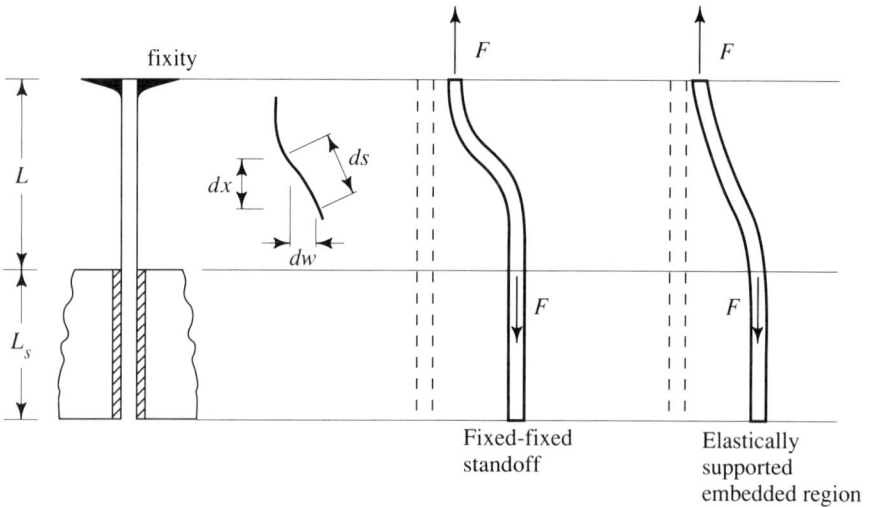

FIGURE 5.15. Axial force F developed by bent pin.

will have an elastic line

$$w(x) = 3ux^2/L^2 - 2ux^3/L^3, \qquad w'(x) = 6u(x/L^2 - x^2/L^3), \qquad (5.44)$$

yielding, after substitution into the integral of Eq. (5.42):

$$\Delta L - 3u^2/5L . \qquad (5.45)$$

Equating this stretch with FL/AE, we get, for a circular pin of area $A = \pi c^2$:

$$F = (3\pi E c^2 u^2)/5L^2 \qquad (5.46)$$

and the pullout shear stress acting on the solder sleeve is

$$\tau = \frac{F}{2\pi c L_s} = \frac{3}{10} \frac{E c u^2}{L^2 L_s} . \qquad (5.47)$$

For a pin elastically supported in the solder joint the elastic line is no longer as simple as that of Eq. (5.44). Now the boundary conditions replacing Eq. (5.43) are

$$x = 0: \qquad w = 0, \qquad w' = 0 \qquad (5.48)$$

$$x = L: \qquad w = u - \Delta, w' = 0 .$$

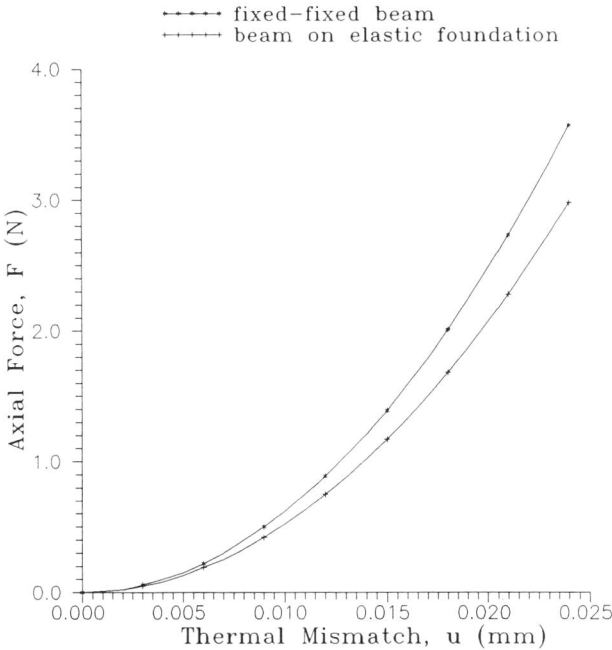

$\cdots\cdots\cdots$ fixed–fixed beam
$\cdots\cdots\cdots$ beam on elastic foundation

FIGURE 5.16. The axial force F versus the maximum mismatch for a fixed-fixed beam and for a beam on an elastic foundation.

The equation of the elastic line is now

$$w(x) = [\theta L - 2(u - \Delta)]\left(\frac{x}{L}\right)^3 + [3(u - \Delta) - \theta L]\left(\frac{x}{L}\right)^2 \tag{5.49}$$

from which the stretch ΔL is computed by Eq. (5.42); the axial force is computed from

$$F = \frac{AE}{2L}\int_0^L \left(\frac{dw}{dx}\right)^2 dx, \tag{5.50}$$

whence

$$F = (AE/30L^2)[18(u - \Delta)^2 - 30\theta L(u - \Delta) + 20^2 L^2] \tag{5.51}$$

Figure 5.16 shows axial force plots for u varied, for the pin and solder dimensions of the example in Section 2. It is evident that a fixed-fixed pin would have higher axial force than the elastically supported one.

If the pin is also plastically deforming over its standoff length, then we have for the slope of the elastic line:

$$\frac{dw}{d\xi} = \frac{M_y}{(L - \xi_e)EI}\left[-(1 + \beta)\frac{\xi^2}{2} + (L + \beta\xi_e)\xi \right.$$
$$\left. + (1 + \beta)\frac{\xi_e^2}{2} - (L + B\xi_e)\xi_e\right]$$
$$- \frac{2\sigma_y\xi_e}{Ec(Bm\alpha - B)}(\sqrt{A - Bm\alpha} - \sqrt{A - B}) \tag{5.52}$$

causing the axial elongation ΔL, numerically calculable from Eq. (5.42).

7. A Magnified-Scale Experiment

The analysis of a single pin described in this chapter took into account only elastic behavior of the solder. A supporting argument might claim that the elastic foundation constant λ of solder is not very sensitive to the elastic foundation modulus k since the former is related to the 1/4-power of the latter. In order to experimentally check the single-pin formulation, a larger-than-life scale experiment was devised.

A circular bar ($c = 5$ mm) of cotton cloth phenolic ($E = 6.2$ GPa, $v = 0.24$, $\sigma_y = 69$–96 MPa) was bonded inside a ring ($b = 10$ mm) of Conathane RN-1511 elastomer (secant modulus at 300% elongation $= 20.7$ MPa, $v = 0.44$, elongation $= 430\%$). The outer surface of the elastomer was adhered to the matching hole of an aluminum plate, both being $L_s = 50$ mm deep. The phenolic bar, simulating the pin, protruded from the rubber-aluminum assembly's plane to a length $L = 96$ mm.

The bar was displaced as a cantilever on top by a transverse force P, and the standoff-deflections measured along L by three dial gages, to check on the

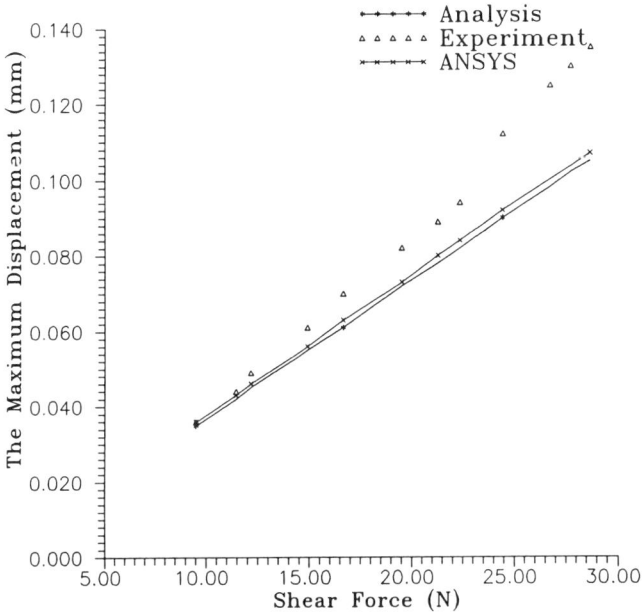

FIGURE 5.17. Base displacements Δ versus transverse force P from the magnified-scale experiment of a phenolic bar bonded to an aluminum plate through rubber lining.

elastic line geometry. In particular, the displacement Δ at the bottom junction was measured. The displacements were also calculated by the elastic foundation method of this chapter. The resulting graph of the experimentally measured and the predicted force P versus displacement Δ curves are shown in Fig. 5.17. The correspondence is good at moderate forces. At increasing force, however, the larger-than-linear deformations of the embedment cause gradual deterioration of the agreement.

This experiment suggests that at large thermal excursions, when pin-plasticity is expected, solder plasticity might also well be taken into account. To the complexity of the preceding problem, the temperature-dependent elastic-plastic properties of solder evidently contribute.

8. Conclusions

The elastic foundation property of the solder embedment was taken into account to calculate the "primary" pin forces: end moments and shear and axial force. When the pins could be considered semi-infinite elastic beams, simple exact solutions were obtained. A calculation of approximate solder joint pressures due to temperature load on a PGA was shown. The analysis was extended to plasticity of the pin. Various computation methods for the foundation constant k were checked through an example. A large-scale experiment showed that at large thermal excursions plastic deformations of the solder foundation can be expected to play a more significant role.

9. Exercises and Questions

1. Enumerate the simplifying assumptions allowing the "primary" analysis of a single pin as the component of a PGA.
2. What material behaviorial (constitutive) laws might be employed to accurately carry out single-pin analysis?
3. Appraise the conservativeness of various solder-foundation models and of the solder pressure calculation.
4. Appraise the conservativeness of elastic pin versus plastic pin analysis.
5. Consider $c = 0.2$-mm-radius Kovar pins of a 36-mm-square PGA inserted into the eutectic Sn/Pb soldered hole ($b = 0.3$ mm) of an $h = 1.25$ mm-thick FR-4 circuit card. Let $\Delta T = 100\,°C$.

    a. Determine k and λ from Engel's Vangheluwe's, and Suhir's expressions. For the latter, let the card dimension a be equal to half the pin spacing, $s/2 = 1.25$ mm.
    b. Determine the maximum pin forces, moments, and displacements for elastic materials.
    c. Determine the maximum pin stresses and the solder pressure.
    d. Calculate the first iteration cycle of the coefficients α and β of a plastic analysis.

References

1. Engel, P.A., Lim, C.K., Toda, M.D., and Gjone, R. (1984), "Thermal Stress Analysis of Soldered Connectors for Complex Electronics Modules," *Computers in Mech. Eng.*, **2**(6), 59–69.
2. Vangheluwe, D.C.L. (1984), "Exact Calculations of the Spring Constant in the Buckling of Optical Fibers," *Appl. Optics*, **23**(13), 2045–2046.
3. Muskhelishvili, N. (1977), *Some Basic Problems of the Theory of Elasticity*, Noordhof, Leyden.
4. Suhir, E. (1988), "Stress in Dual-Coated Optical Fibers," *J. Appl. Mech.*, **55**(4), 822–830.
5. Engel, P.A., and Wu, K.R. (1992), "Thermal Stress Analysis of Pin Grid Array Structures: Pin and Solder Joint Problems," *ASME J. Elec. Packag.*, **114**, 3, 314–321.
6. Engel, P.A. (1976), *Impact Wear of Materials*, Elsevier, New York.

Chapter 6

Thermal Stress in Pin-Grid Arrays: Interaction Between Module and Circuit Card

In the previous chapter we performed thermal stress analysis of a single pin in the pin-grid array (PGA) structure; the analysis was based on the contention that the confining surfaces of the pins, the module above and printed circuit card below, were rigid, i.e., had an infinite modulus of elasticity. In reality, however, these delimiting surfaces act rather like plates and can deform under the moments and forces anchoring the thermally stressed pin.

We may call the M_1, M_2, V, and F pin-end forces, as calculated between rigid plates, the "primary" pin forces (Fig. 6.1). As a "secondary" structural action, we will consider the release forces arising due to the elastic flexure of module and card manifested in transverse displacements. Transverse deflection of the confining plates will result in reducing axial pin forces. In-plane stretch of the plates (described as "tertiary action") will tend to reduce the thermal mismatch, the original cause of all pin and pin-joint stress.

We shall set out to calculate the correction to pin forces and force distributions. Thin-plate theory (Chapter 1, Section 3) will be used for transverse, flexural behavior; for the in-plane stress, the theory of elasticity serves with the two-dimensional Airy stress function method [1].

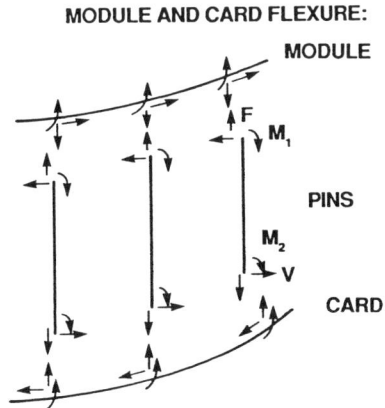

MODULE AND CARD FLEXURE:

FIGURE 6.1. Forces and moments from thermal loading acting at pin ends, and their reactions on the module and card.

Plate analysis is simplified throughout by an equivalent polar symmetrical treatment, analogously to Hall's method [2] shown in Chapter 4 for leadless chip-carrier systems. The pin loads acting on the confining plates are first idealized as a continuous distribution, and the assembly displacement compatibility problem is solved by a collocation (point-matching) scheme.

1. Pin Force Analysis Due to Module and Card Bending

The primary moments M_1, M_2, shear force V, and axial force F anchoring an individual pin to the module above and the circuit card below were given in Chapter 5; Eqs. (5.14)–(5.16) are repeated here:

$$M_1 = \frac{6EI}{L^2} \frac{\gamma^2(\gamma + 1)}{\mathscr{D}} u \tag{6.1}$$

$$M_2 = \frac{6EI}{L^2} \frac{\gamma^2(\gamma - 1)}{\mathscr{D}} u \tag{6.2}$$

$$V = \frac{12EI}{L^3} \frac{\gamma^3}{\mathscr{D}} u \tag{6.3}$$

$$F = \frac{3EA}{5L^2} g(\gamma) \cdot u^2 \tag{6.4}$$

where

$$\mathscr{D} \equiv (\gamma + 1)^3 + 2, \qquad \gamma \equiv \lambda L_s .$$

Note that $g(\gamma)$ modifies the axial force expression [Eq. (5.46)] of a fixed-fixed pin to validate it for an elastic foundation embedment at the bottom. A closed form expression (5.51) corresponds to that condition (Fig. 5.16).

Among the parameters occurring in Eqs. (6.1–6.4) E, A, I, and L are pin properties and γ is a combined property of pin and solder, in terms of the elastic foundation model. Only the quantity u contains the distance dimension r along the module, through Eq. (5.2).

In modeling, we first replace the square PGA of size a with an equivalent polar symmetrical circular one of radius $a_m = \sqrt{2a}/2$, in which the pins transmit moments and shears in radial planes. Secondly, we shall "smear out" the pin forces (Fig. 6.2) which, in reality, are spaced in a square pattern at a distance s. Assuming, for example, a constant density $\eta = 1/s^2$ of pins in a fully populated array, the number of pins inhabiting an annulus dr located at r is $m = \eta \, 2\pi r \cdot dr$.

Our goal is to calculate the secondary axial pin forces P resulting from potential transverse plate displacements, W and w, of the module and card, respectively. W and w are both elastic displacements. Without loss of generality, we shall arbitrarily fix the center of the card in space [$w(0) = 0$]. The

FIGURE 6.2. Pin end forces "smeared out" over an annulus db of the circular plate representing module or card.

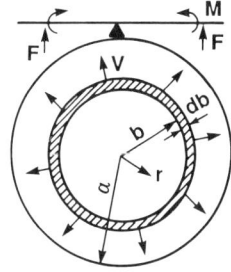

module center, however, will move to a point $W_0 = W(0)$ due to the elastic deformations of the members, and this position parameter is regarded as an additional unknown.

In calculating the secondary pin force distribution [3], three influences are taken into account (Fig. 6.1): that of pin moments (M_1, M_2); primary axial pin forces F; and the secondary axial pin forces P. We shall look at those influences one by one and evaluate them by a technique of collocation or point matching [4]. In doing so, the loaded part of a plate is divided into n equally spaced annuli of width Δr. Over the jth annulus we calculate the total magnitude of the load component of concern (say p_j), and then replace it with a line load at the center of the annulus. Next, by the theory of elasticity, we compute the influence coefficient, i.e., the displacement at the center of lamella i due to the load p_j. At the end, we add the influences using the principle of superposition; collocation will yield the unknown force components (in this case, the secondary axial pin forces) by a set of simultaneous algebraic equations.

2. Influence of Pin-End Moments

Concentrating all the radial moments of an annulus j (Fig. 6.3a), the respective radial plate moments (moment per unit circumference) for the module and plate can be written, by Eqs. (6.1) and (6.2):

$$M_{1j} = \frac{6EI}{L^2} \cdot \Delta T \cdot \Delta \alpha \cdot U_1(\gamma) \eta \Delta r \cdot r_j \tag{6.5}$$

and

$$M_{2j} = \frac{6EI}{L^2} \cdot \Delta T \cdot \Delta \alpha \cdot U_2(\gamma) \eta \Delta r \cdot r_j . \tag{6.6}$$

We shall evaluate the transverse displacement influence at r along both the module and the card, due to the external plate moment M_r applied at $r_j = b$.

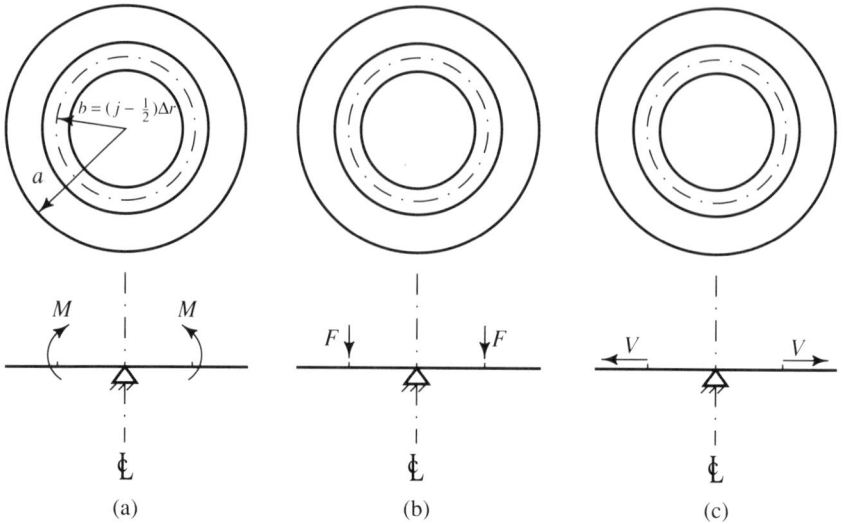

FIGURE 6.3. Discretizing the pin end forces as line forces acting at the midline of n equidistant circular lamellae. (a) Radial moment; (b) transverse force; (c) in-plane force.

The plates will be considered supported at the center point for the computation of elastic influences.

Following Timoshenko and Woinowsky-Krieger [5], the displacement $w(r)$ of a circular plate of radius a, subjected to a radial line moment M_r (N.mm/mm) applied at $r = b$ is derived:

$$w(r) = M_r(b) \cdot \Theta(r, b) \tag{6.7}$$

where $\Theta(r, b)$, an influence function or Green's function, has two distinct realms: $r < b$ and $r > b$:

$$\Theta(r, b) = \frac{r^2}{4D}\left[1 + \frac{1 - v}{1 + v}\frac{b}{a}\right], \quad r \leqslant b \tag{6.8}$$

$$\Theta(r, b) = \frac{b^2}{4D}\left[\frac{r^2 - b^2}{a^2 - b^2}\frac{1 - v}{1 + v}\left\{1 - \left(\frac{b}{a}\right)^2\right\}\right.$$

$$\left. + 2\ln\frac{r}{b} + 1 + \frac{1 - v}{1 + v}\left(\frac{b}{a}\right)^2\right], \quad a \geqslant r \geqslant b \, . \tag{6.9}$$

We have to employ two influence functions: $\Theta_1(r, b)$ for the module, and $\Theta_2(r, b)$ for the card. Θ_1 and Θ_2 contain the respective elastic constants (D_1, v_1) and (D_2, v_2) for the module and the card. The module extends to a radius a_m; the card radius a_c is ordinarily much larger, $a_c \gg a_m$. Equations (6.8) and (6.9) include the case of the "infinite card," $a_c = \infty$. Now the influence coefficients for the respective unit moment loads $M_r = 1$ N.mm/mm

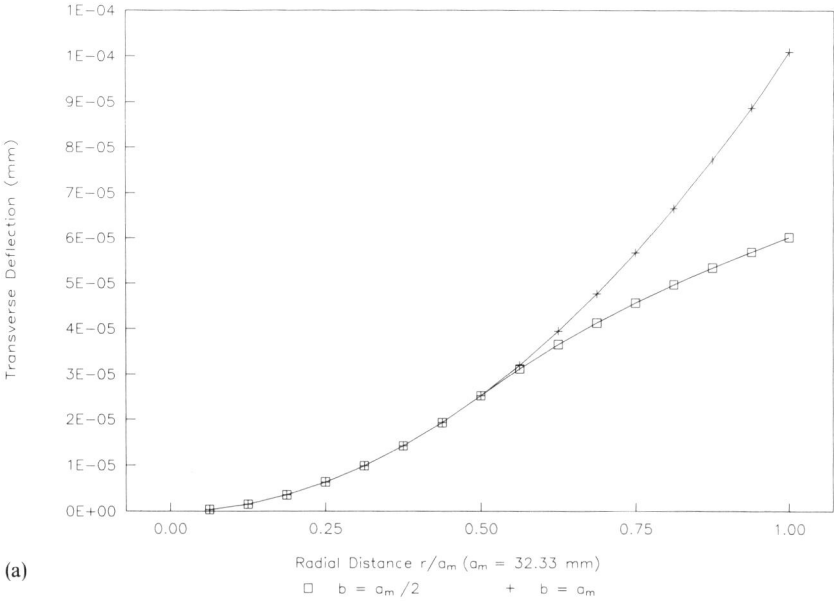

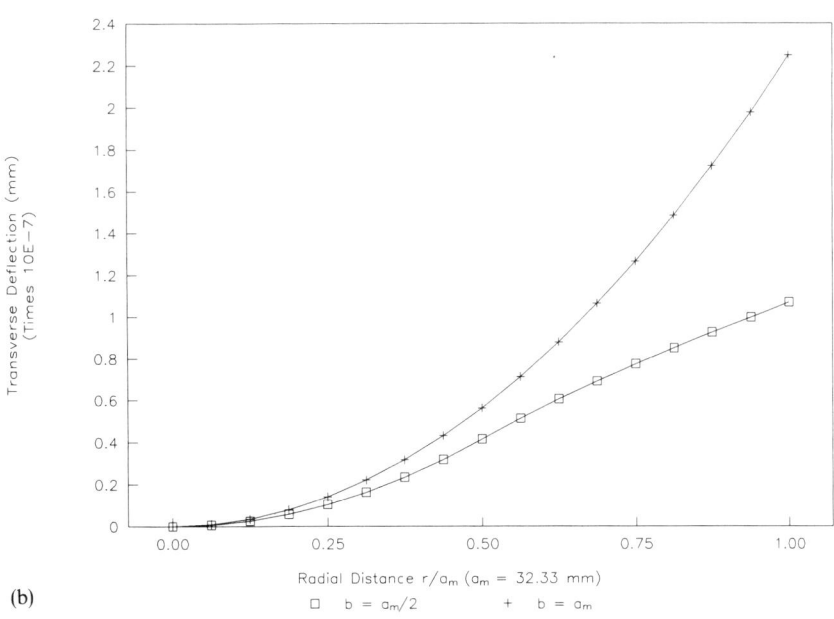

FIGURE 6.4. Transverse displacement influence functions due to unit line plate moment. (a) For FR-4 card described in Example 1; (b) for module described in Example 1.

for the module and the card can be written; both the application of the line loads and the displacements are in the middle of equidistant circular lamellae. We have

$$B_{ij} = \Theta_1 [(i - \tfrac{1}{2})\Delta r, (j - \tfrac{1}{2})\Delta r], \tag{6.10}$$

and

$$b_{ij} = \Theta_2 [(i - \tfrac{1}{2})\Delta r, (j - \tfrac{1}{2})\Delta r], \quad \begin{matrix} i = 1, 2, \ldots n \\ j = 1, 2, \ldots n \end{matrix} . \tag{6.11}$$

Figure 6.4 shows the influence functions $\Theta(r, b)$ for $a_c = \infty$, and for both $b = a_m/2$ and $b = a_m$, computed for the parameters of Example 1 included later at the end of Section 5.

3. Influence of the Primary Axial Pin Forces

The total axial force acting on a plate annulus $(2\pi r \cdot \Delta r)$ (Fig. 6.3b) has a dependence on u^3; we may write its value by Eq. (6.4):

$$f_j = \frac{3EA}{5L^2} g(\gamma)(\Delta T \cdot \Delta \alpha)^2 \cdot \eta 2\pi \Delta r \cdot r_j^3 . \tag{6.12}$$

The transverse plate displacement at r due to a line force $f(b)$ at b is written

$$w(r) = f(b) \cdot \Psi(r, b) \tag{6.13}$$

where, by thin-plate theory:

$$\Psi(r, b) = \frac{r^2}{16\pi D} \left[\frac{1 - v}{1 + v} \left(\frac{b}{a} \right)^2 - 2\ln \frac{r}{b} + 2 \right], \quad r \leqslant b \tag{6.14}$$

$$\Psi(r, b) = \frac{b^2}{16\pi D} \left[\frac{1 - v}{1 + v} \left(\frac{r}{a} \right)^2 + 2\ln \frac{r}{b} + 2 \right], \quad a \geqslant r \geqslant b . \tag{6.15}$$

Now the influence coefficient for primary axial pin force can be written for the module and the card, respectively:

$$Q_{ij} = \Psi_1 [(i - \tfrac{1}{2})\Delta r, (j - \tfrac{1}{2})\Delta r] \tag{6.16}$$

$$q_{ij} = \Psi_2 [(i - \tfrac{1}{2})\Delta r, (j - \tfrac{1}{2})\Delta r], \quad \begin{matrix} i = 1, 2, \ldots n \\ j = 1, 2, \ldots, n \end{matrix} . \tag{6.17}$$

Figure 6.5 shows the influence functions $\Psi(r, b)$ for both $b = a_m/2$ and $b = a_m$.

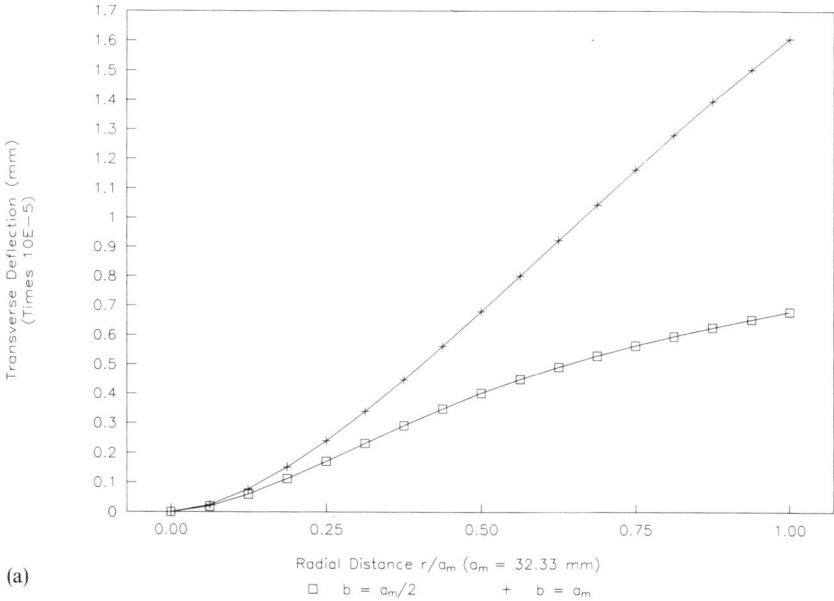

(a)

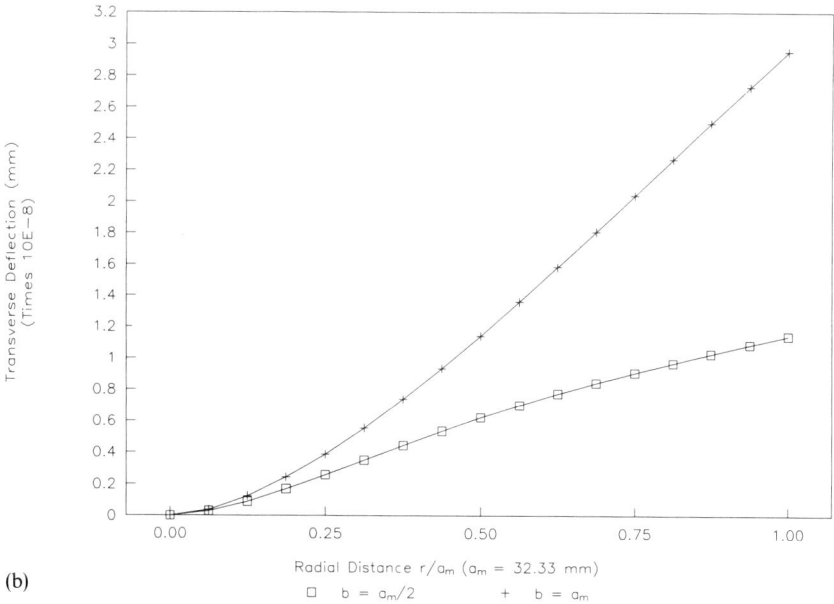

(b)

FIGURE 6.5. Transverse displacement influence function due to a unit transverse line load. (a) For FR-4 card described in Example 1; (b) for module described in Example 1.

4. Influence of Secondary Axial Pin Forces

To appraise the influence of a transverse line force $p(b)$, we write, analogously to Eq. (6.13):

$$w(r) = p(b) \cdot \Psi(r, b) . \tag{6.18}$$

We shall denote by p_j the line load obtained at r_j from summing the distributed load $p(r)$ over a lamella of width Δr:

$$p_j = p(r_j) 2\pi r_j \cdot \Delta r . \tag{6.19}$$

In reality, the P forces are transmitted by pins that have an individual axial stiffness K. These individual pins, at a circle of radius r, are added parallel, forming a spring of stiffness

$$k(r) = K \cdot \eta(2\pi r) \Delta r \tag{6.20}$$

or

$$k_i = 2\pi K \eta r_i \Delta r . \tag{6.21}$$

The pin forces will then be related, by definition, to the difference of module and card displacements as

$$W(r) - w(r) = [p(r) + f(r)]/k(r) . \tag{6.22}$$

5. Solution of the System of Equations

The compatibility of pin displacements with those of the plates delineating them may be stated as follows:

$$W(r_i) - W(0) - w(r_i) = \sum_{j=1}^{n} p_j(Q_{ij} + q_{ij}) + \sum_{j=1}^{n} (M_j B_{ij} - m_j b_{ij})$$

$$+ \sum_{j=1}^{n} f_j(Q_{ij} + q_{ij}), \quad (i = 1, 2, \ldots, n) . \tag{6.23}$$

This may be written in the form of a set of n linear algebraic equations in the secondary pin forces p_j and the unknown displacement position $W_0 \equiv W(0)$ of the module:

$$W_0 + \sum_{j=1}^{n} p_j(Q_{ij} + q_{ij}) + \frac{p_i}{2\pi K \eta \Delta r^2 \cdot i}$$

$$= \sum_{j=1}^{n} (M_j B_{ij} - m_j b_{ij}) - \sum_{j=1}^{n} f_j(Q_{ij} + q_{ij}) - \frac{f_i}{2\pi K \eta \Delta r^2 \cdot i}, \quad (i = 1, 2, \ldots n)$$

$$\tag{6.24}$$

with the additional condition of force equilibrium for the module subjected to f_j and p_j forces bringing up the number of equations to $n + 1$:

$$2\pi\eta\Delta r \sum_{j=1}^{n} j(p_j + f_j) = 0 .$$ (6.25)

The system of equations [(6.24) and (6.25)] can also be written in a matrix form:

$$\{W_0\} + \left([Q] + [q] + \frac{[J]^{-1}}{\omega}\right)\{p\} = [B]\{M\} - [b]\{m\}$$

$$- \left([Q] + [q] + \frac{[J]^{-1}}{\omega}\right)\{f\}$$ (6.26)

subject to

$$[J](\{p\} + \{f\}) = 0$$ (6.27)

where $\{W_0\}$ is a column matrix with each of its members equaling W_0; also

$$[J] = \begin{bmatrix} 1 & & & 0 \\ & 2 & & \\ & & \ddots & \\ 0 & & & n \end{bmatrix}$$ (6.28)

and

$$\omega = 2\pi K\eta\Delta r^2 .$$ (6.29)

It is remarked that the relation between discretized moments M_i and individual pin-end moments M_{ind} is

$$M_i = \eta\,\Delta r M_{ind}$$ (6.30)

while the discretized axial forces are related to the individual ones F_{ind} and P_{ind} as

$$f_i = 2\pi r_i\,\Delta r\,\eta\,F_{ind}(r_i)$$ (6.31)

$$p_i = 2\pi r_i\,\Delta r\eta P_{ind}(r_i) .$$ (6.32)

Example 1

The analysis will be applied and compared to the experimental results of Kelly et al. [6], who studied a 50-mm square PGA, cooled during manufacturing by a 180 °C temperature step. They had the following data:

Ceramic module: $E_m = 345$ GPa (50 Mpsi)

$\alpha = 6.0 \times 10^{-6}/°C$

$v = 0.3$

$h_m = 3.84$ mm (0.151 in.)

$s = 2.54$ mm (0.100 in.)

Epoxy glass card: $E_c = 13.8$ GPa (2 Mpsi)

$\alpha = 15.0 \times 10^{-6}/°C$

$v = 0.3$

$h_c = L_s = 1.27$ mm (0.050 in.)

Eutectic Sn-Pb solder: $E_s = 31$ GPa (4.5 Mpsi)

Kovar pin: $E = 138$ GPa (20 Mpsi)

$L = 1.57$ mm (0.062 in.)

$2c = 0.203$ mm (0.008 in.)

with hole clearance: $h = 0.191$ mm (0.0075 in.)

For the thermal mismatch, $u = 180$ ($[15-6] \times 10^{-6}) \cdot r$ may be used. The elastic foundation has, using k^* for the foundation constant (to avoid confusing the symbols k):

$$k^* \cong 4E_s c/h = 1.32 \times 10^8 \text{ N/mm}^2 \text{ [from Eq. (5.3)]}$$

$$\lambda = (k^*/4EI)^{1/4} = 3.66 \text{ mm}^{-1}$$

$$\gamma = \lambda L_s = 4.65 > 3, \text{ semi-infinite beam range}.$$

This results, by the notation of Eqs. (5.14–5.19), in

$$U_1 = 0.670$$

$$U_2 = 0.433$$

$$U_v = 0.551.$$

The module was idealized as a circular one of $a_m = (45.72/\cos 45°)/2 = 32.3$-mm radius; this was divided into $n = 16$ equally wide annuli, $\Delta r = 2.02$ mm. The card was considered to have a size $d = 200$ mm ($\rho_c \gg \theta_m$). The pin density was $\eta = 1/s^2 = 0.155$ mm^{-2}. The maximum primary pin forces were calculated as: $M_1 = 17$ N/mm (0.15 in/lb), $M_2 = 12$ N/mm (0.11 in/lb), and $F = 12.0$ N (2.68 lb).

The primary moment and shear force variations are shown in Figs. 6.6a and 6.7a, respectively. Note that the moments and shears have a sign (plus or minus) depending on the sign of the temperature increment, ΔT. However, the primary axial pin forces, F, are always positive, i.e., tensile, regardless of whether heating or cooling of the PGA takes place, since they depend on the absolute value of ΔT. Accordingly, the secondary axial forces P can be expected to turn out differently for cooling and heating.

Figure 6.8a shows the results for individual pin-force variation along the radial dimension of the PGA, for cooling of the module by $\Delta T = -180°C$. Primary (F), secondary (P) and total combined ($F + P$) pin forces are plotted separately. The primary force variation is a parabola of tensile values. The

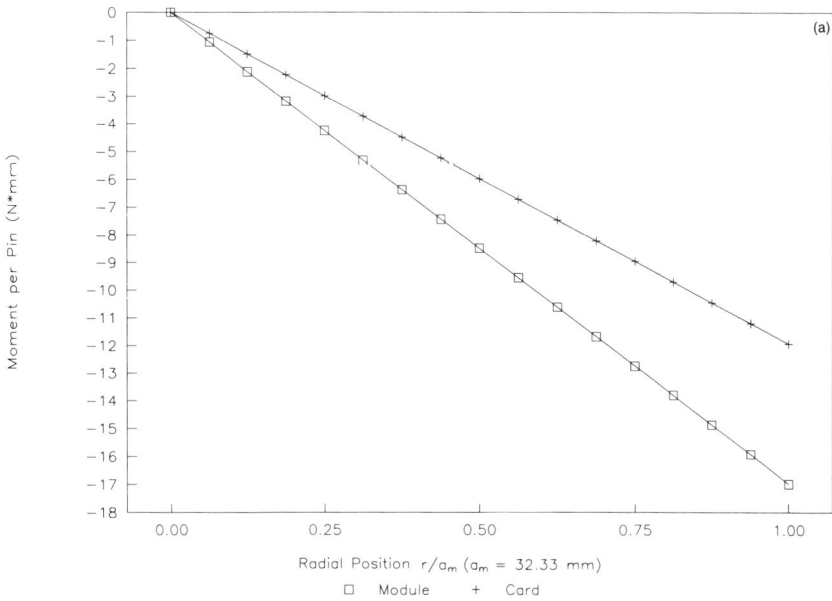

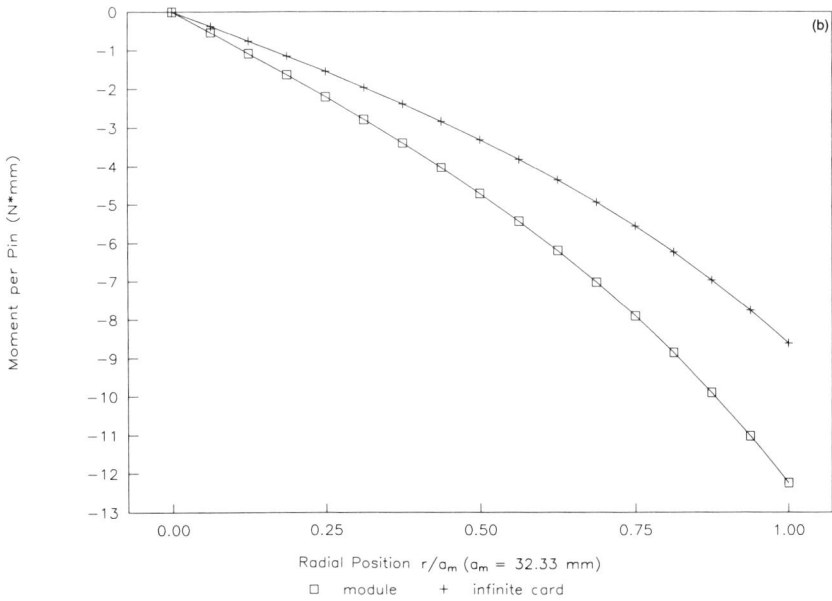

FIGURE 6.6. Moments induced in the pins of Example 1. (a) Primary and secondary effect analysis; (b) tertiary effect analysis – infinite card.

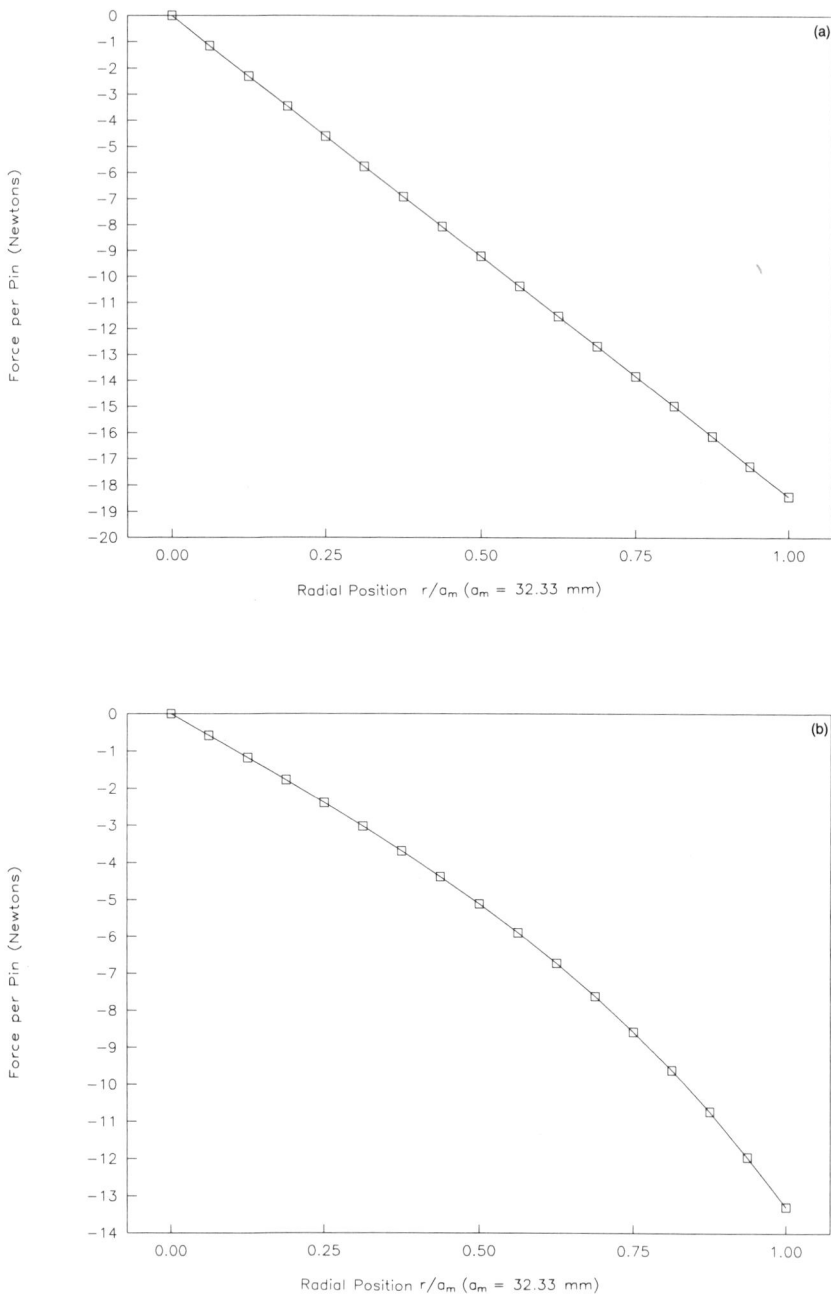

FIGURE 6.7. Shear forces induced in the pins of Example 1. (a) Primary and secondary effect analysis; (b) tertiary effect analysis – infinite card.

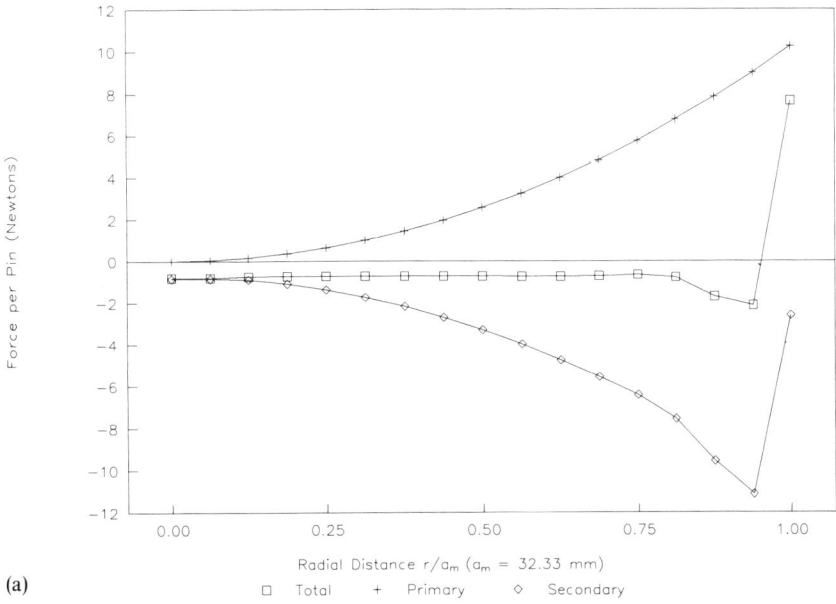

(a)

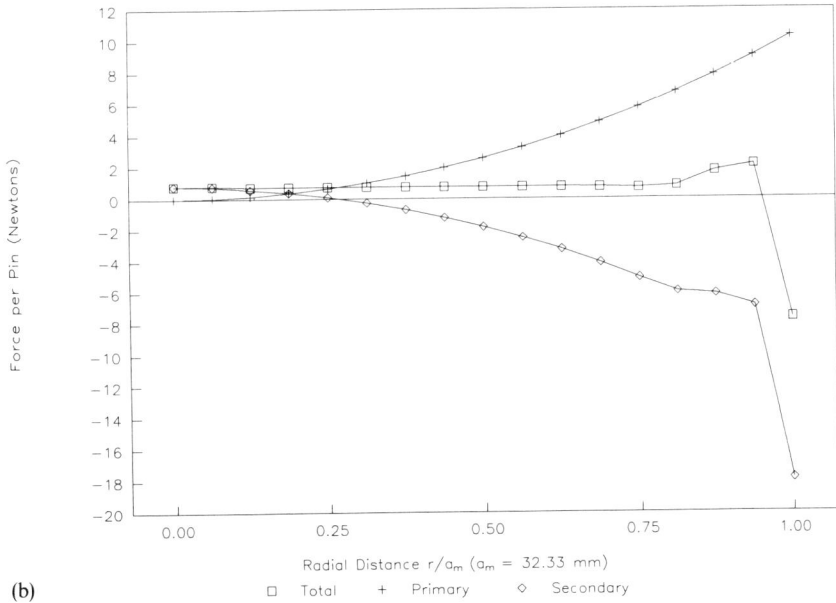

(b)

FIGURE 6.8. Primary, secondary, and resultant axial pin force variation for Example 1. (a) For $\Delta T = 180\,^{\circ}\text{C}$ (heating); (b) for $\Delta T = -180\,^{\circ}\text{C}$ (cooling).

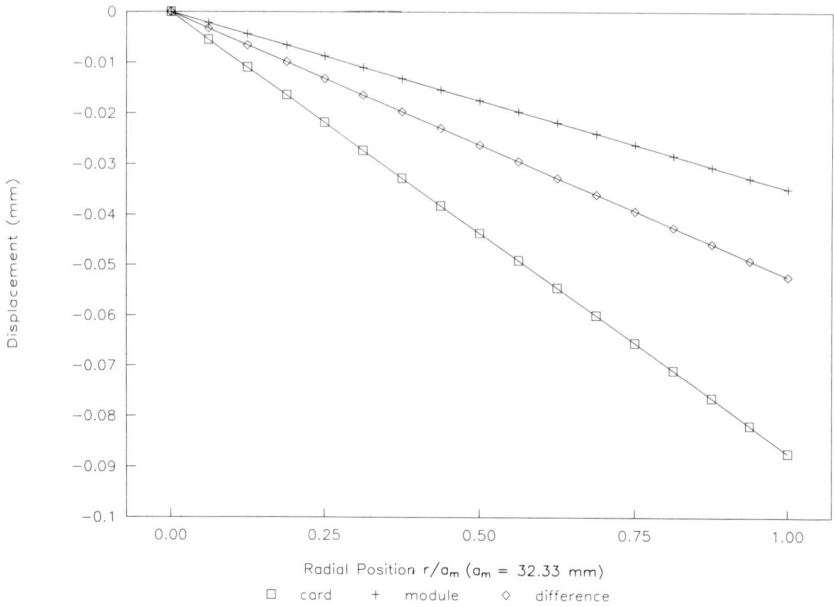

FIGURE 6.9. In-plane free thermal displacements of module and card.

secondary pin force starts as a monotonically increasing compression but is sharply reduced at the periphery. The resulting combined pin force is a low-grade compression over most of the module, with a large tensile ordinate at the periphery.

When the PGA is heated, however, the secondary pin forces start with a shallow tensile distribution at the origin, which gives way to a monotonically increasing trend of compression all the way to the periphery (Fig. 6.8b). Thus the resultant pin-force distribution is a low-grade tension in the central region of the module, while the peripheral pins are under sizable compression. (This compression value is numerically the same as the tension for the corresponding cooling to ΔT.) Good agreement of the maximum net pin force $F + P$ was found with respect to the experimental values of Kelly et al.

The free in-plane module and card displacements are shown in Fig. 6.9. These are not changed by the secondary pin forces P, but are diminished by the pin shear. This "tertiary" effect will next be analyzed.

6. Module and Card Stretch Due to Pin Shear

The shear forces V at the pin holes will cause an in-plane radial stretch v of the module and card, ignored in the primary analysis; the stretch tends to relieve the thermal stress in the pin. This effect can be modeled by assuming a continuous distribution of in-plane force instead of discrete ones at the pins. Writing the radial in-plane displacements $v(r)$ of an elastic plate, due to an applied radial line load R at $r = b$, the theory of elasticity [1] yields

$$v(r) = R(b) \cdot \Phi(r, b) \tag{6.33}$$

with

$$\Phi(r, b) = \frac{r}{2Eh} \left[(1 - v^2) \left\{ \left(\frac{b}{a} \right)^2 + 1 \right\} \right], \quad (r \leqslant b) \tag{6.34}$$

and

$$\Phi(r, b) = \frac{r}{2Eh} \left[(1 - v^2) \left\{ \left(\frac{b}{a} \right)^2 + \left(\frac{b}{r} \right)^2 \right\} \right], \quad (r \geqslant b) . \tag{6.35}$$

Figure 6.10 shows the in-plane displacement influence functions due to an in-plane line load. Proceeding as before to calculate the elastic stretch due to an in-plane radial line force contributed by primary shear, we first get an expression for the latter, from Eq. (6.3):

$$t_j = \frac{12EI}{L^3} \frac{\gamma^3}{\mathscr{D}} \Delta T \cdot \Delta \alpha \cdot 2\pi \Delta r \cdot \eta r_j^2 . \tag{6.36}$$

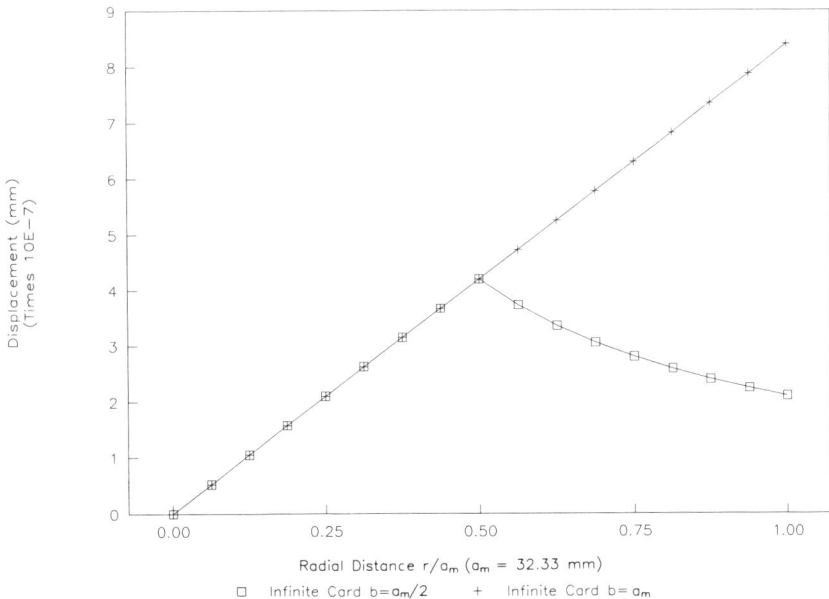

FIGURE 6.10. In-plane displacement influence function due to a unit line force, for an FR-4 card described in Example 1.

The in-plane stretch influence coefficient for shear applied over lamella j is then obtained from Eqs. (6.34) and (6.35) for module and card, respectively:

$$C_{ij} = \Phi_1[(i - \tfrac{1}{2})\Delta r, (j - \tfrac{1}{2})\Delta r] \tag{6.37}$$

$$c_{ij} = \Phi_2[(i - \tfrac{1}{2})\Delta r, (j - \tfrac{1}{2})\Delta r]. \tag{6.38}$$

The correction $v(r)$ to the original mismatch $u(r)$ is thus obtained numerically, by summing the influences of each lamella:

$$v_1(r_i) = \sum_{j=1}^{n} t_j C_{ij}, \tag{6.39}$$

$$v_2(r_i) = \sum_{j=1}^{n} t_j c_{ij}. \tag{6.40}$$

For the data of Example 1, Fig. 6.9 shows the primary thermal displacements $u(r)$ of the card and module. The stretch $v(r)$ and the resultant in-plane displacement $u - v$ for both the module and the card of the PGA are shown in Fig. 6.11. The radial moments corrected by the tertiary effect are shown in Fig. 6.6b, and the corrected shears in Fig. 6.7b.

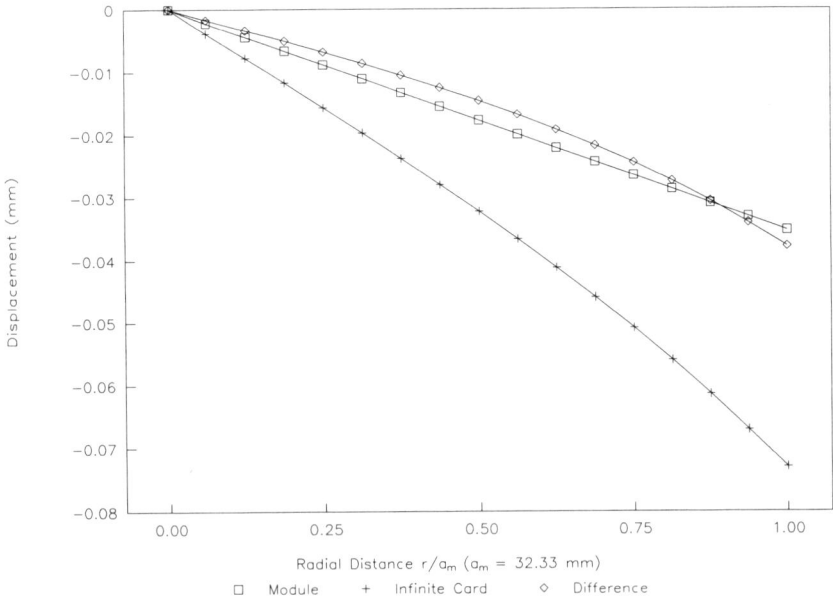

FIGURE 6.11. In-plane end displacements of pins of Example 1. The thermal displacements are corrected by the in-plane stretch effect.

7. System Reduction Factor

A numerical indicator of the effects of module and card flexibility on pin stress versus the primary value of the latter can be expressed in a "system reduction factor," f. Defining

$$f = \text{(total stress)/(primary stress)}, \tag{6.41}$$

several configurations of PGAs were computed in Ref. [7], using finite element results for the total combined stress. The primary pin stresses were calculated along the methods of Chapter 5. The module size a ranged from 25 mm up to $a = 50$ mm; a pin spacing of $s = 2.5$ mm (0.1 in.) was considered for a full area-array of Kovar pins. Pin radii of both $c = 0.4$ and $c = 0.5$ mm (16 and 20 mil) were considered.

The resulting f versus a plots (Fig. 6.12) show f declining with increasing module size and increasing pin stiffness; this is a visual proof of the conservativeness of a mere "primary" pin analysis.

Another source of thermal stress reduction is the partial cracking of the solder joint. One may think of the standoff length L of the pin having been effectively increased by the formation of a longitudinal solder crack. Such solder cracks halfway along the circumference and substantially into the

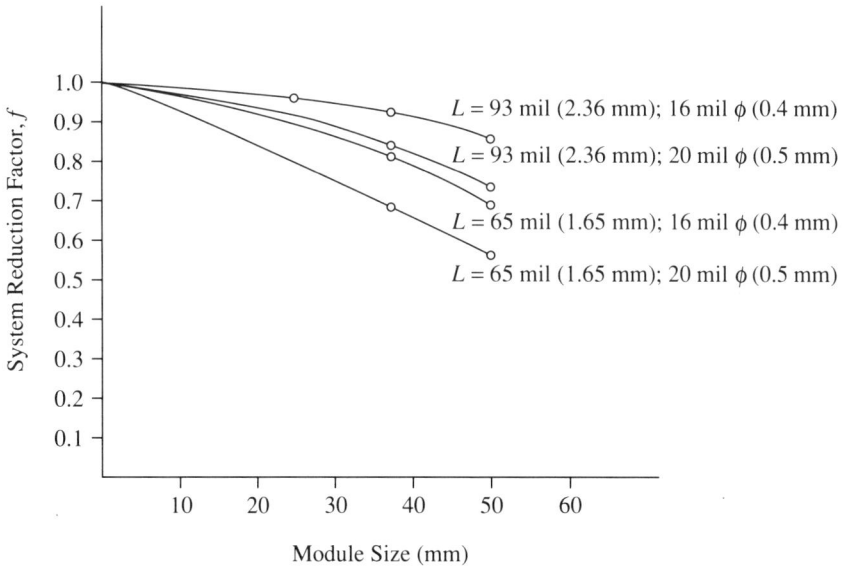

FIGURE 6.12. System reduction factors versus module size, for area-array PGAs. The composition is ceramic module plus FR-4 card, with a Kovar pin spacing of $s = 2.5$ mm.

depth have been accordingly permitted by some specifications, based on the strain relief concept.

8. Conclusions

In Chapters 5 and 6 the hierarchy of thermal stress contributions in a PGA was viewed. We distinguished four stages: (1) Fixed-ended pin analysis; (2) modification due to elastically supported beam effect (pin-in-hole relief in the solder joint); (3) secondary pin forces calculated from transverse bending of module and card; and (4) in-plane stretch of module and card, relieving thermal mismatch (tertiary contribution).

The third and fourth stages were calculated in the present chapter. It is clear that further improvement of the analytical results could be obtained by recomputing the pin forces and pin moments by the new values of mismatch, improved by the calculated stretch. Such further iterations, however, would seem to create only small improvement.

It is not so much for the accuracy of stress calculation (which is greatly influenced by input data) as for obtaining a relatively quick computational method for the prediction of design parameter effects that the preceding procedures may be used for maximum benefit. The larger the module can be made, the more technological interest is attached to it, considering especially multichip modules. A beneficial stress-relieving role derived from the bending of a module and card pair under temperature is therefore an important piece of design information. Finite element calculations of this effect by Engel et al. [7] showed a possible 50 percent stress reduction for 50-mm modules using 0.5-mm (0.020-in.) diameter pins on an $L = 2.5$ mm (0.064 in.) standoff length (Fig. 6.12). By the present approach a similar magnitude was calculated on a comparable structure, by analytical means. By the present method, good agreement with the experimental measurements of Kelly et al. [6] was achieved.

9. Exercises and Questions

1. Derive expressions (6.14) and (6.15) from the elastic theory of thin plates [5].
2. Show that Eqs. (6.14) and (6.15) and Eqs. (6.34) and (6.35) conform to Maxwell's theorem of reciprocal deflections. How would you proceed showing that Eqs. (6.8) and (6.9) conform to the reciprocal theorem?
3. Consider all the geometric and material parameters of a PGA: e.g., α (for module and card); D (for module and card); h (card); c, L, and E (of pin); and E and b (of solder joint). List their effect (increase or decrease) on the thermal stresses of pins and solder joints.
4. Suppose the length ΔL of a longitudinal solder crack can be considered to lengthen the pin standoff. What is the resulting percentage change of an "allowable module size" based on pin stress considerations alone?

References

1. Timoshenko, S.P., and Goodier, J.N. (1951), *Theory of Elasticity*, 2d ed., McGraw-Hill, New York.
2. Hall, P.M. (1984), "Forces, Moments and Displacements During Thermal Chamber Cycling of Leadless Ceramic Chip Carriers Soldered to Printed Wiring Boards," *IEEE Trans. on CHMT*, **CHMT-7**(4), 314–327.
3. Engel, P.A., and Webb, J.R. (1992), "Thermal Stress Analysis of Pin Grid Array Structures: Module and Card Interactions," ASME J. Elec. Packag., **114**(3), 322–328.
4. Conway, H.D., and Engel, P.A. (1969), "Contact Stresses in Slabs Due to Round, Rough Indenters," *Int'l. J. Mech. Sci.*, **11**(11), 709–722.
5. Timoshenko, S.P., and Woinowsky-Krieger, S. (1959), *Theory of Plates and Shells*, 2d ed., McGraw-Hill, New York.
6. Kelly, J.H., Lim, C.K., and Chen, W.T. (1984), "Optimization of Interconnections Between Packaging Levels," *IBM J. Res. Dev.*, **28**(6), 719–726.
7. Engel, P.A., Toda, M.D., and Trivedi, A.K. (1983), "Design Guide for Solder Cracking in Module-Card Assemblies," *IBM Tech. Rep. 01.2678*, Endicott, N.Y.

Chapter 7

Compliant Leaded Systems: The Local Assembly

Compliant leads, introduced in surface-mounted circuit card systems, help reduce stress arising in the interconnecting elements due to thermal mismatch between the module and card. The peripheral SMT joints, however, are also subject to mechanical stress owing to handling, vibration, and shock. The flexural and torsional deformations (called "flexing" of the card in general) lead to a maximum tie force at the module corners; this force has a predominant component in the z-direction, as discussed in Chapter 3. While the axial lead stretch is contributed overwhelmingly by bending action along its curved length, a simplified modeling of leads as axial springs with a spring constant K may account for this flexural deformation; K values have also been measured for various leads (see Table 3.5).

The problems that confront the mechanical analyst and designer of circuit card assemblies include stiffness and lead force distribution in the module-card sandwich. This, for the total circuit card, is generally a complex task, considering the variegated nature of layouts and types of modules. There is much information to be gained, however, from studying a smaller building block. A "local assembly" will be defined as a cutaway structure around a particular module, with a slightly larger piece of the circuit card just including the lead junctures. This chapter is devoted to this special circuit card system.

1. Experimental Studies

One of the simplest local assemblies is made with an SOIC shaped (such as SOJ or SOT) module that has two parallel rows of leads (Fig. 7.1). Its elongated shape warrants beam action. This assembly, simply supported on the card and with equidistant gull wing leads, was tested under two types of load: (1) a central concentrated force P (a line load across) applied to the card; and (2) P applied to the module. In either case, the load-deflection

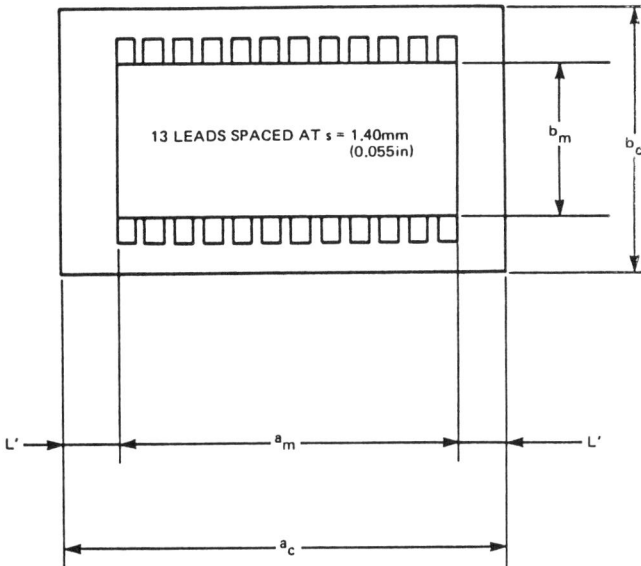

FIGURE 7.1. Dimensions of a local assembly.

characteristics in the linear range were checked as P versus w and P versus W, respectively, where w means card displacement and W is module displacement. The measured stiffness values $S = P/w_{max}$ or P/W_{max} of these local assemblies were entered as the first entries of Table 7.1.

These assembly stiffnesses are greater than the individual stiffnesses of a corresponding card or module element; the assembly is a "coupled beam" structure – two beams reinforcing one another by an interposed medium, the leads. The assemblies may also be varied for a study of coupled beams (Fig. 7.2): the card portion may be made longer, the loading may be end moments, the modules may be of the "stacked" variant, or they may be "double-sided" meaning identical modules surface soldered on both sides of the card. Such load versus deflection measurements [1] were made indeed, and the resulting stiffnesses added to Table 7.1.

What is apparent from the results of Table 7.1 is that the coupled beam system is not, by far, behaving as a composite beam. The latter would, by definition, satisfy the "planes remain plane" beam criterion. In fact, the coupled beams of a local assembly may have less total stiffness than the individual members (card and module) added together; this is due to the role leads play. The true structural behavior of a local assembly will be assessed in the next chapter, when we compare experimental data with those supplied by an analytical model.

TABLE 7.1. Local assembly (single-module) stiffness. Lengths are shown in mm (in.).

Card	Module	Leads	Stiffness N/mm (lb/in)	
			Measured	Analytical
C_1, FR4	M_1, SOJ	J, 13 pairs	Load on card	
$a = 17.5$ (0.69 in.)	14.0 (0.55 in.)		420	427
$b = 13.2$ (0.52 in.)	7.6 (0.30 in.)	$s = 1.4$ mm	(2400)	(2440)
$R = 28,220$ N mm^2	54,230	(0.055 in.)		
(9.83 lb in^2.)	(18.89)			
C_2, FR4	M_2, SOJ	Gull, 13 pairs	Load on card	
$a = 17.5$ (0.69 in.)	14.0 (0.55 in.)		499	596
$b = 14.5$ (0.57 in.)	7.6 (0.300 in.)	$s = 1.4$ mm	(2850)	(3400)
$R = 41,660$ N.mm^2	67,670	(0.055 in.)		
(14.51 lb in^2.)	(23.57)			
C_3, FR4	M_3, SOIC	Gull, 13 pairs	Load on card	
$a = 18.8$ (0.74 in.)	17.3 (0.68)		318	298
$b = 10.9$ (0.43 in.)	7.6 (0.300)	$s = 1.4$ mm	(1814)	(1699)
$R = 26,380$ N.mm^2	79,580	(0.055 in)	Load on module	
(9.19 lb in^2.)	(27.72)		445	365
			(2540)	(2083)
	M_4, (SOIC)		556	
	$a = 17.3$ (0.68)		(3175)	–
	$b = 7.6$ (0.300)			
	$R = 59,720$			
	(20.80)			
	M_5, stacked		636	
	$a = 17.3$ (0.68)		(3628)	–
	$b = 7.6$ (0.300)			
	$R = 59,720$			
	(20.80)			
	$\phi = 1.14$			
C_3	M_4 double-sided	Gull, 13 pairs $s = 1.4$ mm (0.055 in.)	470	512
			(2681)	(2921)
			Load on card	
C_3	M_5 stacked	Gull, 13 pairs $s = 1.4$ mm (0.057 in.)	342	297
			(1954)	(1695)
			Load on module	
			445	357
			(2540)	(2040)

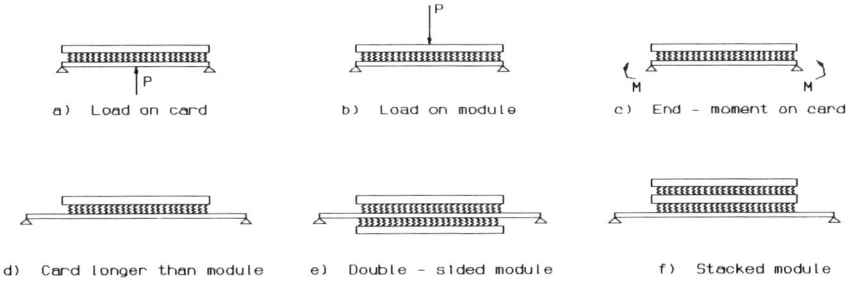

FIGURE 7.2. Various configurations of module-lead-card local assemblies. For symmetric loading, the x axis originates at mid-span.

2. Analytical Model

The two parallel rows of n leads in the local assembly of Fig. 7.1 (and in the schematic representations of Fig. 7.2) may be assumed to act as parallel independent elastic springs constituting a Winkler foundation of constant k (N/mm^2): from the lead spacing s and individual axial lead spring K we may write

$$k = 2K/s \approx 2nK/a . \tag{7.1}$$

The module (subscripted m) and card (subscripted c) have beam rigidities $R_m = D_m b_m$ and $R_c = D_c b_c$. The distributed load, of intensity q, is the averaged transverse force transmitted through the leads: a tensile q acting upward on the card and downward on the module. Thus the individual beam equations for module and card are written, using the convention of Fig. 1.1:

$$W^{iv} = - q/R_m \tag{7.2}$$

and

$$w^{iv} = q/R_c . \tag{7.3}$$

Introducing a new variable $\eta = W - w$, and noting that by definition

$$q = k\eta , \tag{7.4}$$

we may combine Eqs. (7.2) and (7.3) into a single ordinary linear constant-coefficient fourth-order differential equation:

$$\eta^{iv} + 4\lambda^4 \eta = 0 \tag{7.5}$$

where

$$R = (1/R_m + 1/R_c)^{-1} \tag{7.6}$$

and

$$\lambda = (k/4R)^{1/4} . \tag{7.7}$$

We recognize Eq. (7.5) as the "elastic foundation" equation [Eq. (1.10)], which, in terms of symmetrical loading on the beam assembly (the x-axis originated at mid span) yields [2], a result analogous to Eq. (1.11):

$$\eta = C_1 \cosh \lambda x \cdot \cos \lambda x + C_4 \sinh \lambda x \cdot \sin \lambda x . \tag{7.8}$$

The two coefficients C_1 and C_4 are determined from two boundary conditions dictated by the loading. The latter are shown in Table 7.2, for the three loading configurations: (1) concentrated force on the card, (2) concentrated force on the module, and (3) end moments. Having obtained η, the preceding procedure has now resulted in an exact solution for the interbeam pressure q by Eq. (7.4) and Table 7.3. Should we find it necessary to calculate shears, moments, slopes, or displacements for the module and card, it can be done by successive integration of Eqs. (7.2) and (7.3), respectively. We note that great simplifications in the equations occur if the value $\rho \equiv \lambda a/2 > 3$.

In the following development, we shall refer to beam stiffness under concentrated load ($S = P/w$ or P/W) or end-moments. For the end-moment loading, the "stiffness" may be defined as the ratio of end-moment to end slope, $S = M_0/\theta$.

In addition to the stiffness, we are interested in the lead forces. The end leads will be maximally stressed in either load configuration, and these maximum lead forces are important, especially in the case of moment loading

TABLE 7.2a. Boundary conditions for η.

	Pure moment		Concentrated central force P	
x	M	x	On card	On module
0	$\eta' = 0$	0	$\eta' = 0$	$\eta' = 0$
$a/2$	$\eta'' = -\dfrac{M'}{E_c I_c}$	$a/2$	$\eta'' = 0$	$\eta'' = 0$
0	$\eta''' = 0$	0	$\eta''' = \dfrac{P}{2R_c}$	$\eta''' = -\dfrac{P}{2R_m}$
$a/2$	$\eta''' = 0$		$\displaystyle\int_0^{a/2} \eta\, dx = 0$	$\displaystyle\int_0^{a/2} \eta\, dx = -\dfrac{P}{2k}$

TABLE 7.2b. Boundary conditions for w and W.

	Pure moment,	Simply supported card, concentrated force	
x	on card	Card loaded	Module loaded
0	$w = 0$	$-$	$-$
$a/2$	$-$	$w = 0$	$w = 0$
0	$w' = 0$	$w' = 0$	$w' = 0$
$a/2$	$-$	$-$	$-$
0	$-$	$-$	$-$
$a/2$	$w'' = -M/E_c I_c$	$w'' = 0$	$w'' = 0$
0	$w''' = 0$	$w''' = -P/2R_c$	$-$
$a/2$	$-$	$-$	$W''' = P/2R_m$

TABLE 7.3. Solutions for three symmetric loading cases:
$\eta(x) = C_1 \cosh \lambda x \cdot \cos \lambda x + C_4 \sinh \lambda x \cdot \sin \lambda x$.

| | P concentrated central force. | | |
	On card	On module	M edge moment
Coefficients			
C_1	$-\dfrac{P}{4\lambda^3 R_c} \cdot \dfrac{c \cdot ch}{ch \cdot sh + c \cdot s}$	$-\dfrac{P\lambda}{k} \cdot \dfrac{c \cdot ch}{ch \cdot sh + c \cdot s}$	$\dfrac{M}{2\lambda^2 R_c} \cdot \dfrac{ch \cdot s - sh \cdot c}{ch \cdot sh + c \cdot s}$
C_4	$-\dfrac{P}{4\lambda^3 R_c} \cdot \dfrac{s \cdot sh}{ch \cdot sh + c \cdot s}$	$-\dfrac{P\lambda}{k} \cdot \dfrac{s \cdot sh}{ch \cdot sh + c \cdot s}$	$-\dfrac{M}{2\lambda^2 R_c} \cdot \dfrac{ch \cdot s + sh \cdot c}{ch \cdot sh + c \cdot s}$

Abbreviations: $s = \sin \rho$, $c = \cos \rho$, $sh = \sinh \rho$, $ch = \cosh \rho$, $\rho \equiv \lambda a/2$.

that corresponds to the natural loading configuration of a module mounted on and bent along with a circuit card. For the last lead, we may integrate the pressure distribution over the end spacing, and get for the lead-force:

$$F = (1/2) \int_{(n-2)s/2}^{(n-1)s/2} q(x)\, dx \ . \tag{7.9}$$

The formula resulting for end-moment load M_0 is, simplified by the conservative assumption of constant pressure over the end-spacing,

$$F = akM_0/8nR_c\lambda^2 \ . \tag{7.10}$$

Example 1

The SOIC module of a local assembly depicted in Fig. 7.1 has the dimensions $a_m = 17.3$ mm (0.68 in.), $b_m = 7.6$ mm (0.30 in.), $h_m = 3$ mm (0.118 in.); its equivalent modulus is $E_m = 3.91$ GPa (567 Kpsi) and Poisson ratio $v_m = 0.4$. The FR-4 card, $b_c = 12.75$ mm (0.5 in.) wide and $h_c = 1.25$ mm (0.050 in.) thick, is of matching span a_m, and its properties are $E_c = 13.8$ GPa (2 Mpsi), $v_c = 0.3$. The $n = 13$ pairs of gull wings have individual stiffness $K = 630$ N/mm (3600 lb/in.) and are spaced at a distance $s = 1.25$ mm (0.05 in.) apart. Find the equations of interbeam pressure q under midspan concentrated loadings and due to end moments. Also calculate the approximate end-lead force.

We calculate the rigidities:

$$R_m = E_m b_m h_m^3/[12(1 - v_m^2)] = (3910)(7.6)(3^3)/[12(1 - 0.4^2)]$$

$$= 79{,}600 \ \text{N.mm}^2$$

$$R_c = (13{,}800)(12.75)(1.25^3)/[12(1 - 0.3^2)] = 31{,}500 \ \text{N.mm}^2 \ .$$

By Eq. (7.6)

$$R = (1/79{,}600 + 1/31{,}500)^{-1} = 22{,}570 \text{ N.mm}^2 \ .$$

Now the foundation modulus computed from 13 pairs of leads is

$$k = (630)(2)(13)/17.3 = 947 \text{ N/mm}^2$$

so that the foundation constant by Eq. (7.7) is

$$\lambda = (947/[(4)(22{,}570)])^{1/4} = 0.32 \text{ mm}^{-1}$$

and the dimensionless foundation parameter is

$$\rho = \lambda a/2 = (0.32)(17.3)/2 = 2.77 \ ,$$

which is smaller than 3. The assembly may not pass for a semi-infinite beam treatment.

Three sets of the coefficients C_1 and C_4 are calculated from Table 7.3. The trigonometric and hyperbolic functions evaluated at $\rho = \lambda a/2$ are

$$s = \sin(2.77) = 0.363, \qquad sh = \sinh(2.77) = 7.948$$

$$c = \cos(2.77) = -0.932, \qquad ch = \cosh(2.77) = 8.011 \ .$$

Now from load P on the card, we get

$$C_1 = -\frac{(-0.932)(8.011)P}{(4)(0.32^3)(31{,}500)} \frac{1}{(8.011)(7.948) - (0.363)(0.932)}$$

$$= 28.55 \times 10^{-6} \cdot P \ ,$$

and similarly,

$$C_4 = -11.03 \times 10^{-6} \cdot P \ .$$

For a central load P on the module we get: $C_1 = -39.84 \times 10^{-6} \cdot P$ and $C_4 = -15.39 \times 10^{-6} \cdot P$. For end moments M there result $C_1 = 25.25 \times 10^{-6} \cdot M$; $C_4 = -11.03 \times 10^{-6} \cdot M$.

Note that C_1 has the physical meaning of midspan gap displacement η_0. It is interesting that η_0 is (28%) greater for module loading by P than for card loading by P. The opposite is true, however, when one evaluates the total displacements, W_0 and w_0.

For end-lead force under moment M_0 load we get, from Eq. (7.10):

$$F_{max} = \frac{(17.3)(947)M_0}{(8)(13)(31{,}500)(0.32^2)} = 0.04885\,M_0 \ .$$

3. Properties of Simple Local Assemblies

We shall check the effect of some basic parameters on the stiffness and lead force of simple local assemblies, ones consisting of a module and card of equal span like that in Example 1.

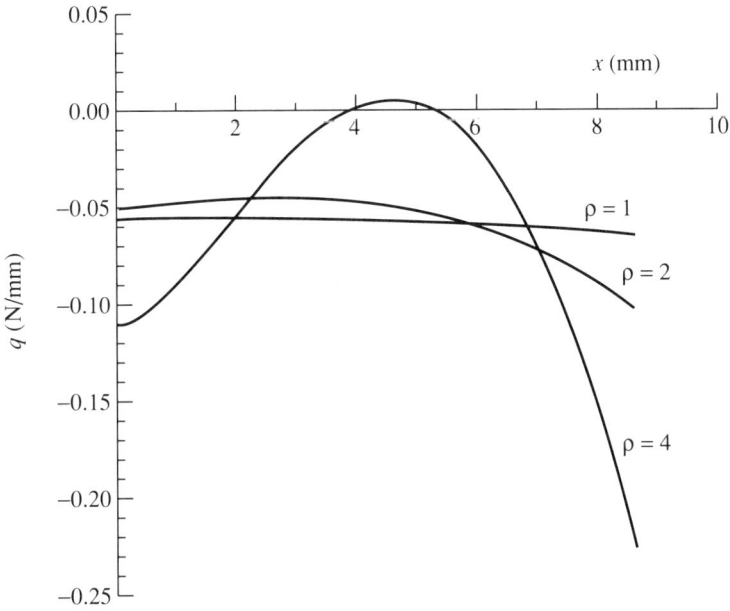

(a)

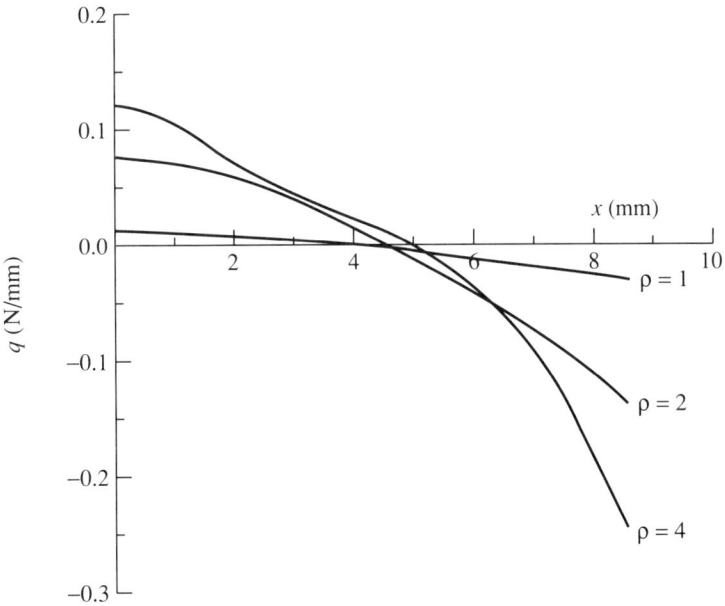

(b)

FIGURE 7.3. Interbeam pressure distribution for concentrated central load. (a) Load on module; (b) load on card.

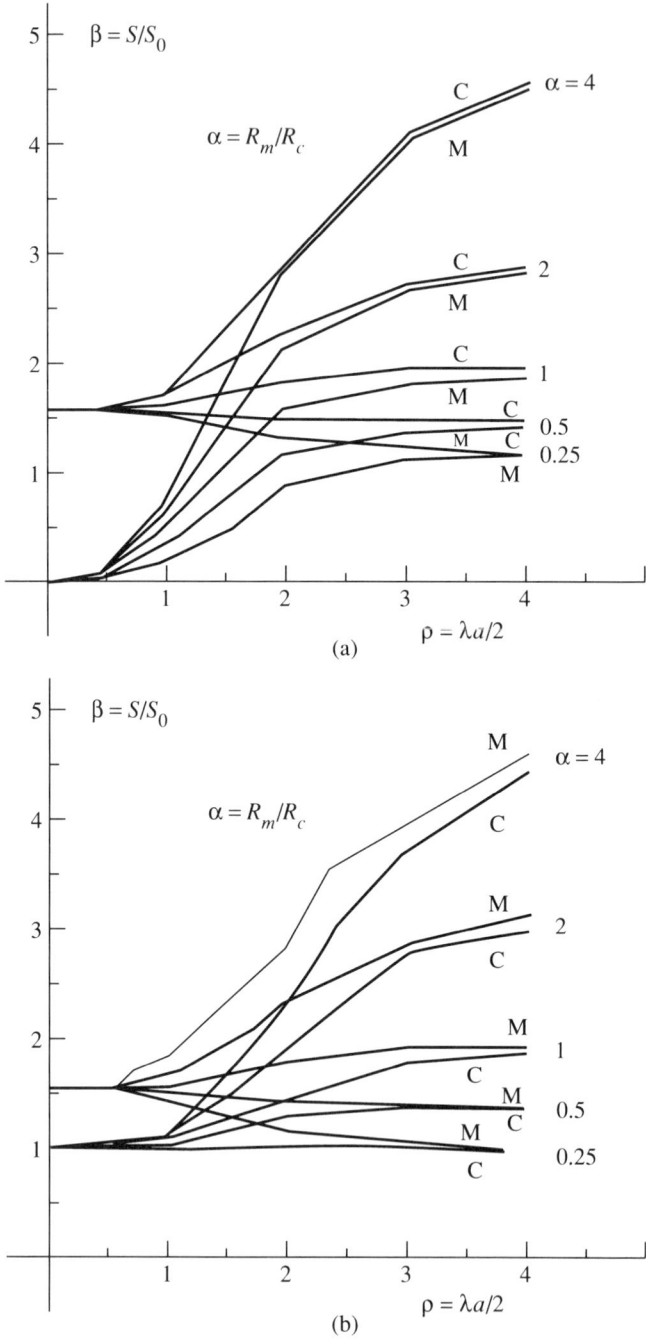

FIGURE 7.4. Stiffness of local assembly for concentrated central load. (a) Load on module; (b) load on card. Note: "M" refers to module stiffness, and "C" to card stiffness.

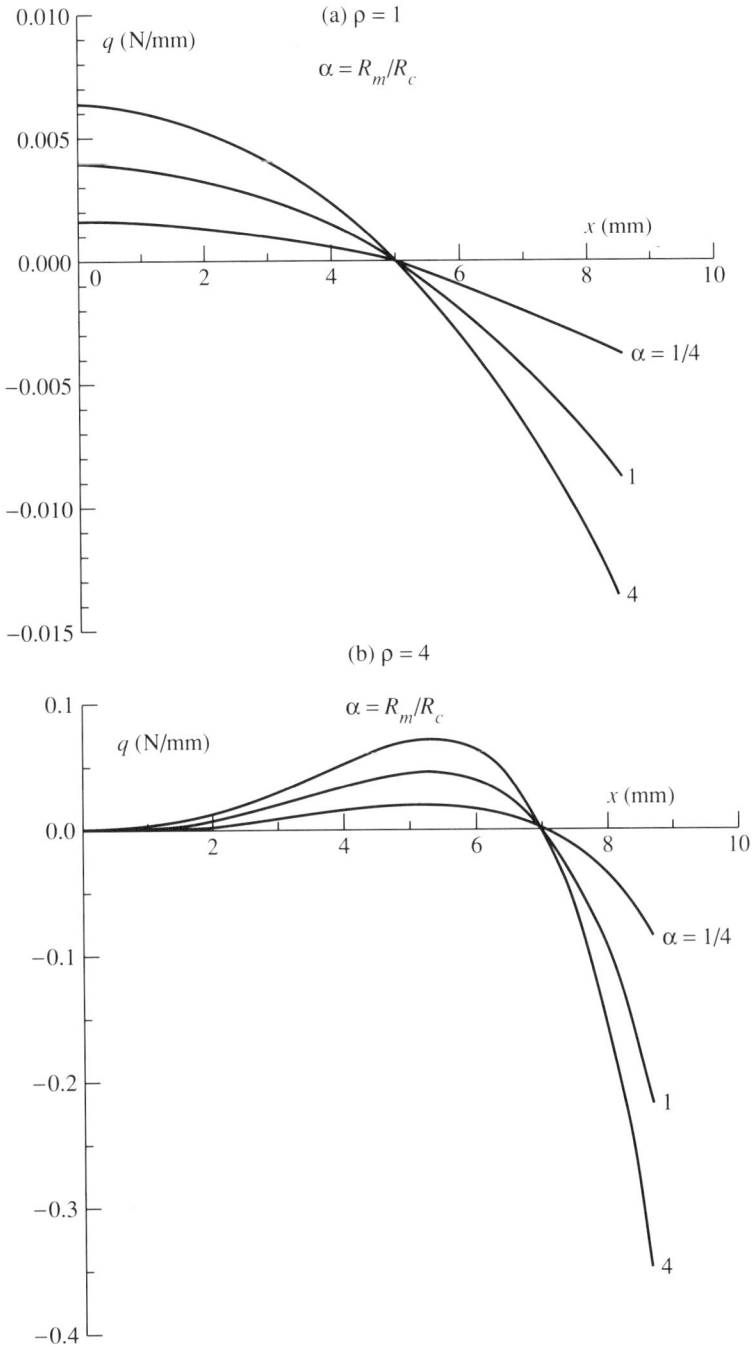

FIGURE 7.5. Interbeam pressure distribution for end-moment load. (a) $\rho = 1$; (b) $\rho = 4$.

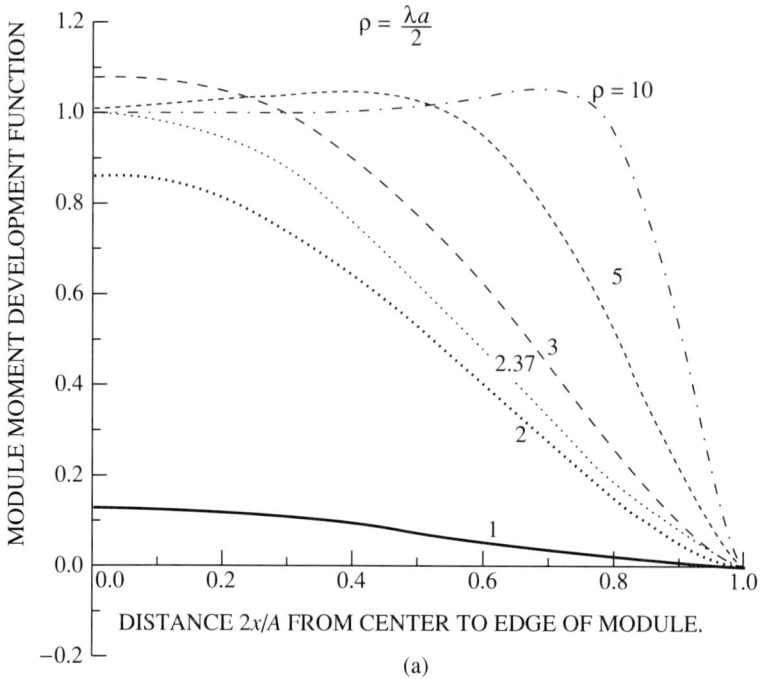

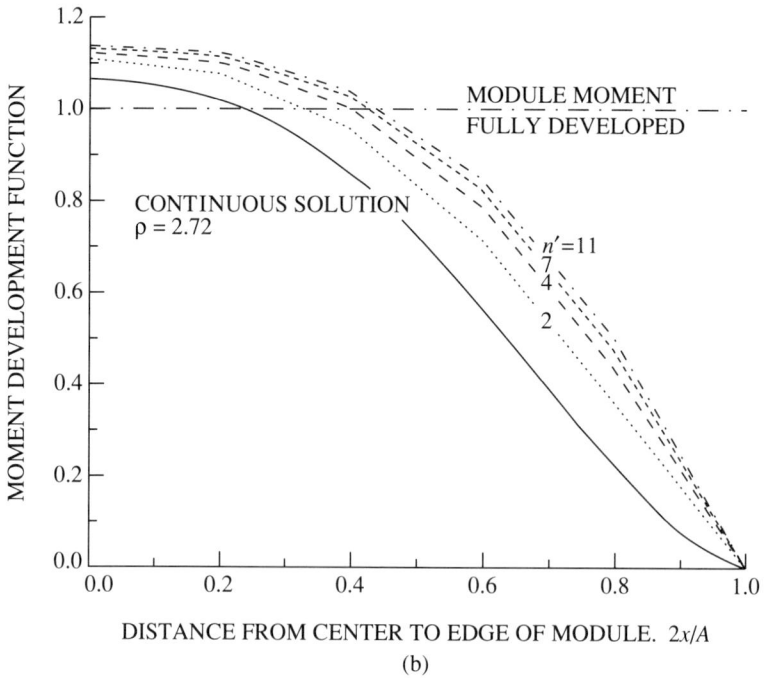

FIGURE 7.6. Module-moment development function versus lead stiffness. (a) Continuous lead distribution; (b) discrete lead distribution.

By the continuous model of coupled beams, the pressure distribution under concentrated load P on the module (Fig. 7.2b) is shown in Fig. 7.3a, and for P applied to the card (Fig. 7.2a), in Fig. 7.3b. Both diagrams have the card with the same properties as in Example 1, and the module parameters equaling those of the card. This corresponds to a nondimensional parameter $\alpha \equiv R_m/R_c$ chosen as 1.

The nondimensional parameter $\rho = \lambda a/2$ stands for the stiffness of the leads for constant a. Varying both ρ and α, but keeping $R_c = 31500$ N.mm^2 as in Example 1, the respective system stiffnesses S were then calculated and are shown in Figs. 7.4a and 7.4b for P on the module and the card, respectively. The nondimensional system stiffness is $\beta = S/S_0$, where S_0 stands for the stiffness corresponding to the card alone. The curves marked "M" refer to effective stiffness of the module and those marked "C" to the card stiffness. An interesting feature of low lead-stiffness configurations ($\rho \approx 0$) is the distribution of the interbeam pressure, tending to inverse the overall beam stiffness. For end-moment loading applied to the card (Fig. 7.2c), the interbeam pressure distribution q for soft leads ($\rho = 1$) is shown in Fig. 7.5a, and for stiffer leads ($\rho = 4$) in Fig. 7.5b. It is instructive to establish the gradual moment sharing from the card ends to the module [3]. Soft leads ($\rho < 2.37$) never manage to transfer the appropriate portion of module moment M_{m0} according to the ratio of module rigidity R_m to total rigidity $R_T = R_m + R_c$, which would result in $M_{m0} = (R_m/R_T)M_0$. The buildup of moment in the module through progressively stiffer leads ($1 < \rho < 10$) is shown in Fig. 7.6a. For stiffer leads ($\rho > 2.37$), the module moment reaches the plateau $M_m = M_{m0}$ progressively closer to the ends, $x = \pm a/2$.

4. Discrete Local Assembly

The previous model of leads connecting module and card as a continuous elastic medium becomes impractical when we deal with a square module. Even if we continue assuming beam action in, say, the x-direction, the leads marching in the perpendicular y-direction at a module's ends destroy the "uniformity" of the elastic foundation. We now will have n' leads in the first and last interval of the beam, and only two in the intermediate intervals. The practical implication is that, in view of Figs. 7.5a, 7.5b, the maximum lead forces at low lead stiffness would materialize at the center, not at the ends.

This problem can be treated numerically, by assuming a finite number of leads along the local assembly. More relevantly, we shall call for intermediate springs of intensity k and end springs of intensity $n' \cdot k_i$ (n' not necessarily equal to n, which would be the square-module condition). Thus, in the numerical treatment, we can establish the influence of n'.

Let the module and card be acted upon by equal and opposite lead forces F_i (Fig. 7.7). We shall consider even configurations with N unknown lead forces. (In general, $N = n/2$). We can write the displacements W_i and w_i,

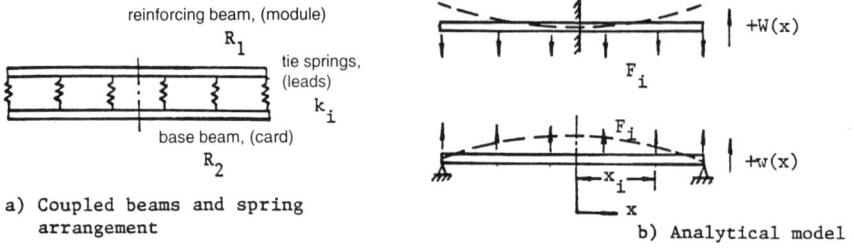

a) Coupled beams and spring
   arrangement

b) Analytical model

FIGURE 7.7. Discrete coupled-beam arrangement.

respectively, at discrete points x_i of lead junctures, using the proper influence functions $C_i(x)$ and $c_i(x)$; these express displacements at x due to unit load $F = 1$ at x_i. For the module, we have

$$C_i(x) = x^2(3x_i - x)/6R_m, \quad 0 < x < x_i \tag{7.11a}$$

$$C_i(x) = x_i^2(3x - x_i)/6R_m, \quad x_i < x < a/2 \tag{7.11b}$$

and for the card

$$c_i(x) = (a^3 - 6ax_i^2 + 4x_i^3 - 6ax^2 + 12x_i x^2)/24R_c, \quad 0 < x < x_i \tag{7.12a}$$

$$c_i(x) = (a^3 - 6ax^2 + 4x^3 - 6ax_i^2 + 12xx_i^2)/24R_c, \quad x_i < x < a/2. \tag{7.12b}$$

Thus, for the simply supported card:

$$w(x) = \sum F_i \cdot c_i(x) + y(x) \tag{7.13}$$

where $y(x)$ is the deflection due to external loads on the card. For the cantilevered module, which has an unknown rigid body displacement W_0 at $x = 0$:

$$W(x) = W_0 - \sum F_i \cdot C_i(x) + Y(x), \tag{7.14}$$

where $Y(x)$ is the displacement due to external loads P_m on the module.

The axial elongation of leads is expressed in

$$W(x_i) - w(x_i) = F_i/k_i . \tag{7.15}$$

Finally, $N + 1$ linear algebraic equations in the N lead forces and W_0 are obtained. To the equation of displacement compatibility

$$\sum F_i[C_i(x_j) + c_i(x_j)] + F_j/k_j - W_0 = Y(x_j) - y(x_j) \tag{7.16}$$

we add the criterion of equilibrium

$$\sum F_i = -P_m . \tag{7.17}$$

Example 1 can now be checked by the discrete method. It is also of interest to compute the response of the discrete assembly assuming an increasing number n' of end leads. In Fig. 7.6b, the module moment development function is graphed, corresponding to n' end leads; this can be compared with the continuous lead-system of Fig. 7.6a. In Fig. 7.8, the maximum lead force

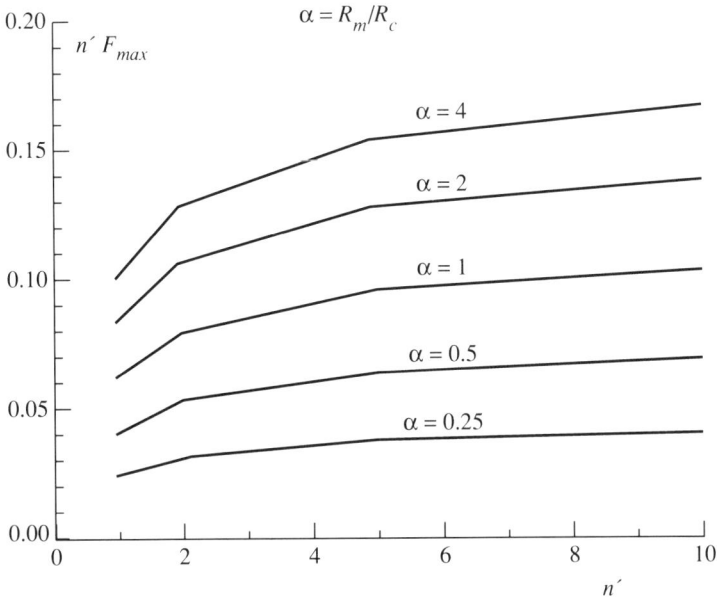

FIGURE 7.8. Maximum lead force versus end-spring number n', for discrete coupled-beam assembly under edge moment $M = 1$ N. mm applied to the card. Parameters are $R_c = 31,500$ N.mm, $a = 17.3$ mm, $\rho = 2$.

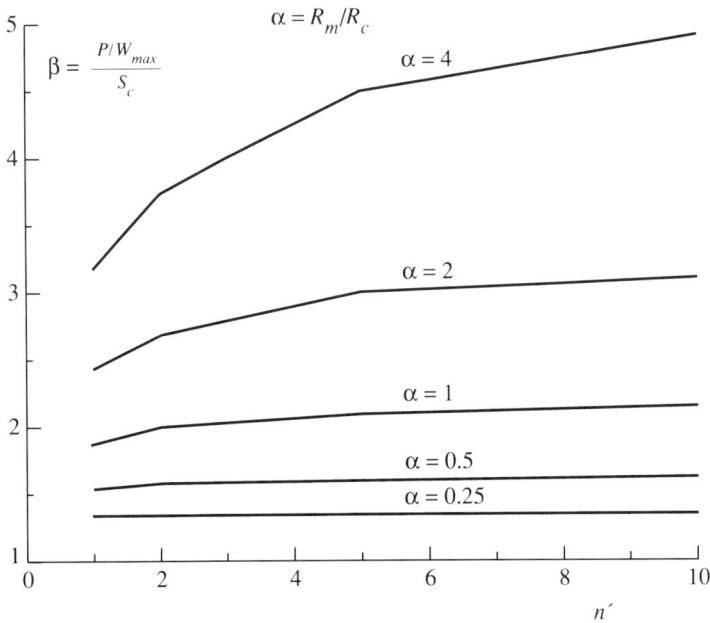

FIGURE 7.9. System stiffness versus number of end springs n' for different rigidity ratios $\alpha = R_m/R_c$; $\rho = 2$.

F_{max} (located over the ends) in end-moment loading is illustrated: five values of stiffness ratio α: 0.25, 0.5, 1, 2, and 4 were used, and n varied between 1 and 10. The parameter ρ was kept at a value of 2. We can compare the value of $F_{max} = 0.049M_0$ obtained in Example 1 (for $\rho = 2.77$) with a value obtained from Fig. 7.8 by interpolating with respect to α, but for $\rho = 2$; namely, $F_{max} = (0.11/2)M_0$. The increase of system stiffness S versus n' for concentrated force (P-load) on the module, and again using the α and ρ values given earlier, is demonstrated in Fig. 7.9.

5. Built-up (Multiple-Module) Local Assemblies

Figure 7.2e shows a "double-sided" local assembly, and Fig. 7.2f one of stacked modules. We shall call these and other possible variants "multiple-module" local assemblies; the structure has but one module seen in plan view, unlike "module groups."

5.1. Stacked-Module Arrangement

A two-story module is put together by an extra tier of leads, but we will be concerned only with lead forces of the first tier, the ones to be surface-soldered to the card. The extra rigidity added by the second story may be expressed by a factor ϕ larger than 1 in the beam equation (7.2) of the stacked module:

$$W^{iv} = - q/\phi R_m , \qquad (7.18)$$

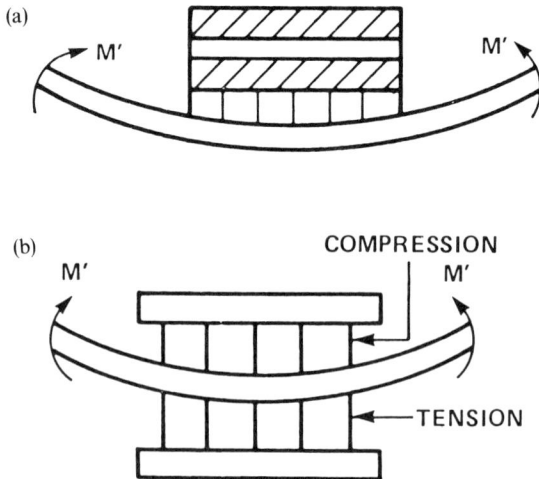

FIGURE 7.10. Flexural deformation in multiple local assemblies. (a) stacked-module arrangement; (b) double-sided module arrangement.

which makes the equivalent rigidity of the local assembly, by Eq. (7.6):

$$R = (1/R_c + 1/\phi R_m)^{-1} . \tag{7.19}$$

Thus, the rigidity of the original structure has been somewhat increased (Fig. 7.10a) and so is the assembly stiffness; a corresponding increase in lead forces is also expected by our theory over those of the simple-module assembly.

5.2. Double-Sided Module Assemblies

Consider an identical module attached to the card segment above and below (Fig. 7.10b). Now any external loading $f(x)$ of the assembly can be expressed as the superposition of a symmetric and an antisymmetric part (Fig. 7.11). The symmetric part produces only direct compression upon the structure, while the antisymmetric part produces bending. We can calculate the lead forces separately for the two cases, but the antisymmetric part is found to dominate over the purely compressive contribution, and we may neglect the latter for simplicity.

From the antisymmetric contribution we will then have equal distributed forces acting in the same direction with respect to the card, and Eq. (7.3) becomes

$$w^{iv} = 2q/R_c \tag{7.20}$$

from which, by Eqs. (7.2) and (7.4) the equivalent rigidity becomes

$$R = (2/R_c + 1/R_m)^{-1} . \tag{7.21}$$

When the loads such as those of the three-point bending test are specified for the local assembly, the lead forces may be determined as before; they will be found smaller due to the decreased equivalent rigidity of the system. This should be expected since the leads from both sides help to carry the load. Further treatment of the stiffness conditions of various multiple module configurations will be given in Chapter 8 (e.g., Fig. 8.11).

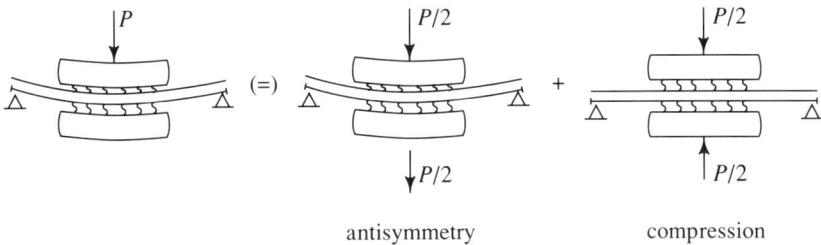

FIGURE 7.11. The loading on a double-sided module, superposed from an antisymmetrical (bending) and a symmetrical (compressive) system.

5.3. Card Longer Than Module

When the card segment of a local assembly is appreciably longer than the module (Fig. 7.2d), the assembly stiffness tends to substantially decrease. Such a case can also be analytically appraised with ease. Let the card length be $a_c = a_m + 2L'$. If the total central displacements of an equal-length ($L' = 0$) card and module assembly caused by a concentrated central load P are w_0 and W_0, then the additional displacements have two corrections: (1) Due to an additional moment caused by $PL'/2 = \Delta M$ on the card we get Δw_1 and ΔW_1; and (2) due to a contribution of bending moment $Px/2$ over the edge region we get Δw_2 and ΔW_2:

$$w = w_0 + \Delta w_1 + \Delta w_2$$

and

$$W = W_0 + \Delta W_1 + \Delta W_2 . \qquad (7.22)$$

5.4. Experimental Results

Local assemblies of all the configurations of Fig. 7.2 were subjected to two-point bending tests, and the measured assembly stiffnesses are shown in Table 7.1. These tests served to confirm the theory expounded and will be used in the next chapter to deduce the actual support condition of solder joints. Test results on double-sided modules indicated a stiffness increase from simple to corresponding double-sided modules. For stacked modules, stiffness increase was experimentally found only when the load was applied to the stacked side.

6. Orthotropy of Local Assemblies

The structure of modules greatly influences their directional properties; a DIP module or its SMT equivalent of the SOIC type will have drastically different equivalent moduli in two perpendicular directions, along the leads (x-direction) and perpendicular to them (y-direction): see Fig. 3.16. This anisotropy is also reflected in the local assemblies made with these modules.

The general procedure to measure the orthotropic properties of local assemblies is to subject them to beam tests: one bending test ("three-point bending test") in the xz-, and one in the yz-plane, yielding E_x and E_y, respectively. Beam-torsion tests, twisting the card while holding the ends in a straight line, will yield the Poisson ratios v_x and v_y, respectively.

An improved means of obtaining the elastic constants of local assemblies from measurements was suggested by Chang and Magrab [4]; according to their estimate, two three-point bending tests and a four-point torsion test would yield some improvement in the accuracy of the orthotropic material properties.

7. Module Group Assemblies

"Module group" is a term we will use to indicate more than one module attached to a card segment in the plan view. Out of a great variety of module groups tested in the three-point bending configuration, Fig. 7.12 shows two comparable ones: Fig. 7.12a is the picture of a 2×4 matrix of SOIC modules, single-sided over a 51.56- $\times$ 44.96-mm card segment. In Fig. 7.12b, the same-size module group in a double-sided version is shown.

The beam-test results for the two-directional elastic properties are shown as follows together with the equivalent mass distribution, useful for dynamic analysis:

single-sided modules:

$$E_x = 17{,}030 \text{ MPa}, \qquad E_y = 20{,}615 \text{ MPa}, \qquad \gamma = 0.047 \text{ N/cm}^3$$
$$2.47 \text{ Mpsi} \qquad\qquad 2.99 \text{ Mpsi} \qquad\qquad 0.174 \text{ lb/in}^3 .$$

Double-sided modules:

$$E_x = 18{,}960 \text{ MPa}, \qquad E_y = 23{,}718 \text{ MPa}, \qquad \gamma = 0.072 \text{ N/cm}^3$$
$$2.75 \text{ Mpsi} \qquad\qquad 3.44 \text{ Mpsi} \qquad\qquad 0.266 \text{ lb/in}^3 .$$

8. Conclusions

The local assembly as a building block of the module/lead/card system was analyzed for bending. From the simplest of local assemblies we progressed to various multiple-module configurations, such as stacked and double sided assemblies. Experimental assembly stiffnesses were obtained from beam tests. A theory of "coupled beams" was fashioned for computing the displacements and internal force distributions of such local assemblies. Orthotropy and module groups were discussed.

In the next chapter we will make use of the information gathered here, toward analysis of a full circuit card system.

9. Exercises and Questions

1. Enumerate the benefits of studying a local assembly in bending.
2. Under what conditions should a module/lead/card assembly be modeled as a continuous or discrete coupled beam arrangement?
3. Consider the SOIC and FR-4 sandwich of Example 1. Under a unit end-moment load $M = 1$ N.mm, calculate the stiffness and maximum lead force for a stacked ($\phi = 1.14$) and for a double-sided structure.
4. If instead of the SOIC a square module of the same lead spacing and plate rigidity was used, what lead forces would result upon end-moment loading?

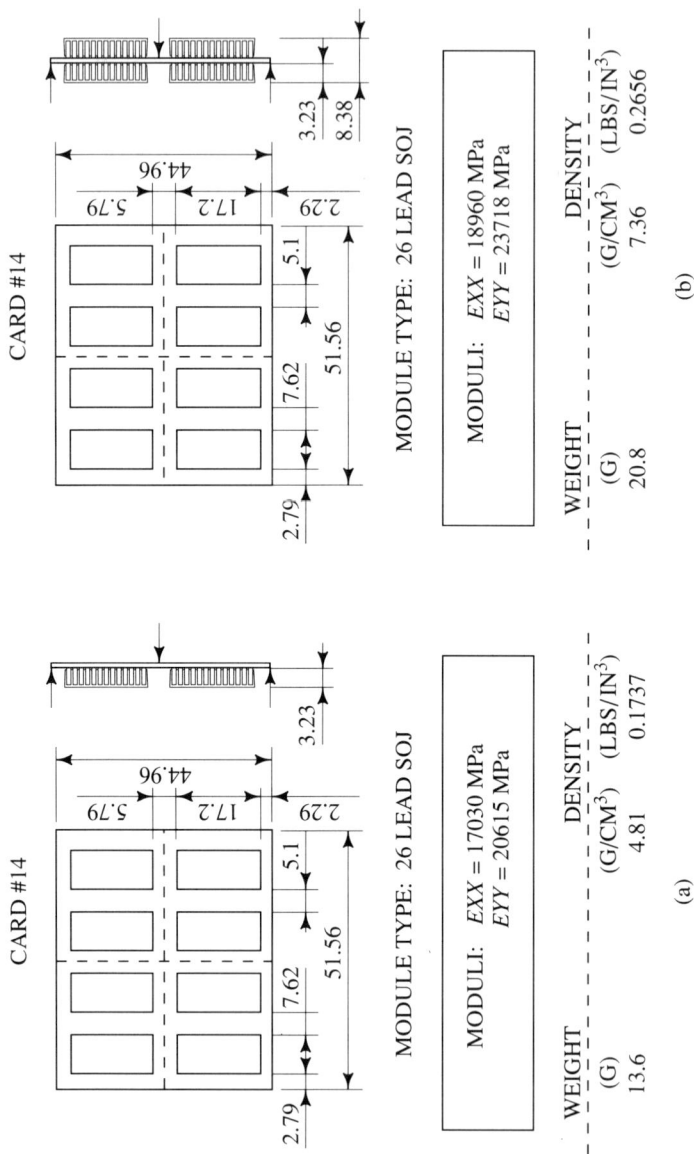

FIGURE 7.12. Module groups tested to establish an equivalent set of elastic constants, E_x and E_y. (a) Single-sided arrangement; (b) double-sided arrangement.

5. For a unit central concentrated load applied to the SOIC module in Example 1, calculate the stiffness and maximum lead force.

References

1. Engel, P.A., Caletka, D.V., and Palmer, M.R. (1991), "Stiffness and Fatigue Study for Surface Mounted Module/Lead/Card Systems," *ASME J. Elec. Packag.* **113**(2), 129–137.
2. Suhir, E. (1988), "On a Paradoxical Phenomenon Related to Beams on Elastic Foundation," *J. Appl. Mech.*, **55**(4), 818–821.
3. Engel, P.A. (1990), "Structural Analysis for Circuit Card Systems Subjected to Bending," *ASME J. Elec. Packag.* **112**(1), 2–10.
4. Chang, T.S., and Magrab, E.B. (1991), "An Improved Procedure for the Determination of the Elastic Constants of Component-Lead-Board Assemblies," *ASME J. Elec. Packag.* **113**(4), 427–430.

Chapter 8

Bending in Compliant Leaded Systems

In the previous chapter we set out to analyze module-populated circuit card (or board) systems subjected to mechanical loading. A small unit of such a system, the local assembly, was introduced. We shall now turn our attention to larger and more complex structures, and distinguish two basic approaches toward their solution: (1) the total-structure approach and (2) the building block approach. In the first, the total structure is analyzed at once, fitting together its components by the principles of compatible displacements. The second approach uses "substructures" such as the local assembly to attain the desired analytical objectives.

Before discussing these computational procedures for flexurally loaded assemblies, we shall first determine the role played by leads. This is facilitated by our previous study of local assemblies.

1. The Role of Leads

Let us consider five configurations of local assemblies such as the ones depicted in Fig. 7.2: simple assembly with load on module (*m*); simple assembly with load on card (*c*); stacked assembly with load on module (*sm*); stacked assembly with load on card (*sc*); and double-sided assembly (*ds*). Experimental measurements of system stiffness S ($= P/w, P/W$) for five such configurations were made [1]. In order to match the measured S values with analytical ones, the right spring stiffness had to be used. In Chapter 3 it was explained that leads act overwhelmingly in an axial mode under flexing of a circuit card; however, the spring constant K is a function of the spring attachment boundary condition at the solder joint. We shall check under what conditions of lead support those five sets of experimental data would be fitted by analytical data.

K values for individual leads were measured (Table 3.5) while holding both their ends (card base plate and module juncture) rigid in the tester fixtures. This would correspond to a fixed-fixed support condition for the lead; other possible boundary conditions for a lead loaded in the *yz*-plane of Fig. 3.16 are depicted in Fig. 8.1: fixed-free would model the lead as cantilevered from the module, free on the solder side; fixed-hinged would allow rotation but no sideways translation in the solder; fixed-rollers would allow sideways translation but no rotation in the solder; fixed-fixed would allow neither rotation nor translation in the y-direction at either joint.

It is possible to calculate the K constant for various support conditions, based on a known lead geometry. The fixed-fixed spring constant K_{zz} could be checked from load test measurements, but the other members of a lead stiffness matrix $[K]$ would be more difficult to measure. Calculations by finite elements or use of statically indeterminate structural analysis principles yield good results; for the statically determinate fixed-free condition, Kotlowitz [2] derived formulas for compliance under a wide variety of lead geometries. Figure 8.2 shows the schematics of some leads.

The compliance of the statically determinate lead, unsupported at the base, is expressed in the flexibility matrix $[C]$, which is defined, in conjunction with

CARD END CONDITIONS ON A LEAD

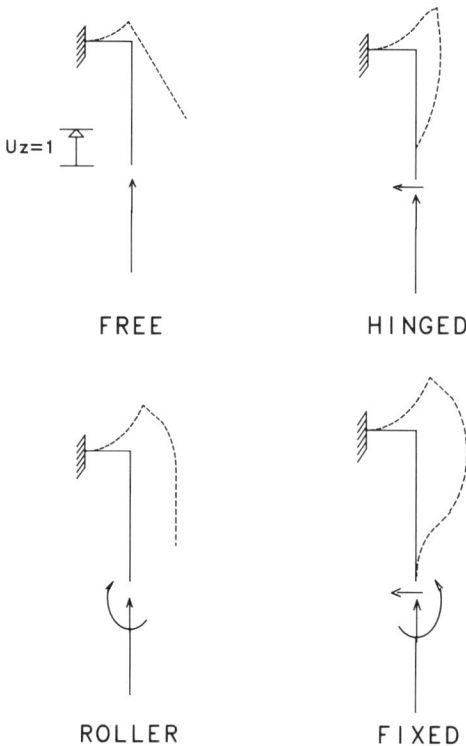

FIGURE 8.1. Four kinds of boundary conditions of a compliant lead at the solder joint. The module juncture is considered fixed in each case.

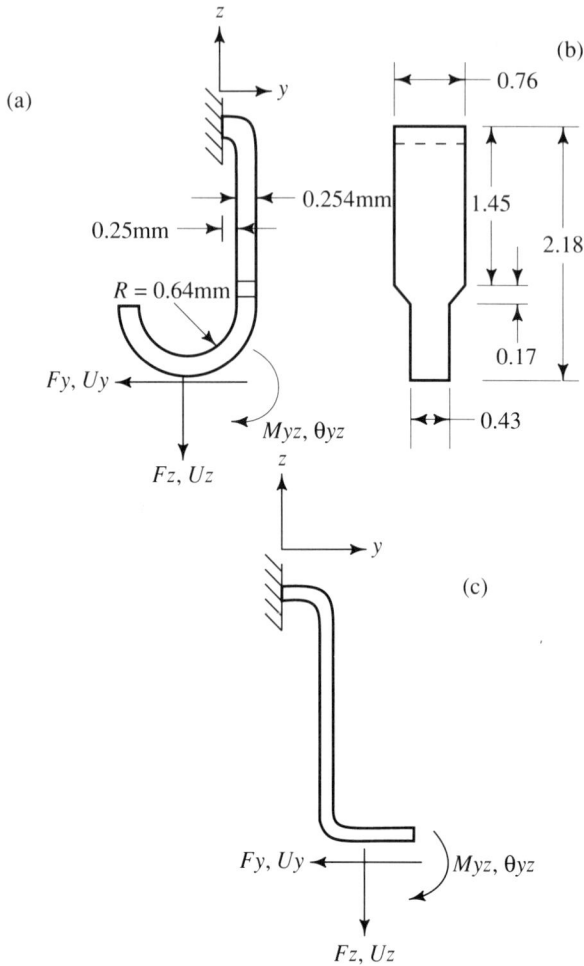

FIGURE 8.2. Nodal loads (F_y, F_z, or M_{yz}) applied at the solder joint for construction of the flexibility matrix. The corresponding displacement degrees of freedom are u_y, u_z, and θ_{yz}. (a) J-lead, yz cross section; (b) J-lead, front view; (c) Gull wing lead.

Fig. 8.1 and 8.2 as follows:

$$
\begin{bmatrix}
C_{yy} & C_{yz} & C_{y_\theta} \\
C_{yz} & C_{zz} & C_{z_\theta} \\
C_{y_\theta} & C_{z_\theta} & C_{\theta_\theta}
\end{bmatrix}
\begin{pmatrix}
F_y \\
F_z \\
M_{yz}
\end{pmatrix}
=
\begin{pmatrix}
u_y \\
u_z \\
\theta_{yz}
\end{pmatrix}.
\tag{8.1}
$$

The three degrees of freedom of a cantilever lead (Fig. 8.2) are u_y, u_z, and θ_{yz}; each member of $[C_{ij}]$ is, accordingly, the displacement of the degree of freedom i due to the unit force applied in the sense of the degree of freedom j. The use of this matrix is totally interchangeable with that of the stiffness matrix $[K]$, its inverse:

$$[K] = [C]^{-1} . \tag{8.2}$$

The K_{ij} element of the $[K]$ matrix is, as usual, defined as the force at i when dof j is unity, all other dofs being zero.

We define a lead spring constant as $K = F_z/u_z$. For a free-ended lead $F_y = M_{yz} = 0$, whereas $F_z = 1$; from the second row of Eq. (8.1), we get $C_{zz} = u_z$, and thus the fixed-free spring constant is

$$K_f = 1/C_{zz} . \tag{8.3}$$

When the solder-joint base is hinged, $M_{yz} = u_y = 0$, so that the first two lines of the matrix equation (8.1) yield

$$C_{yy} F_y + C_{yz} F_z = 0$$
$$C_{yz} F_y + C_{zz} F_z = 1 .$$

Eliminating the force F_y from the two equations, the fixed-hinged spring constant $K_h = F_z/1$ is computed:

$$K_h = \frac{C_{yy} C_{zz} K_f}{C_{yy} C_{zz} - C_{yz}^2} . \tag{8.4}$$

Similarly, the roller support condition, $F_y = \theta_{yz} = 0$ yields the spring constant for the fixed-roller support condition:

$$K_r = \frac{C_{\theta\theta}}{C_{zz} C_{\theta\theta} - C_{z\theta}^2} . \tag{8.5}$$

The calculated $[C]$ matrix for the gull wing leads of Type 602 listed in Table 3.5 can be used to identify a K-value for any of the four boundary conditions listed earlier. The compliance $[C]$ and stiffness $[K]$ matrices were computed in the (N, mm) system:

$$[C] = \begin{bmatrix} 0.14851 & -0.060787 & -0.10809 \\ -0.060787 & 0.033500 & 0.063826 \\ -0.10809 & 0.063826 & 0.13577 \end{bmatrix}$$

$$[K] = [C]^{-1} = \begin{bmatrix} 29.410 & 83.919 & -16.036 \\ 83.919 & 525.59 & -180.27 \\ -16.036 & -180.27 & 79.344 \end{bmatrix} .$$

From these, we get

$K_f = 29.85$ $K_h = 116.0$ $K_r = 286.1$ $K_{zz} = 525.6 \text{ N/mm}$

(170.4)   (662.3)   (1633)   (3000 lb/in.)

$k_f = 44.93$ $k_h = 174.6$ $k_r = 430.6$ $k_{zz} = 791.1 \text{ N/mm}^2.$

To any K-value, there corresponds a foundation constant k, which for an SOIC (two lead-row) type of local assembly, is calculated by $k = 2nK/a =$

(2) (13) $K/17.3 = 1.505 \, K$. Then through the whole region of interest for K, k can be evaluated, and from k, λ obtained by Eq. (7.7), using the combined rigidity R of the local assembly of the card plus the SOIC module. Next, for each of the five local assembly configurations tested, we can evaluate a system stiffness S, meaning central load per displacement. We will have S_m (simple assembly, load on module), S_c (simple assembly, load on card), S_{sm} (stacked assembly, load on module), S_{sc} (stacked assembly, load on card), and S_{ds} (double-sided assembly).

Let us construct a diagram of S versus k, computed for each of the five configurations. Figure 8.3 shows the five corresponding curves, in double logarithmic representation. In the meantime, a horizontal line can be drawn with the measured experimental assembly stiffness value S of each configuration. Where the horizontal line intersects the corresponding curve, there lies the abscissa for the correct k, revealing the true support condition for the leads.

In Fig. 8.3 each of the five intersections lies about midway between the free and hinged lead support conditions. This means that the solder joint is far from being a fixity; rather, it offers elastic support to the lead.

Having found experimental evidence as to the behavior of solder joints, the lead forces F can also be inferred from the computed values, corresponding to the right k constant. Figure 8.4 shows maximum lead forces obtained for all five local assembly configurations tested. It is evident that the stacked configuration yields the highest lead forces, and the double-sided scheme the lowest ones.

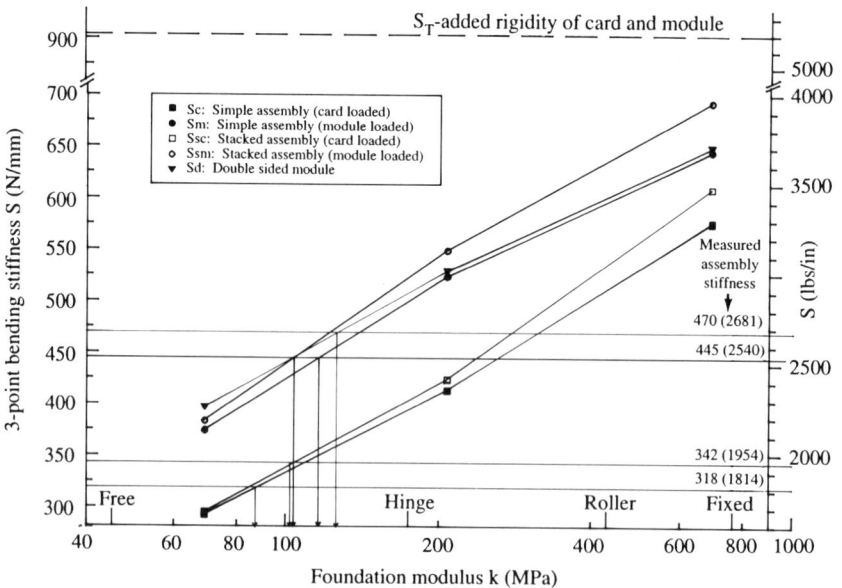

FIGURE 8.3. System stiffness variation for five local assemblies in terms of the foundation constant k.

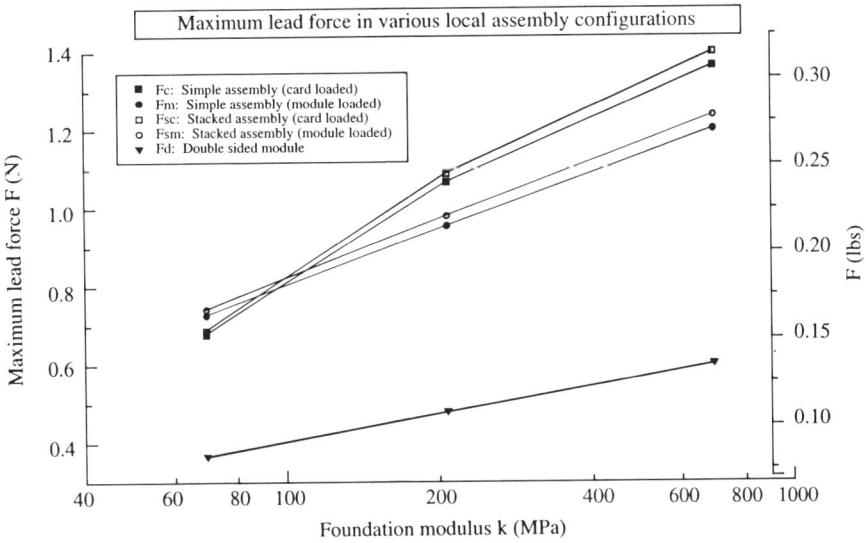

FIGURE 8.4. Maximum lead force variation for five local assemblies in terms of the foundation constant k.

2. Application of the Finite Element Method

The finite element method offers an expedient tool when relatively small and slightly populated circuit card systems are to be analyzed. As already explained, the number of elements required by everyday circuit card systems is excessive. The finite element procedure can, however, also be used to calculate "building-block" properties, and to check out results obtained by other methods. Plate behavior is fully taken into account.

In Ref. [3] solutions by the ANSYS computer software program are described; for FR-4 circuit cards of $L = 12.5 \times 12.5$ cm (5 in.) square size and $h_c = 1.25$ mm (0.05 in.) thickness, STIF63 plate and membrane-stress elements were used, and STIF63 elements stood for peripheral-leaded PLCC-type modules as well. J-leads were fashioned as beams (STIF4) connecting the module and card. Module and lead dimensions were varied; the control dimensions of the module were $a = 25$ mm (1 in.), a thickness of $h_m = 3$ mm (0.12 in.); the isotropic modulus $E = 3.24$ GPa (470,000 psi) and $v = 0.4$ were kept constant. The leads had a length of 3 mm (0.12 in.) and modulus $E = 124$ GPa (18,000 Kpsi); the equivalent cross-sectional area $A_{eq} = 0.01$ mm^2 (1.65×10^{-5} in.^{2}) yielding K = 543 N/mm (3100 lb/in.). Near-zero beam-moments of inertia I, and a spacing $s = 2.5$ mm (0.1 in.), were other control variables for the leads.

The boundary conditions were simple (knife-edge) supports at $y = \pm L/2$, and free edges at $x = \pm L/2$. A symmetrical displacement constraint of $\theta = 10$

degrees was input over the supports. This displacement was only used as a round figure to be scaled down in small-displacement elastic analysis; no geometrically nonlinear action was assigned. There was two-fold symmetry, allowing four-fold element reduction (Fig. 8.5). The maximum lead forces

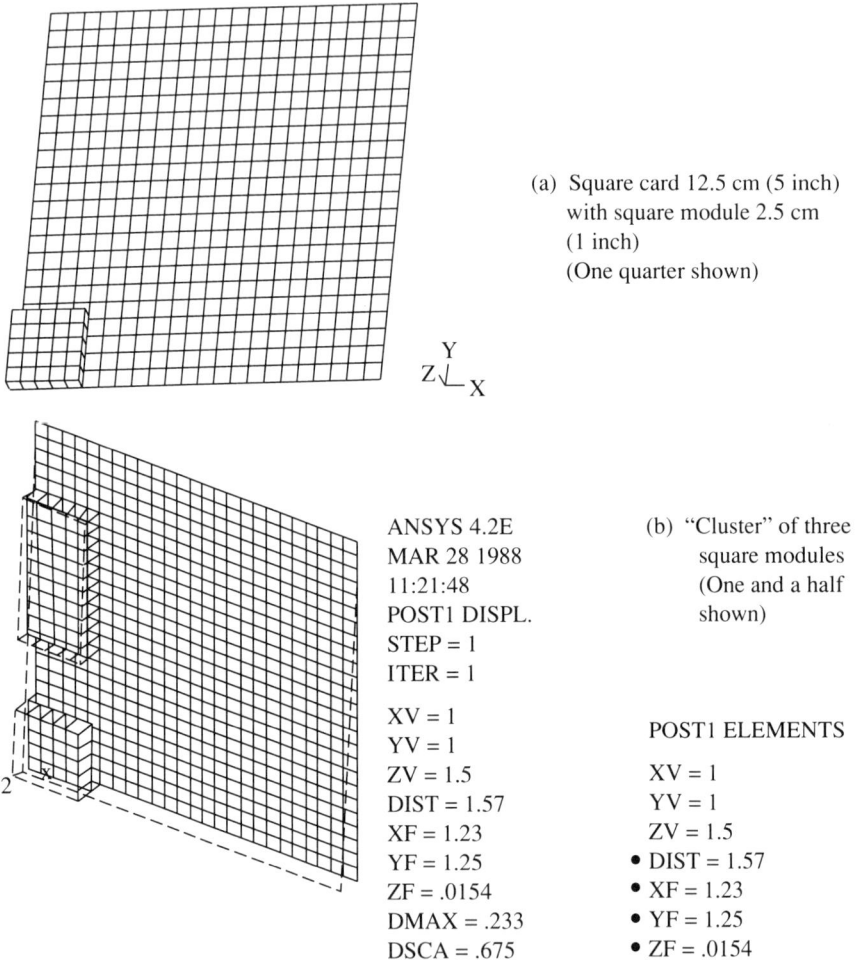

(a) Square card 12.5 cm (5 inch) with square module 2.5 cm (1 inch) (One quarter shown)

Y
Z⤵X

ANSYS 4.2E   (b) "Cluster" of three
MAR 28 1988   square modules
11:21:48   (One and a half
POST1 DISPL.   shown)
STEP = 1
ITER = 1

XV = 1   POST1 ELEMENTS
YV = 1
ZV = 1.5   XV = 1
DIST = 1.57   YV = 1
XF = 1.23   ZV = 1.5
YF = 1.25   • DIST = 1.57
ZF = .0154   • XF = 1.23
DMAX = .233   • YF = 1.25
DSCA = .675   • ZF = .0154

FIGURE 8.5. Finite element (ANSYS) model for various module/card arrangements; note the two-fold symmetry.

FIGURE 8.6. Lead forces (lb) for bending of a square 12.5-cm card with square 12.5-mm modules attached, to $\theta_x = 10$ degrees of edge rotation parallel to the x-axis: elastic ANSYS analysis. (a) Four modules at card corners; (b) five modules along each edge $x = L/2$; (c) five modules along each edge of the card.

TABLE 8.1. Dependencies studied. Const. Card: $E_c = 13.79$ GPa $(2 \times 10^6$ psi), $t = 1.25$ mm (0.05 in.); Module: $E_m = 3.24$ GPa (470,000 psi), $t = 3$ mm (0.121 in.); Dependent Variable: Lead Force, F_c.

Independent Variables:	Control	Variations
1) Loading Modes:	M_x	T
	Bending Moment	Torque
2) Lead size:	$A_{eq} = 0.01$ mm² $(1.65 \times 10^5$ in.²)	$\times 10^{-1}$, $\times 10^1$
3) Lead Inertia:	$I_x = I_y = 0$	$I_x \cdot I_y = 4.16 \times 10^{-5}$ mm⁴
		$I_x = I_y = (10^{-10}$ in.⁴)
4) Square Module Size:	$a = 25$ mm $\times 25$ mm (1 in. $\times$ 1 in.)	50×50 mm, (2in. $\times$ 2in.)
		75×75 mm, (3in. $\times$ 3in.)
5) Lead Spacing:	$s = 2.5$ mm (0.1 in.)	5 mm, 1.25 mm
		(0.2 in., 0.05 in.)
6) Module Shape:	$a \times b = 25$ mm $\times 25$ mm (1in. $\times$ 1in.)	25×50 mm (1in. $\times$ 2in.)
		50×25 mm (2in. $\times$ 1in.)
7) Module Cluster:	2.25 mm $\times$ 25 mm	Module back to back
	(1in. $\times$ 1in.)	
8) Card Size:	$L_x \times L_y = 12.5 \times 12.5$ cm	18.75×12.5 cm (7.5 $\times$ 5in.)
	(5in. $\times$ 5in.)	8.33×12.5 cm (3.33 $\times$ 5in.)
9) Multiple Modules:	One module	Stacked
		Double-sided

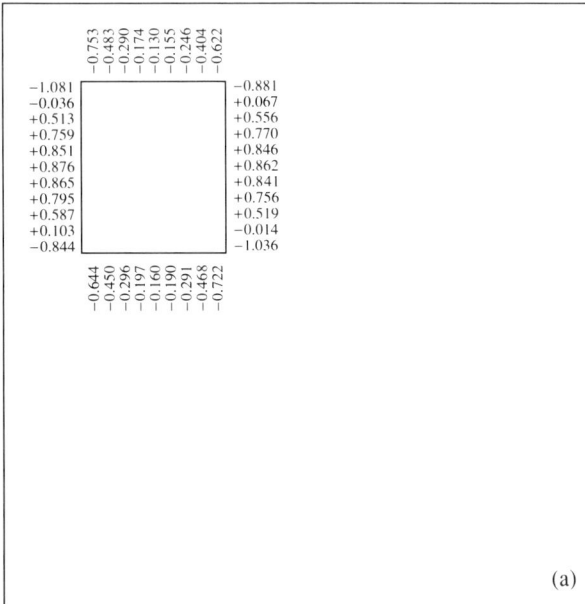

(a)

Top lead values (vertical): −0.787, −0.481, −0.268, −0.143, −0.097, −0.126, −0.229, −0.409, −0.663

First module — left / right:

−1.174	−0.980
−0.067	+0.035
+0.520	+0.564
+0.788	+0.800
+0.892	+0.888
+0.923	+0.910
+0.911	+0.889
+0.828	+0.791
+0.586	+0.523
+0.029	−0.078
−1.051	−1.222

Between modules (vertical): −0.695, −0.414, −0.222, −0.113, −0.078, −0.116, −0.235, −0.450, −0.779

(vertical): −0.803, −0.482, −0.261, −0.130, −0.078, −0.101, −0.203, −0.396, −0.695

Second module — left / right:

−1.212	−1.097
−0.055	−0.006
+0.552	+0.561
+0.825	+0.813
+0.926	+0.904
+0.951	+0.924
+0.932	+0.900
+0.839	+0.800
+0.581	+0.531
−0.001	−0.068
−1.121	−1.209

Between modules (vertical): −0.741, −0.442, −0.235, −0.115, −0.072, −0.104, −0.219, −0.432, −0.761

(vertical): −0.773, −0.463, −0.249, −0.123, −0.075, −0.101, −0.208, −0.410, −0.724

Third module — left / right:

−1.167	−1.152
−0.051	−0.060
+0.514	+0.492
+0.740	+0.714
+0.791	+0.765
+0.782	+0.757

(b)

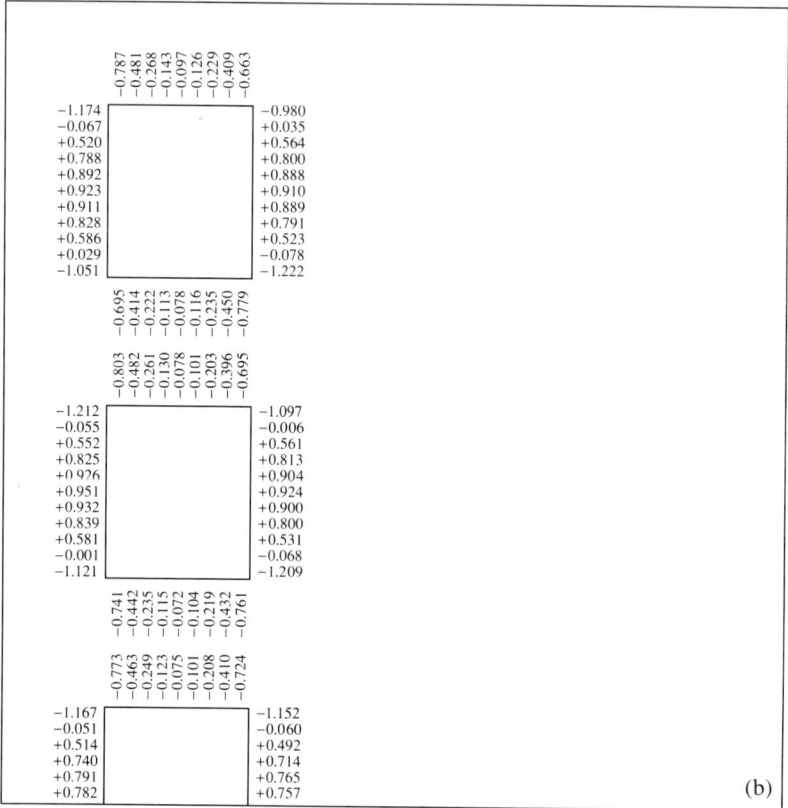

FIGURE 8.6. Continued.

F_{max} occurred at the module corners, and they were calculated for each set of parameters (See Table 8.1). Note that on each module there was a lead placed in the corner position in order to serve as comparison with other methods.

It was found that the biggest lead force components were the axial ones, and this hardly changed when lead inertias I_x, I_y were beefed up to 4.16×10^{-3} mm^4 (10^{-8} in.^{4}). A survey of parameter dependences is given in Table 8.2, comparing finite element results obtained by the "strip-method" see Section 3.

Various modules and module clusters were also analyzed by Vogelmann [4]; his finite element lead forces, shown in Figs. 8.6a–8.6c were for 12.5 mm (0.5 in.) square PLCC modules with a lead spacing of s = 1.25 mm (0.05 in.). The feature of a single corner lead in each module corner was consistently used, for reasons to be explained later. The cluster is gradually increased from

Figure data (Fig. 8.6c):

Top (vertical headers, first box): -0.751, -0.463, -0.263, -0.145, -0.101, -0.123, -0.204, -0.340, -0.518
Top (vertical headers, middle box): -0.599, -0.392, -0.241, -0.151, -0.120, -0.147, -0.227, -0.355, -0.516
Top (vertical headers, right box): -0.532, -0.360, -0.225, -0.138, -0.106

First	Middle		Right	
-1.116	-0.712	-0.840	-0.681	-0.711
-0.056	+0.127	+0.045	+0.125	+0.103
+0.504	+0.506	+0.459	+0.487	+0.473
+0.759	+0.628	+0.608	+0.602	+0.593
+0.857	+0.645	+0.645	+0.615	+0.608
+0.885	+0.639	+0.658	+0.608	+0.603
+0.872	+0.637	+0.677	+0.607	+0.603
+0.791	+0.608	+0.674	+0.587	+0.583
+0.556	+0.456	+0.559	+0.465	+0.464
+0.016	+0.021	+0.175	+0.100	+0.103
-1.032	-0.917	-0.695	-0.698	-0.686

Bottom (vertical headers, first box): -0.687, -0.412, -0.223, -0.112, -0.072, -0.098, -0.192, -0.361, -0.608
Bottom (vertical headers, middle box): -0.564, -0.418, -0.293, -0.208, -0.173, -0.191, -0.263, -0.384, -0.539
Bottom (vertical headers, right box): -0.523, -0.361, -0.231, -0.146, -0.115

Top (vertical headers, second section): -0.751, -0.461, -0.261, -0.147, -0.111, -0.151, -0.270, -0.475, -0.771

First	Middle
-1.115	-1.147
-0.012	-0.003
+0.565	+0.592
+0.825	+0.855
+0.924	+0.950
+0.952	+0.970
+0.936	+0.946
+0.846	+0.846
+0.585	+0.573
-0.010	-0.037
-1.159	-1.203

Bottom (vertical headers, second section): -0.773, -0.466, -0.254, -0.131, -0.087, -0.119, -0.234, -0.446, -0.770

Top (vertical headers, third section): -0.788, -0.477, -0.262, -0.136, -0.089, -0.119, -0.231, -0.439, -0.758

First	Middle
-1.181	-1.187
-0.043	-0.054
+0.532	+0.519
+0.761	+0.749
+0.812	+0.800
+0.801	+0.791

(c)

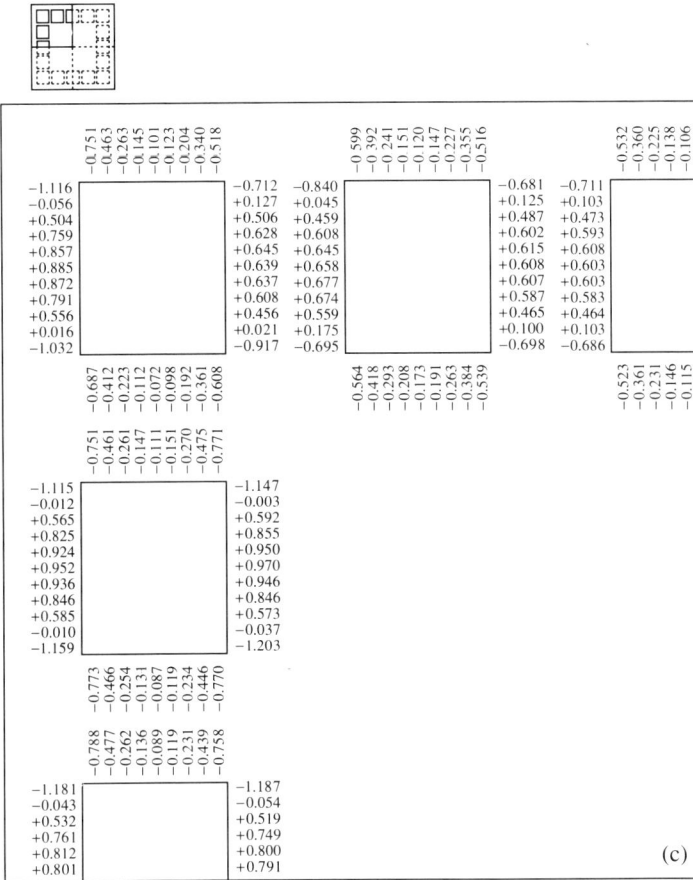

FIGURE 8.6. Continued.

four corner modules (Fig. 8.6a) to two parallel rows (Fig. 8.6b), and finally to a round assembly of peripheral modules (Fig. 8.6c). The maximum (corner) lead forces increase only slightly as the stiffness of the populated circuit card increases with the addition of modules.

3. Strip Method

A local assembly could be used directly to find approximate stiffness and lead force conditions; however, the local assemblies described in the previous chapter were beam systems, while in most circuit card systems plate action, i.e., the presence of double curvature, is strong. The computational method [3] described as follows attempts to take care of this by cutting a strip that contains the module in two perpendicular (first in the y-, then x-) directions. Figure 8.7, finite element modeled by Fig. 8.5a, is for a rectangular module ($a \times b$) attached to the center of a rectangular card $L_x \times L_y$.

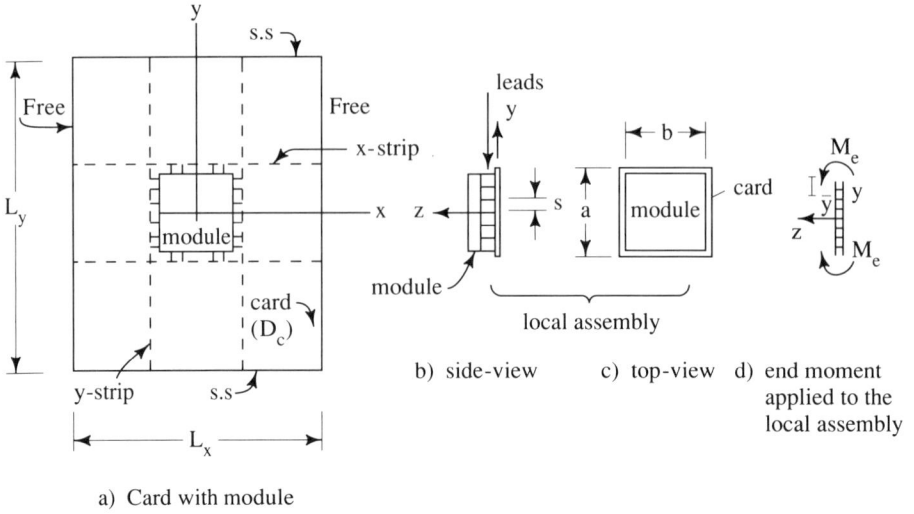

a) Card with module

FIGURE 8.7. Modeling of a card with central module attachment for the strip method.

We first consider bending of the y-strip; it is subjected to an external support rotation, say $\theta = 10$ degrees, caused by plate moments M'_0; clearly this displacement is only meant as a round number subject to scaling purposes, since it exceeds linearly elastic plate behavior. A cut ABCD (cut 1 in Fig. 8.8c) is made, the plane BC slicing across the first row of leads, but not touching the module. We can write the sum of moments acting on ABCD as

$$- M'_0 \cdot b + s S^*_x + M'_c \cdot b + 2M'_{xy}([L_y - a]/2) = 0 \qquad (8.6)$$

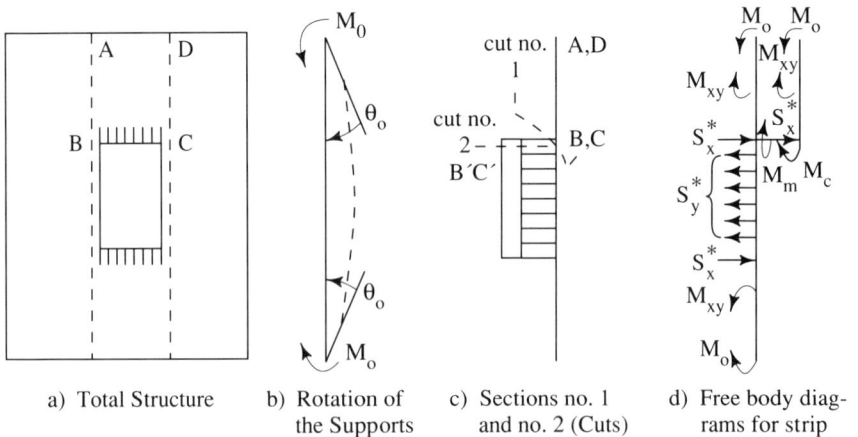

a) Total Structure   b) Rotation of the Supports   c) Sections no. 1 and no. 2 (Cuts)   d) Free body diagrams for strip

FIGURE 8.8. Flexural displacement constraint θ of a circuit card with central module.

where S_x^* is the sum of pin forces along the edge $y = a/2$, seen as compressive from positive card curvature $\partial^2 w/\partial y^2$. The single asterisk ($*$) refers to the y-strip, while the x-strip data, later, will be denoted by two asterisks ($**$). M_{xy} is the average twisting moment along AB and CD. Primes denote plate moments (moment per unit length).

A second cut AB'C'D (cut 2 in Fig. 8.8c) has B'C' in a vertical plane across module and card, not intersecting leads. Invoking equilibrium, we write

$$- M_0' \cdot b + M_c' \cdot b + 2M_{xy}'([L_y - a]/2) + M_m' \cdot b = 0 . \tag{8.7}$$

We may relate the module moment M_m' to the external applied moment M_0' by using the local assembly relation

$$M_m' = (D_m/[D_m + D_c]) g\{[a/2] - s\} \cdot M_0' \tag{8.8}$$

where the $g\{[a/2] - s\}$ term is the moment-development function depicted in Fig. 7.6a or 7.6b, evaluated at the first lead past the corner lead. The value of this function reaches 1 at $y = a/2 - \bar{y}$ as shown in Fig. 8.7d. The non-dimensional g-function is a fast-rising one, its maximum ordinate being 1.0 if the leads are stiff and numerous, and if the number of cross leads n' is large. Writing the total plate rigidity $D_T = D_c + D_m$, at $g = 1$ we get

$$M_m'/M_0' = D_m/D_T \tag{8.9}$$

$$M_c'/M_0' = D_c/D_T . \tag{8.10}$$

We now obtain the sum of lead forces along $y = a/2$:

$$S_x^* = (D_m/D_T)(b/s)M_0' \cdot g\{[a/2] - s\} \tag{8.11}$$

along with a uniform lead force distribution F_x^* and a "parabolic" one F_y^*, according to the local assembly under end moment, say, Fig. 7.5.

Cutting a strip in the x-direction now, a similar development will be augmented by the argument of a lesser stiffness of the strip at the sides $(x > b/2)$ than in the module area $(0 < x < b/2)$. Making an approximation for a constant curvature $\partial^2 w/\partial x^2$ in the edge region, the sum of edge forces S_y^{**} is now calculated as

$$S_y^{**} = \left(\frac{D_m}{D_T}\right)^2 g\{[b/2] - s\} \cdot \frac{a(2L_y - a)}{(L_x - b)^2} \tag{8.12}$$

containing the card/module shape constant, depicted in Fig. 8.9:

$$C = \frac{a(2L_y - a)}{(L_x - b)^2} . \tag{8.13}$$

The lead force distributions resulting from the two strips can now be added.

The corner forces $F_c = F_c^* + F_c^{**}$ are obtained from the superposition of two sets of edge-force distributions that reinforce one another at the corners as shown in Fig. 8.10. We get approximately

$$F_c \cong (1/n')S_x^* + (1/n)S_y^* , \tag{8.14}$$

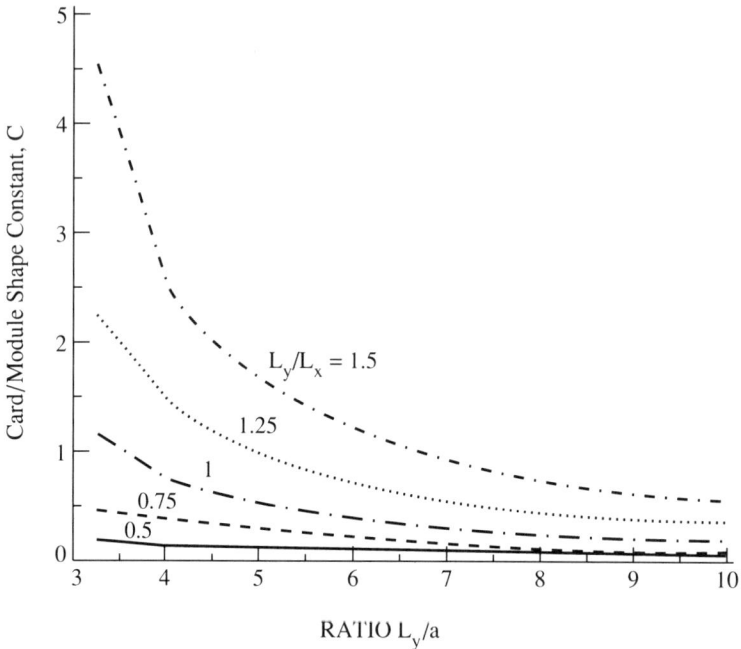

FIGURE 8.9. Variation of shape-factor C versus L_y/a and L_y/L_x for square module, $a = b$.

and for the case of a square module $g(a/2 - s) = g \; (b/2 - s) = g$, there results:

$$F_c \cong g \cdot \left(1 + \frac{D_m}{D_T} \cdot C\right) \left(\frac{D_m}{D_T}\right) M'_0 . \tag{8.15}$$

The more general expression, for arbitrary boundary conditions on the card is

$$F_c \cong \left(g_y + \frac{D_m}{D_T} C g_x\right) \left(\frac{D_m}{D_T}\right) M'_0 . \tag{8.16}$$

For stacked and double-sided modules, cuts 2 to be used in deriving S^* are made vertically through the modules. Cut 1 for double-sided modules slices through the leads on both sides of the module. The rigidity-ratio terms like D_m/D_T will be as follows for four cases:

Simple	Stacked	Double-sided	Stacked, double-sided
$\dfrac{D_m}{D_m + D_c}$	$\dfrac{D_m(1 + \phi)}{D_m(1 + \phi) + D_c}$	$\dfrac{D_m}{2D_m + D_c}$	$\dfrac{D_m(1 + \phi)}{2D_m(1 + \phi) + D_c}$

The nondimensional lead forces for each case, depending on the D_m/D_c ratio, with $\phi = 2$, are shown in Fig. 8.11.

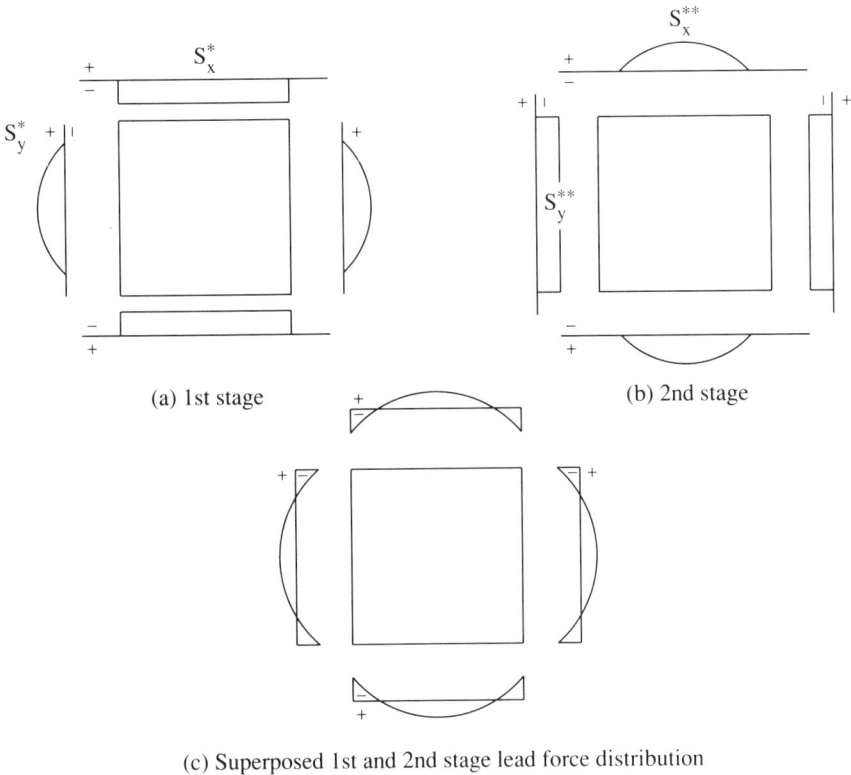

(a) 1st stage

(b) 2nd stage

(c) Superposed 1st and 2nd stage lead force distribution

FIGURE 8.10. Schematic of lead force distribution due to bending.

Table 8.2 shows a comparison between lead forces obtained by the strip method and those by finite elements and the trends for the relations describing lead forces in terms of the system parameters. It appears that the finite element method gives about twice the magnitude of F_c than the strip method; however, recall that the finite element results are based on *single* corner leads, unlike the two near-corner leads always used in practice. This will be discussed further for the case of twisted circuit card systems in the next two chapters.

Equations (8.15) and (8.16) show the basic dependences of the maximum (corner) lead force F_c, on all the three constituent elements of the system: card, module, and leads. For an infinitely flexible module ($D_m = 0$) there would be no transmission of load to the module, making $F_c = 0$. If the card were very flexible, yet still able to transmit the moment M'_0, lead forces could materialize so that the lead force is proportional to the local bending moments in the card. Finally, the stiffness of leads is manifested in the steepness of the rise of the module-moment development function g; its maximum value is unity.

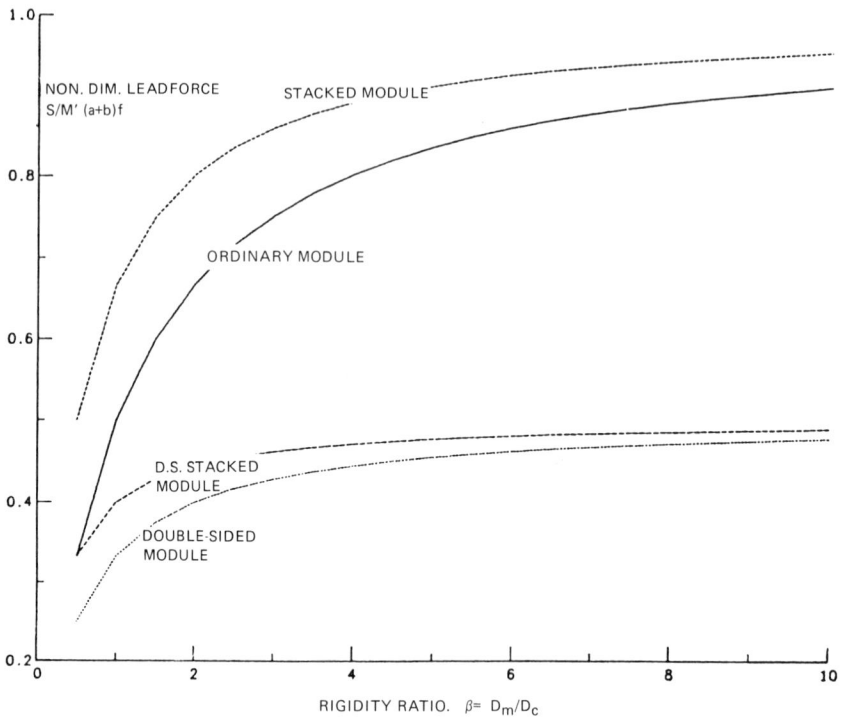

FIGURE 8.11. Nondimensional lead force variation versus rigidity ratio D_m/D_c for various module configurations.

4. Building Block Method

So far we considered and analyzed local assemblies as essentially beam structures subjected to beam loading. This treatment need not be exclusive; the local assembly of any module may be subjected to plate moments of bending components M_x and M_y and a twisting moment component M_{xy} (Fig. 8.12). The stiffness and lead-force conditions of the local assembly can then be calculated, be it by analytical or numerical (finite element) methods.

When one looks at a circuit card under external load, the distribution of bending moments in any region ($x_1 < x < x_1 + b$; $y_1 < y < y_1 + a$) is usually continuous and would be rather smooth, if not for the presence of a module of size $a \times b$ attached within that region. If, however, the leads transmit mostly transverse force to the card and only minimal moment, then the smoothness of the moment variation is essentially undisturbed. Thus the lead forces of the local assembly may be calculated from the average bending

TABLE 8.2. Comparison of parametric influences of engineering analysis method vs. finite element method.

	By engineering analysis	By finite elements
Lead stiffness, A_{eq} (mm^2)	$F(s) = F_0(s)\left(\dfrac{A}{A_0}\right)^\alpha$ $\alpha = 0.365$	$F = F_0\left(\dfrac{A}{A_0}\right)^\alpha$ $\alpha = 0.343$
Lead spacing s (mm)	$F(s) = F_0(s_0)\left(\dfrac{s}{s_0}\right)^\beta$ $\beta = 1.212$	$F = F_0\left(\dfrac{s}{s_0}\right)^\beta$ $\beta = 0.204$
Square module size a (mm)		$F = F_0\left(\dfrac{a}{a_0}\right)^2$ $\gamma = 0.274$
Module shape a, b (mm)		$F = F_0\left(\dfrac{a}{a_0}\right)^\alpha\left(\dfrac{b}{b_0}\right)^\delta$ $\alpha = \delta = 0.274$
Module cluster 25-mm square modules back-to-back	$F_{\text{front module}} = F_{\text{center module}}$	$F_{\text{front module}} = F_{\text{center module}}$
Card size	$M' \cong M_0'\dfrac{L_{yo}}{L_y}$	
18.75 × 12.5 cm (flat) (tall)	$F/F_0 = 1$ $F/F_0 = 1$	$F/F_0 = 1.007$ $F/F_0 = 0.973$
8.33 × 12.5 cm		
Multiple modules	$\dfrac{E_m I_m}{E_c I_c} \equiv \beta \Rightarrow (1.331)$	
Double-sided 25-mm square	$\dfrac{F}{F_0} = \dfrac{1 + 1/\beta}{2 + 1/\beta} \Rightarrow (0.571)$	$\dfrac{F}{F_0} = 0.781$
Stacked 25-mm square	$\dfrac{F}{F_0} = \dfrac{1 + 1/\beta}{1 + 1/2\beta} \Rightarrow (1.142)$	$\dfrac{F}{F_0} = 1.197$

Based on parameters of card and module as noted in Table 8.1; Load: θ_x symmetric rotation of supports. Subscript "0" means control configuration.

moment on the card in the module area, essentially ignoring the influence of the module attachment.

This chain of thinking suggests a simple approximate procedure for lead stress analysis: first evaluate the lead forces of a local assembly in question, due to unit plate moments $M_x' = 1$, $M_y' = 1$, $M_{xy}' = 1$, separately; each of these load systems is in equilibrium by itself. Let these lead forces F_i (biggest ones expected in the corners) be called $F_{i,x}$, $F_{i,y}$, and $F_{i,xy}$, respectively, indicating the loading system as the source of the load. Next, compute the components of local bending moment on the card (in the region of the module attachment) from gross plate analysis: M_x, M_y, M_{xy}. For regular-shape cards and ordinary simple loading cases, these may be available from handbooks [5]. Finally, calculate the lead forces from

$$F_i = F_{i,x} \cdot M_x + F_{i,y} \cdot M_y + F_{i,xy} \cdot M_{xy} . \tag{8.17}$$

The accuracy of this procedure may be improved if the module is taken into account in the "gross" plate analysis; but to do that, we have to abandon simple computation of the card moments M_x, M_y, and M_{xy} and do this work

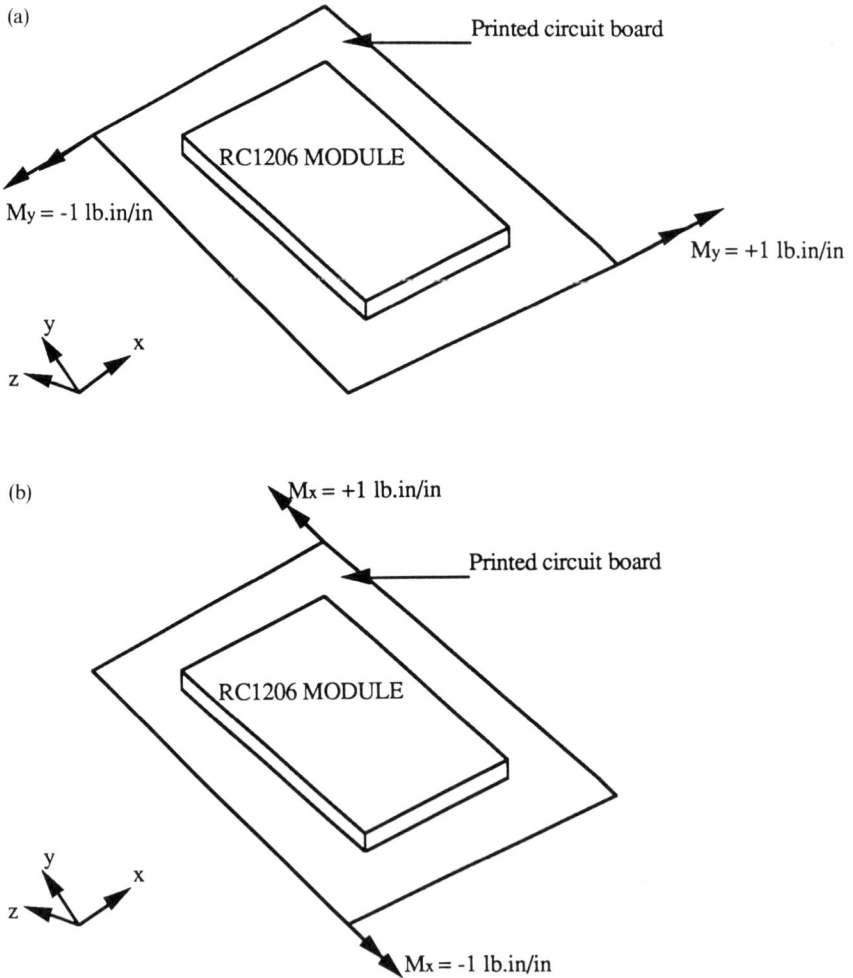

(a)

Printed circuit board

RC1206 MODULE

$M_y = -1$ lb.in/in

$M_y = +1$ lb.in/in

y

x

z

(b)

$M_x = +1$ lb.in/in

Printed circuit board

RC1206 MODULE

y

x

z

$M_x = -1$ lb.in/in

FIGURE 8.12. Building block method applied to a local assembly consisting of RC1206 module and circuit card of commensurate size. (a) Load case I; unit M_y; (b) load case II; unit M_x; (c) load case III; unit M_{xy}.

(c) $M_{yx} = -1$ lb.in/in

Printed circuit board

$M_{xy} = +1$ lb.in/in

RC1206 MODULE

$M_{xy} = -1$ lb.in/in

y

x

z

$M_{yx} = +1$ lb.in/in

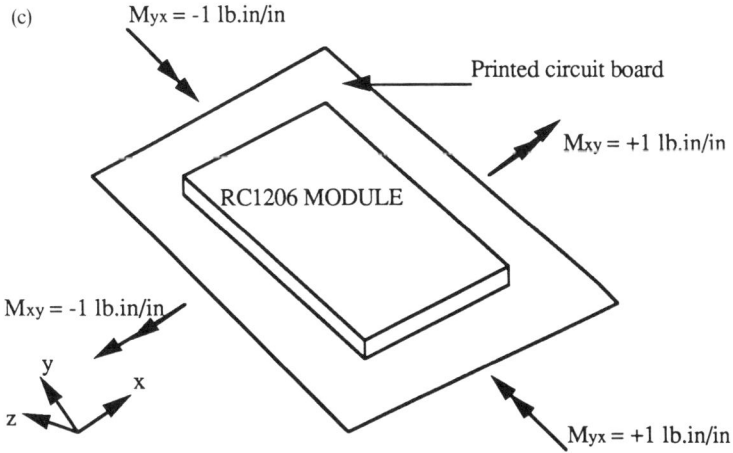

FIGURE 8.12. (Continued)

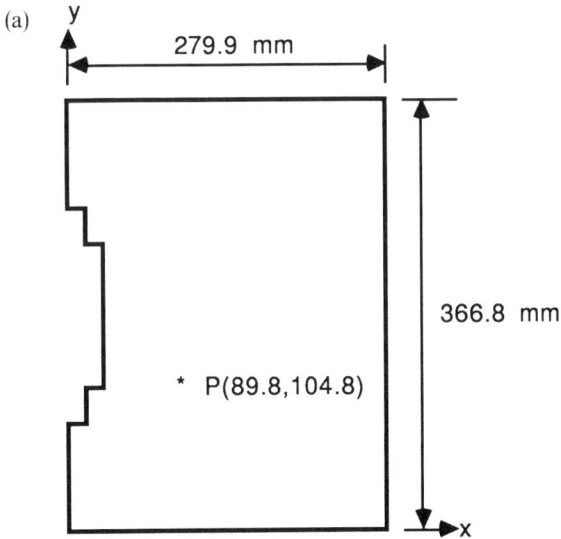

(a)   y

279.9 mm

366.8 mm

* P(89.8,104.8)

x

FIGURE 8.13. Circuit card problem with glued-on module. (a) Circuit card. The z-axis comes out of the paper. Thickness of printed circuit board $= 2.032$ mm. Imposed deflection at point P (89.88 mm, 104.8 mm) is $w = 0.762$ mm. The four corners of the printed circuit board were simply supported; (b) Finite elements of the card; (c) 3×3 array of glue-spring elements.

(b)

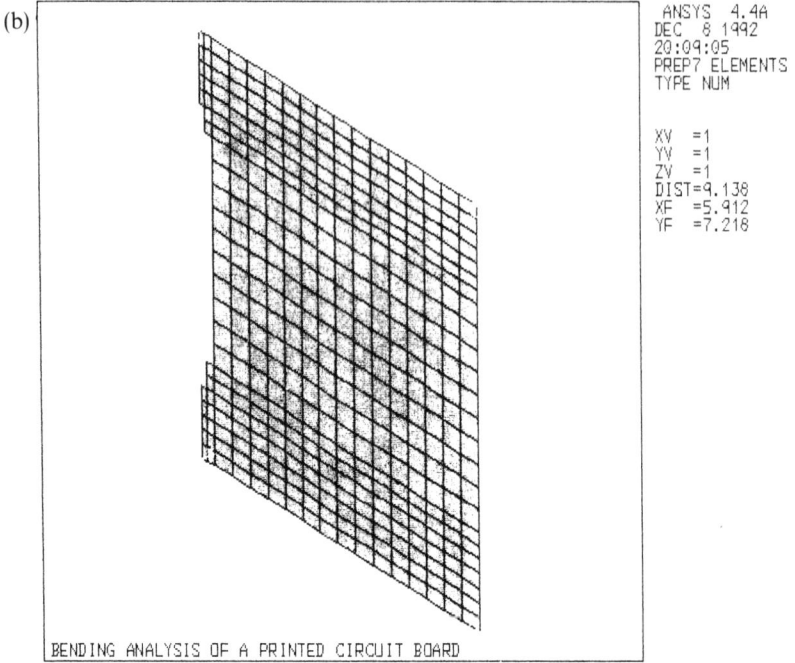

```
ANSYS 4.4A
DEC  8 1992
20:09:05
PREP7 ELEMENTS
TYPE NUM

XV  =1
YV  =1
ZV  =1
DIST=9.138
XF  =5.912
YF  =7.218
```

BENDING ANALYSIS OF A PRINTED CIRCUIT BOARD

(c)

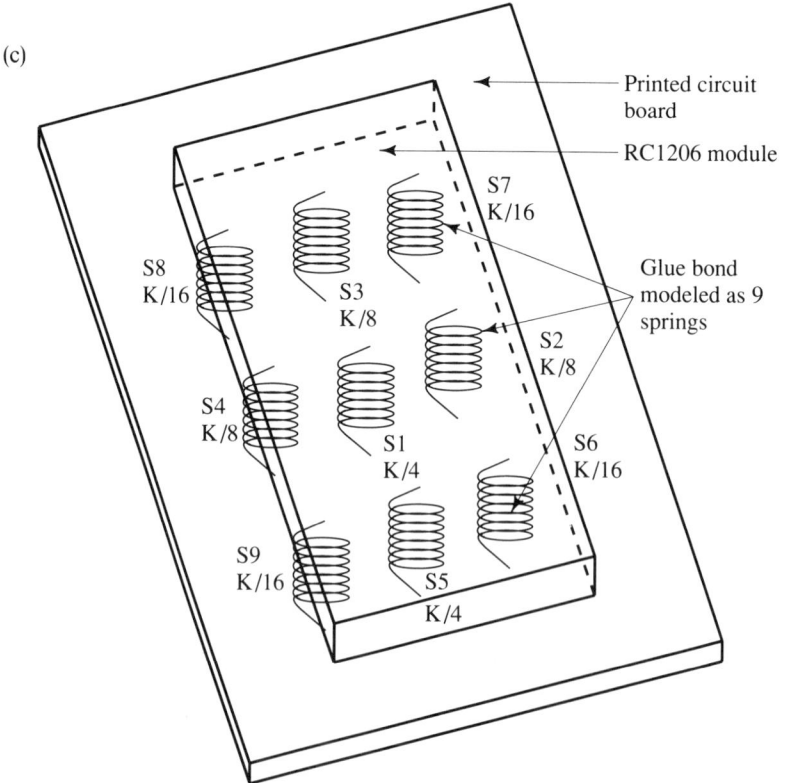

Printed circuit board

RC1206 module

S7 K/16

Glue bond modeled as 9 springs

S8 K/16

S3 K/8

S2 K/8

S4 K/8

S1 K/4

S6 K/16

S9 K/16

S5 K/4

FIGURE 8.13. Continued.

TABLE 8.3. Adhesive bond analysis for RC1206 module attached to a circuit board at P shown in Fig. 8.13a. Discretized springs are used. The thickness of the board is 2 mm (0.08 in). Axial forces in springs: $F_i = M_y \times F_{i,y} + M_x \times F_{i,x} + M_{xy} \times F_{i,xy}$. Plate moments at point P for $W_p = 0.762$ mm: $M_y = 0.7874$ N; $M_x = 0.9947$ N; $M_{xy} = 0.1662$ N.

Spring #, i	Influence coefficients for spring force			Spring force, F_i (N)
	$F_{i,y}$	$F_{i,x}$	$F_{i,xy}$	
1	− 0.00293	− 0.00096	0	− 0.00326
2	− 0.00179	0.000801	0	− 0.00062
3	0.001770	− 0.00084	0	0.000549
4	− 0.00179	0.000801	0	− 0.00062
5	0.001770	− 0.00084	0	0.000549
6	0.000747	0.000264	0.002757	0.000393
7	0.000747	0.000264	− 0.00275	0.001310
8	0.000747	0.000264	0.002757	0.000393
9	0.000747	0.000264	− 0.00275	0.001310

in finite elements as well. If the latter choice is attractive, the area of the module is best inserted with a rigidity corresponding to the equivalent rigidity R of its local assembly; $R_m + R_c$ would be somewhat high (see Fig. 8.3), but admissible for the purposes of the finite element treatment.

As an example for the application of the building block method, Fig. 8.13 shows a circuit board to which flat-bottomed elements (RC1206 modules) were glued temporarily while the board was handled in the manufacturing cycle [6]. The bonded area was modeled as a 3×3 array of axial glue springs. Any glued component would be combined into a local assembly with the circuit board and its glue springs. After the spring constant of glue springs was determined from load test on a bond of known dimensions, the force influences $F_{i,x}$, $F_{i,y}$, and $F_{i,xy}$ were evaluated. The central glue spring received relatively large contributions from the unit-moment loading configurations; see Table 8.3. Upon calculating the plate moments M_x, M_y, and M_{xy} at the location of a glued element, the maximum glue-element force F_i was obtained at the central spring $i = 1$ from Eq. (8.17).

5. Hybrid Experimental/Analytical Method

Wong et al. [7] devised an iterative method of lead force analysis that combines experimental evaluation of load versus displacement characteristics for the individual leads with analytically (or numerically) calculated circuit card deflections (Fig. 8.14). Since by this method the circuit card deflections are assumed to be unaffected by the modules (and, in fact, for simplicity, they may be calculated from beam formulas), this analysis procedure belongs to the "building block ' category.

I.C. Package   Package Model

Nonlinear Springs

Lead/Solder Characteristics   Pull Test

Pull out force (lbs)

Displacement (mils)   N leads

System Response

Load

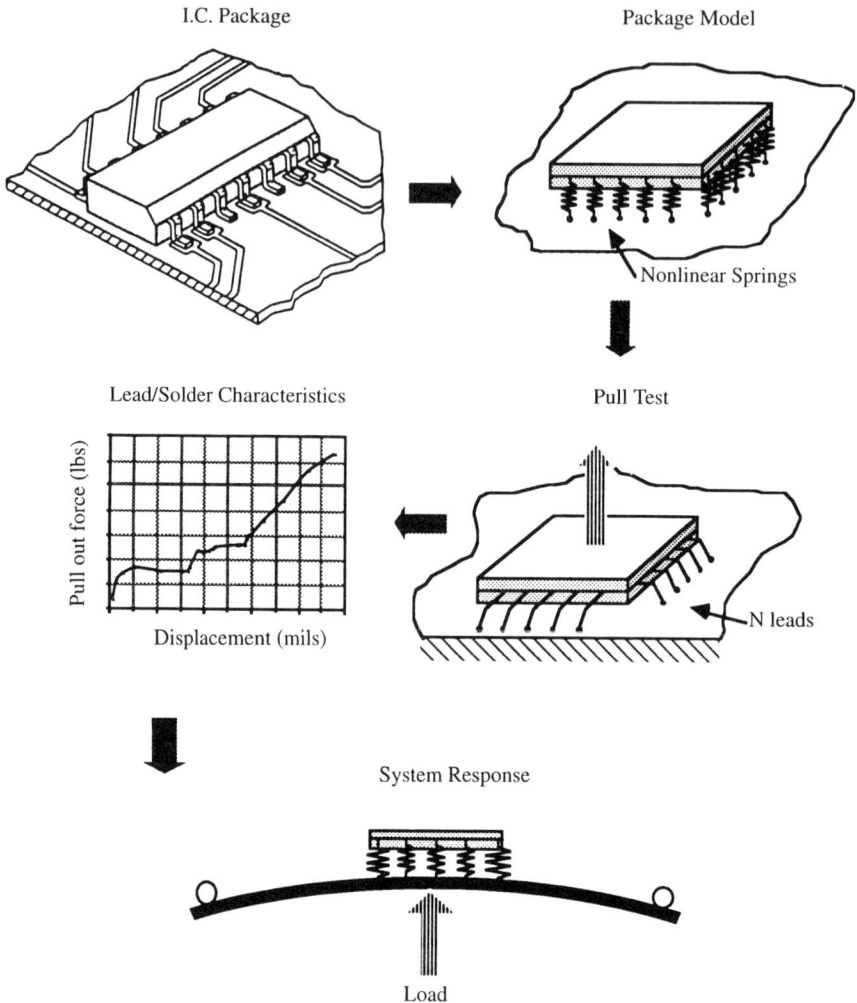

FIGURE 8.14. Hybrid analytical/experimental method.

The load test of individual leads established a much greater stiffness in the axial direction than in the shear direction (Fig. 8.15), and therefore all leads were modeled as axial springs. However, the nonlinear load deformation characteristics and the feature of greater stiffness in the tension direction were included in the computational procedure.

Neglecting module deformations, the analysis would start with a guessed position W of the module, in the vertical sense, with respect to the circuit card position, w. This would determine the lead extensions u, the value of which allows the sum total of vertical lead forces $\sum F$ to be calculated. Since the

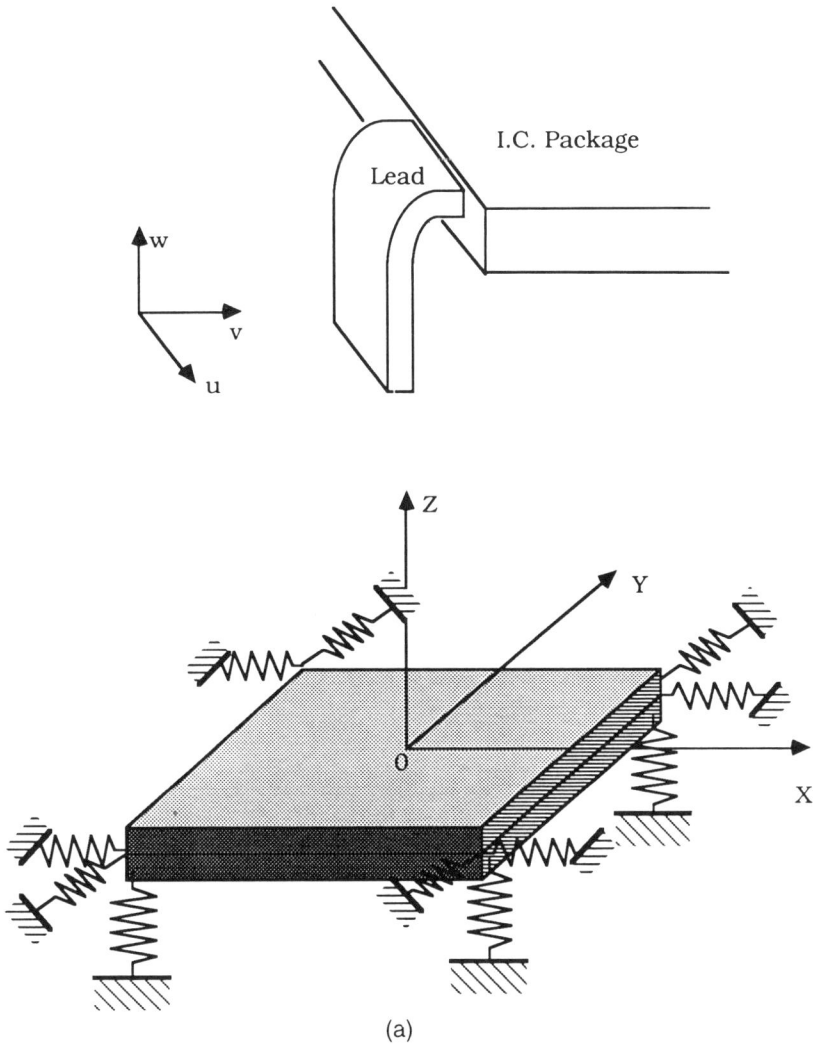

(a)

FIGURE 8.15. Lead deformation. (a) Physical setup and axes; (b) displacement in the v-direction; (c) displacement in the u-direction; (d) displacement in the w-direction. From Wong et al. [7].

latter should add up to a known value P (or zero, for an unloaded module), this supplies a criterion to the convergence of the iteration process, with module position W-w approaching u. Figures 8.16a and 8.16b show respective flowcharts for bending and twisting configurations of the card.

This method was used to evaluate the lead forces arising in assemblies of a single 44-lead PLCC and a single 68-lead PLCC module attached to a 15.24 cm (6-in.) long and 8.89 cm (3.5 in.) wide FR-4 circuit card; the latter

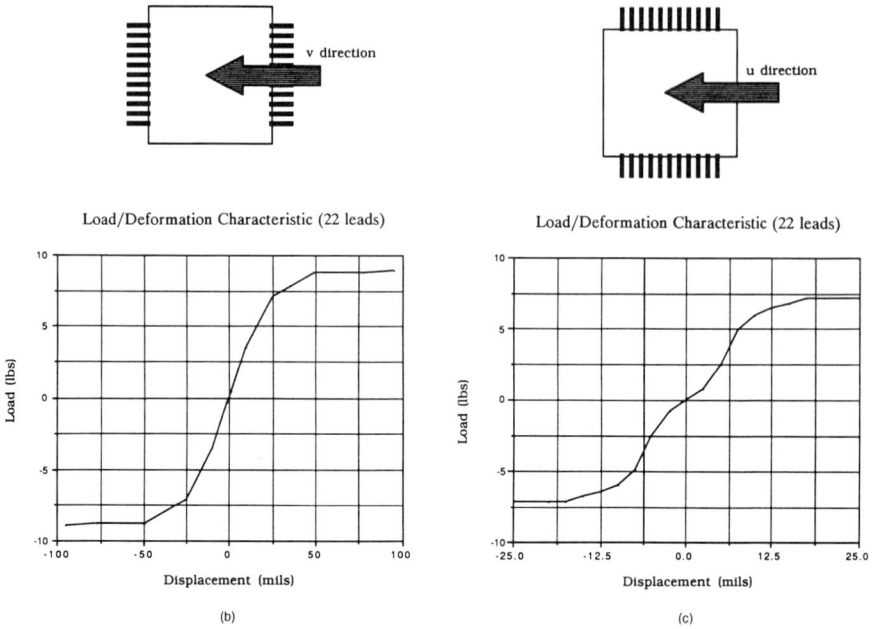

FIGURE 8.15. Continued.

was simply supported at the ends and a central line load P was applied to the card. Two positions of the modules were calculated: a central one and one toward the support. Figure 8.17 shows maximum lead forces with symmetrical module positions at the corners and at the center of a rectangular card. According to their locations both tensile or compressive loads may arise in the leads, as shown in the curves. The nonlinear lead-deformation effect is shown at higher loads. Some maximal lead forces materialized in the central locations of the module, rather than at its corners.

6. Conclusions

Leads of surface-soldered modules (J-leads and gull wings) were found to have elastically supported (nearly hinged) end conditions at the card. Linearly elastic finite element analysis of a typical circuit card, with a single central module and with increasingly dense module clusters, showed maximum corner lead forces. Results and trends for various system parameters were shown and compared with those obtained by the "strip method." A building block method used local assemblies such as the ones introduced in Chapter 7, in addition to plate formulas for card deflection, to perform the lead force analysis. A hybrid experimental/analytical procedure used modules

Load/Deformation Characteristic (44 leads)

Displacement (mils)
(d)

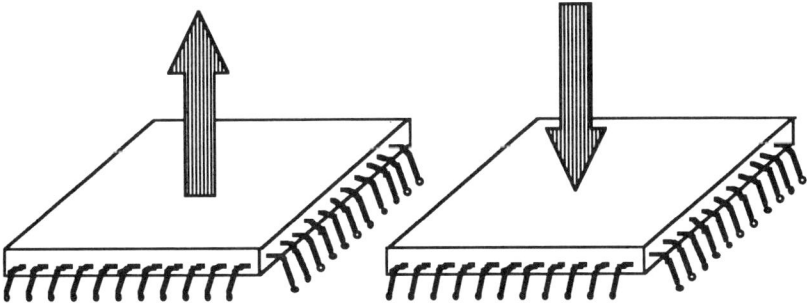

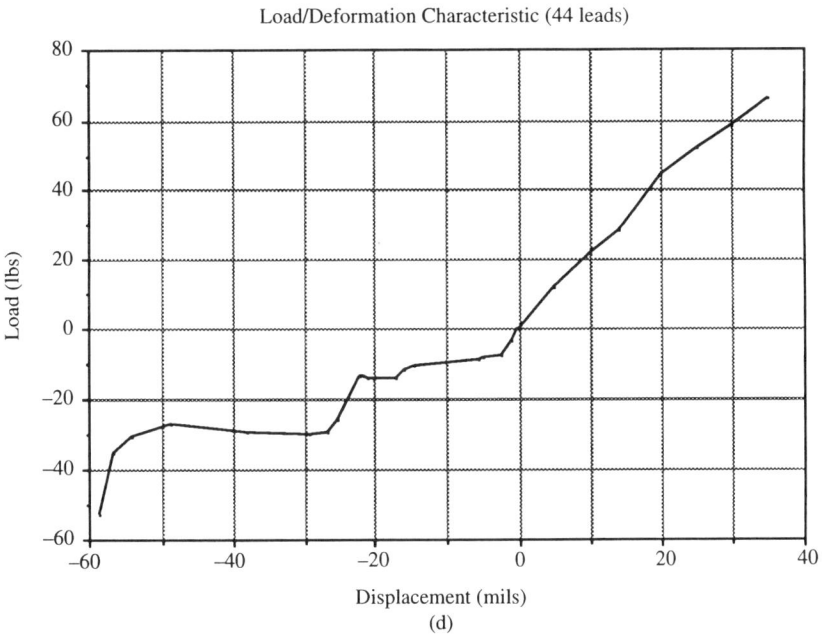

FIGURE 8.15. Continued.

with measured nonlinear lead compliances as building blocks in an iterative scheme.

The general lead force distribution under bending of a module/lead/card system depends a great deal on the lead stiffness. Corner lead forces are likely to be maximal when stiff leads induce a strong elastic foundation effect. When bending a square module with many soft leads sharing end loads, the maximum lead force may arise at the central position.

(a)

Input prescribed force & position of the I.C. Package.

I.C.Package at center ?

No

Input PC board slope of given I.C.package position , Rotate P.C.board about Y-axis

Yes

Guess the vertical position of the I.C. Package

Calculate relative displacement for a particular lead in vertical direction

Use force/displacement chart, find out the lead forces

No

Calculation of the total lead forces

Total Force = 0 ?

Yes

Output displacement of leads

Stop

(b)

Start

Read data from MARC for PCB deflections

Input I.C. Package position

Interpolation from MARC data, calculate PCB deflection at the solder joints.

Initial guess of the delta positions & orientations.

Calculate relative displacements for a particular lead in u, v, & w directions.

Change the delta

Use force/displacement charts, find out the forces in u, v & w directions.

Calculate total force in X, Y, & Z directions and total moments about X, Y, & Z axes.

Total force and total moment = 0 ?

Yes

Stop

No

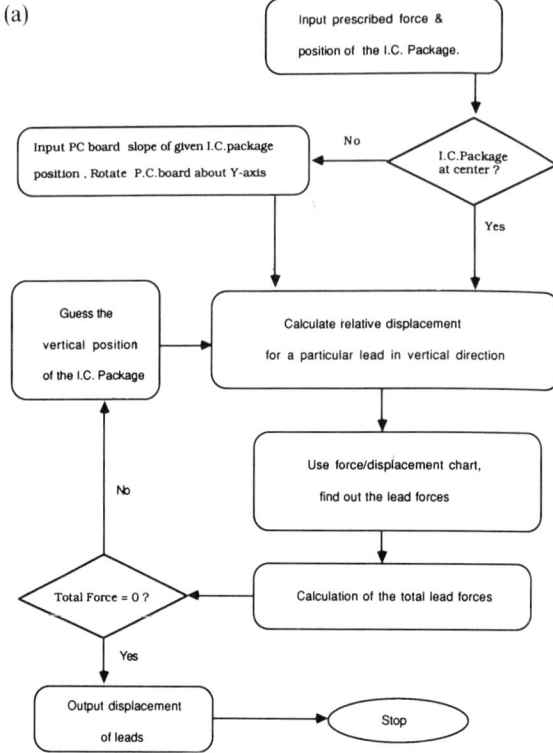

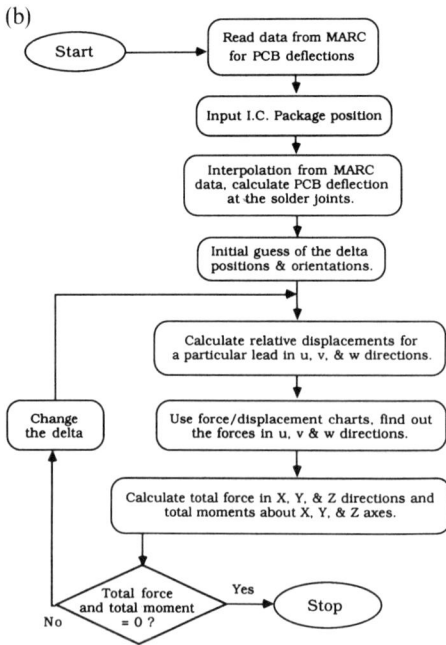

FIGURE 8.16. Calculation of lead force by the hybrid analytical/ experimental method. (a) Bending; (b) twisting. From Wong et al. [7].

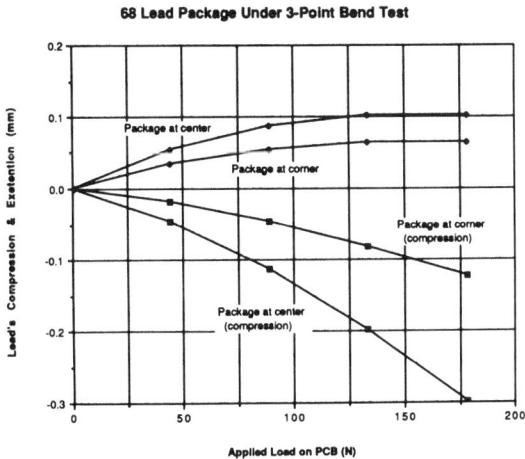

FIGURE 8.17. Lead compression and extension in three-point bending test, involving a 68-lead package. From Wong et al. [7].

7. Exercises and Questions

1. Derive the 3×3 in-plane flexibility influence matrix [C] for a one-segment vertical Amzirc lead of $L = 2.5$ mm length, $b = 1$ mm width, and $h = 0.2$ mm height, fixed at the module, corresponding to Eq. (8.1). Then (a) obtain the stiffness matrix [K], and (b) calculate K_f, K_h, K_r, and K_{zz}.
2. Find the maximum lead forces from the strip method for a 25 mm-square PLCC module centrally attached to a 12.5-cm-square FR-4 card.
3. Use the building block method for determining the lead forces of the same structure.
4. Find the maximum lead force from the strip method for the structure of Fig. 8.6a under a unit plate moment $M_x = 1$ Nmm/mm, applied along the top and bottom with the sides free.

References

1. Engel, P.A., Caletka, D.V., and Palmer, M.R. (1991), "Stiffness and Fatigue Study of Module/Lead/Card Systems Subjected to Bending," *ASME J. Elec. Packag.*, **113**(2), 129–137.
2. Kotlowitz, R.W. (1988), "Comparative Compliance of Generic Lead Designs for Surface Mounted Components," *IEPS J.*, **10**(1), 7–19.
3. Engel, P.A. (1990), "Structural Analysis for Circuit Card Systems Subjected to Bending," *ASME J. Elec. Packag.*, **112**(1), 2–10.
4. Vogelmann, J.T. (1991), M.S. thesis, State University of New York, Binghamton.
5. Roark, R.J., and Young, W.C. (1992), *Formulas for Stress and Strain*, 6th ed., McGraw-Hill, New York.

6. Prakash, V., Engel, P.A., Pitarresi, J.M., Albert, T., and Westby, G. (1991), "Stress Analysis of Component Attachments to Circuit Cards," *Proc. Int'l IEPS Conf.*, San Diego, Calif., Vol. 2, pp. 794–804.
7. Wong, T.L., Stevens, K.K., Wang, J., and Chen, W. (1990), "Strength Analysis of Surface Mounted Assemblies Under Bending and Twisting Loads," *ASME J. Elec. Packag.*, **112**(2), 168–174.

Chapter 9

Approximate Engineering Theory for the Twisting of Compliant Leaded Circuit Card/Module Systems

Twisting is a favorite method of product assurance testing agencies for verification of robustness. The geometry of flexure is simpler, and structural analysis benefits from the symmetries. This chapter will introduce the basic problems involved in the twisting of circuit cards, and an approximate solution will be shown for a single module attached to a card. This approximate treatment, while useful in defining some concepts, also produces acceptable results for relatively widely spaced stiff leads in plastic modules. (The reader may skip to the next chapter for an analytical theory, confirmed by finite element analysis.) The procedure is extended to the analysis of module groups.

1. Fundamental Approach

An engineering analysis for the twisting of a circuit card with a central module attached will be outlined at present. For greater simplicity, the card and module are chosen square at first; thus $b = a$ and $L_x = L_y = L$ are chosen in Fig. 9.1, which shows a card subjected to a linear displacement $w = \pm cx$ along its top and bottom edges. There is complete antisymmetry with respect to the x- and y-axes.

When the circuit card without a module is twisted, the linear deflection is caused by torques T exerted by pairs of concentrated forces $P = T/L$ at the card corners. All over the card the deflection is

$$w = 4w_p xy/L^2 \tag{9.1}$$

where, by St. Venant's theory of torsion,

$$d\phi/dy = T/GJ . \tag{9.2}$$

In Eq. (9.2), $G = E/[2(1 + v)]$ and $J = L_x h^3/3$, and the torque is

$$T = 8(1 - v)Dw_p/L . \tag{9.3}$$

If a square module is attached to the card center, some stress is transferred through the leads to the module. While antisymmetry is preserved, two main

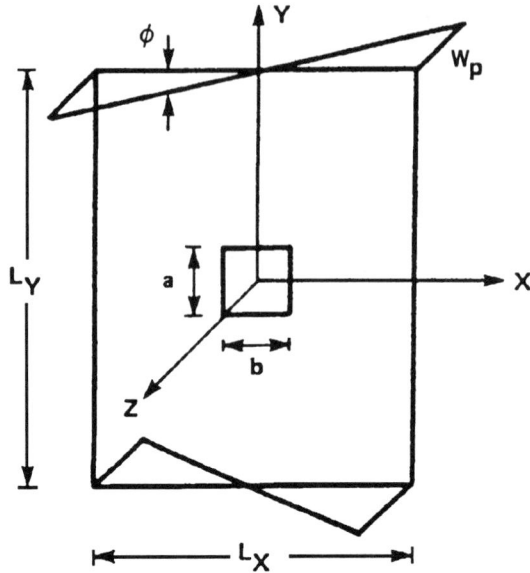

FIGURE 9.1. Twist of a circuit card with module attached at the center. Note the constant twist angle ϕ.

differences arise with respect to the bare card: Eq. (9.1) will no longer be strictly valid, and linear card-edge displacements no longer correspond to pure concentrated forces P at the card corners.

We now make an approximation [1, 2] on the side of conservativism, concentrating the lead-force distribution along the module edges into corner lead forces Q (Fig. 9.2). Of course, the four corner forces Q will be in equilibrium among themselves since no external force acts on the module.

In calculating the corner lead forces Q, we evaluate the deflections of two plates (those of the card, w, and those of the module, W) and stipulate the compatibility of these displacements with the stretch $u = w - W$ of the corner leads.

We make use of the condition of antisymmetry (Fig. 9.3) in formulating deflections through a square plate $L \times L$ of rigidity D, due to a concentrated force F applied transversely at a point r' along the diagonal. The solution can be obtained numerically, by finite elements and by the Ritz method; it is given in Fig. 9.4 and Table 9.1 for five equally spaced points $(x_i = y_i)$ along the diagonal. The figure shows the results as a 5×5 influence matrix $\varLambda(\rho, \rho')$ defined by the equation

$$w(r) = \varLambda(\rho, \rho') \cdot F(r') \cdot L^2/D \ . \tag{9.4}$$

The dimensionless diagonal distance $\rho = r/L^*$ were used in constructing $\varLambda$, so that it is applicable to any square-size plate. The half-diagonal of the plate is $L^* = \sqrt{2}L/2$.

Note that for other than an even 5×5 lead spacing, interpolation can be used for the respective $\varLambda$ values (Table 9.2). In possession of the influence

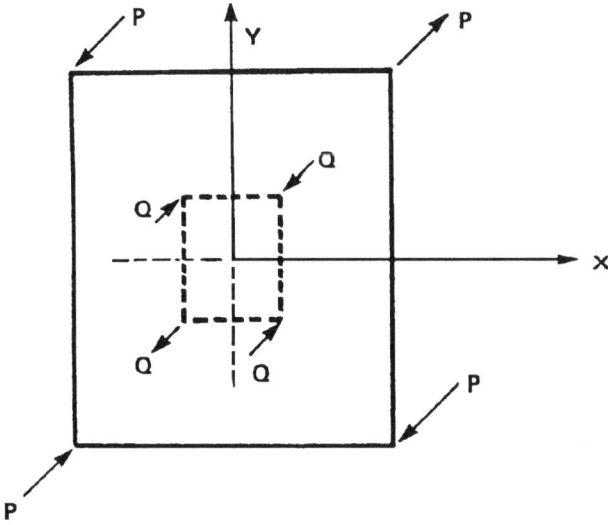

FIGURE 9.2. Forces acting on a card: P ($P \cdot L_x = T$ constituting the torque) and Q (concentrating the lead reactions at the module corners).

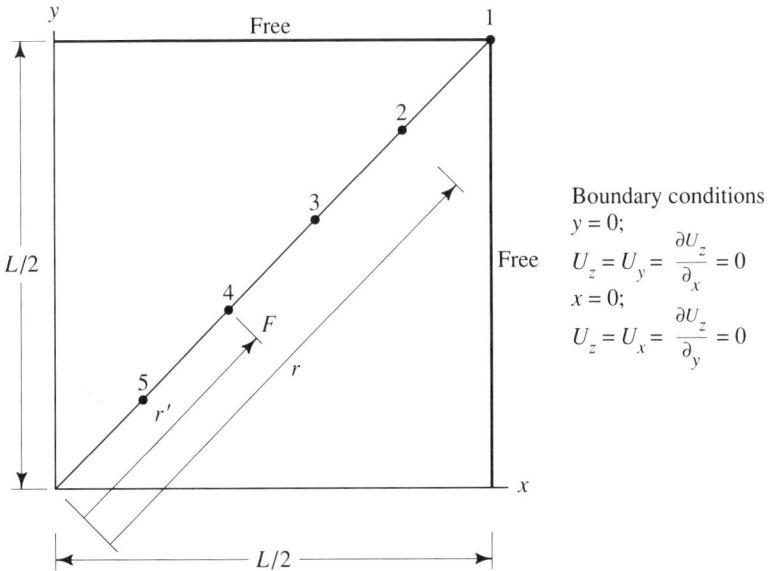

Boundary conditions
$y = 0$;

Free $U_z = U_y = \dfrac{\partial U_z}{\partial_x} = 0$

$x = 0$;

$U_z = U_x = \dfrac{\partial U_z}{\partial_y} = 0$

FIGURE 9.3. Quarter-plate with boundary conditions corresponding to antisymmetry. The diagonal is divided into five equal spaces, and the fifth points are marked 5, 4, 3, 2, 1.

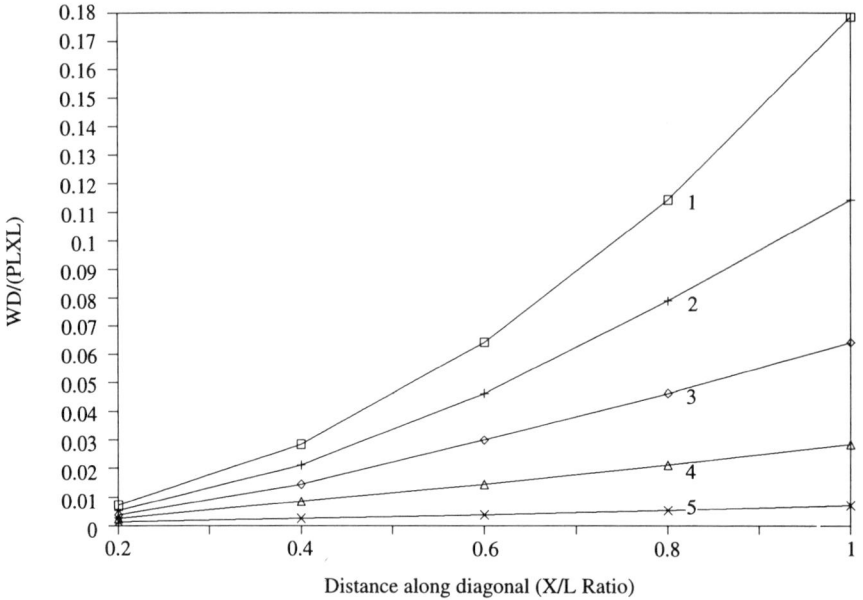

FIGURE 9.4. Antisymmetric displacement influence functions for a square elastic thin plate, loaded by a unit transverse force at fifth points of the diagonal as shown in Fig. 9.3, $v = 0.3$. Computations by finite elements, and by the Ritz method (in parentheses) are shown in Table 9.1.

TABLE 9.2. Coefficients of $\Lambda(\rho; 0.2j)$, fitted by a fifth-degree polynomial.

$$\Lambda(\rho; 0.2j) = \sum_{i=1}^{5} c_i \cdot \rho^i$$

j	c_1	c_2	c_3	c_4	c_5
0	0	0	0	0	0
1	0.0093070	-0.0158244	0.0267818	-0.0167031	0.0035807
2	-0.0132922	0.1970738	-0.4127875	0.3977812	-0.1402083
3	0.0059933	0.0337066	0.2177677	-0.3444167	0.1512240
4	-0.0004975	0.140555	-0.02448021	0.027625	-0.0289323
5	-0.00003	0.1788398	-0.0009651	0.0012552	-0.0005599

function, the following simple computation results in the corner lead force Q. Let the card deflection at Q due to $P =$ unity be w_{QP}, and let the card deflection at Q due to $Q = 1$ be w_{QQ}. Then the card deflection at Q is

$$w_Q = w_{QP} \cdot P - w_{QQ} \cdot Q . \tag{9.5}$$

The module deflection at Q is written in terms of the module deflection W_{QQ} at Q due to a unit force $Q = 1$:

$$W_Q = W_{QQ} \cdot Q . \tag{9.6}$$

The lead has an axial displacement proportional to its axial spring constant K:

$$u = Q/K . \tag{9.7}$$

Compatibility of displacements requires

$$w = W + u . \tag{9.8}$$

Substituting Eqs. (9.5)–(9.7) into Eq. (9.8) and solving for Q in terms of P gives

$$Q = \frac{w_{QP}}{w_{QQ} + W_{QQ} + 1/K} P . \tag{9.9}$$

Example 1: Lead Force Calculation

Consider a 25-mm (1 in.) square PLCC module of 3-mm (0.121-in.) thickness and equivalent modulus $E = 3.24$ GPa (470,000 psi); its plate rigidity is

$$D_m = \frac{E_m h_m^3}{12(1 - v_m^2)} = \frac{(3.24 \times 10^3)(3^3)}{12(1 - 0.4^2)} = 8680 \text{ N.mm (76.80 lb-in.)} .$$

The module is attached to an $L = 12.5$ cm (5 in.) square FR-4 card of thickness $h = 1.25$ mm (0.05 in.) and modulus $E = 13.8$ GPa (2 Mpsi). Thus the card has a plate constant

$$D_c = \frac{(13.8 \times 10^3)(1.25^3)}{12(1 - 0.3^2)} = 2674 \text{ N.mm (23.66 lb-in.)} .$$

Consider the J-leads with a fixed-free spring constant (see Chapter 8) of $K = 26$ N/mm (148 lb/in.), so that $1/K = 0.0385$ mm/N.

Since $a = L/5$, we can find the influence coefficients directly from Fig. 9.4:

$$w_{QP} = \frac{L^2 \cdot \Lambda(0.2; 1)}{D_c} = \frac{(125^2)(0.007142)}{2674} = 0.04173 \text{ mm/N}$$

$$w_{QQ} = \frac{L^2 \cdot \Lambda(0.2; 0.2)}{D_c} = \frac{(125^2)(0.001417)}{2674} = 0.00828 \text{ mm/N}$$

$$W_{QQ} = \frac{a^2 \Lambda(1; 1)}{D_m} = \frac{(25^2)(0.1785)}{8680} = 0.01286 \text{ mm/N} .$$

Substituting into Eq. (9.9), we obtain

$$Q = \frac{0.04173}{0.00828 + 0.01286 + 0.0385} P = 0.700 P .$$

TABLE 9.3. Leadforce and stiffness approximation for the engineering theory. α-values in $Q = \alpha \cdot P$; β-values in $k_T = \beta \cdot D_c$.

Lead B.C. Module size	Fixed hinged $K = 26$ N/mm			Fixed-fixed $K = 543$ N/mm		
	α	β	β/α	α	β	β/α
25 mm	0.700	2.88	4.11	1.816	3.02	1.66
50 mm	1.193	3.46	2.90	1.618	3.78	2.34

Note: $\beta = 2.80$ for the bare card. $Q = \beta/\alpha \cdot (D_c \phi / L)$.

Table 9.3 shows the corresponding relationship between Q and P for twice the module size and for fixed-fixed leads as well. The numerical value of Q, of course, depends on the actual twist.

2. Torsional Stiffness Calculation

We can evaluate the torsional stiffness $k_T = T/\phi$ of a card like that of Example 1, with and without a central square module attached. The procedure is writing the card corner displacement w_P from which, through $\phi = 2w_P/L$ and $T = PL$, we get

$$k_T = PL^2/2w_P . \tag{9.10}$$

For a bare card, we have

$$w_P = \Lambda(1; 1)PL^2/D_c . \tag{9.11}$$

By Eq. (9.10), using $\Lambda(1; 1)$ from Fig. 9.3:

$$k_{T0} = D_c/2\Lambda(1; 1) = D_c/(2)(0.17854) = 2.80 \, D_c . \tag{9.12}$$

For a card with module, the card corner deflection is now $P \cdot w_{PP} - Q \cdot w_{PQ}$ or

$$w_P = (L^2/D_c) \cdot [P \cdot \Lambda(1; 1) - Q \cdot \Lambda(1; 0.2)]$$

and k_T can be obtained from Eq. (9.11). The lead force is then, in general,

$$Q = \frac{2k_T \Lambda(1; 1) - D_c}{2k_T \Lambda(1; a/L)} P . \tag{9.13}$$

For combinations of 25- and 50-mm modules as well as fixed-hinged and fixed-fixed leads, see Table 9.3 for k_T values. Table 9.3 demonstrates that torsional stiffness increases rapidly with module size, while lead force increases with lead stiffness.

3. Rectangular Cards with a Module

To validate the preceding procedures for rectangular cards (L_x by L_y), torsional stiffness is calculated on an equivalent square card of the same size L_y as that of the rectangular card (Fig. 9.5). Then the stiffness k_{re} of the latter (with central module) is approximately related to the stiffness k_{sq} of the corresponding square card with a central module:

$$k_{re} = k_{sq} \cdot L_x/L_y . \tag{9.14}$$

Once the stiffness is known, the lead forces can be calculated for given T or given ϕ based on Q, the lead force calculated on the corresponding square module.

$$\text{For given } \phi: \quad \text{For given } T:$$

$$Q_{re} = Q_{sq} \qquad Q_{re} = Q_{sq} \cdot L_y/L_x \tag{9.15}$$

4. Module Clusters

We may generalize the lead force calculation by obtaining the torsional stiffness for the case of a module cluster, a group of modules arranged in a symmetrical fashion over a card. From Eq. (9.11) the torsional stiffness

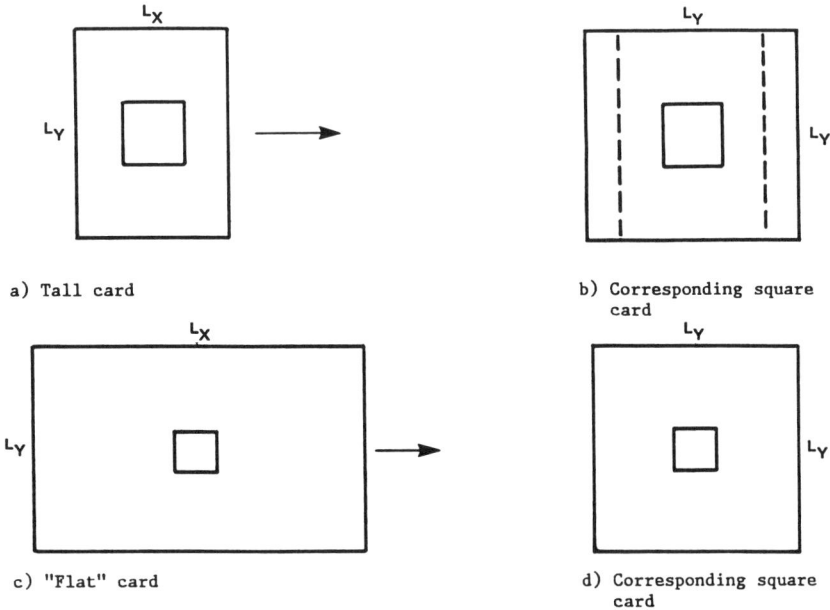

a) Tall card

b) Corresponding square card

c) "Flat" card

d) Corresponding square card

FIGURE 9.5. Rectangular and corresponding "equivalent" square cards.

augmented by a single central module can be written as

$$k_T = \frac{PL^2}{2\left[w_{PP} - w_{PQ}\left(\dfrac{w_{QP}}{w_{QQ} + W_{QQ} + 1/K}\right)\right]P} = \frac{L^2}{2\left[w_{PP} - \dfrac{w_{QP}^2}{w_{QQ} + W_{QQ} + 1/K}\right]} . \tag{9.16}$$

The addition of another module, similar to the first, along the x-axis would increase the spring constant by yet another increment

$$\Delta k_T = k_{T1} - k_{T0} \tag{9.17}$$

where k_{T0}, the spring constant of the bare card, is, based on Eq. (9.2):

$$k_{T0} = 2GL_x h^3 / 3L_y . \tag{9.18}$$

Then, for n module attachments along the x-axis (Fig. 9.6a):

$$k_T \approx k_{T0} + n \cdot \Delta k_T . \tag{9.19}$$

For a given twist ϕ, the value of the torque $T = PL$ is increased as k_T is increased from k_{T0}. But what happens to the lead forces Q of module clusters?

Assuming that the circuit card is twisted into the saddle shape of Eq. (9.1), the prescribed twist ϕ being unchanged, only the torque T and thus the forces P have been increased. When a rectangular plate is subjected to flexure, it must develop anchoring forces at its corners to resist the tendency of curling up, as described by Timoshenko [3]. These anchoring forces are proportional to the twisting amount M_{xy} given in Eq. (1.30). In our case, they constitute the lead forces Q, and their dependence on the saddle curvature $\partial^2 w/\partial x \partial y$ and the module rigidity D_m may be expressed by the aid of a proportionality factor ψ:

$$Q = 2D_m(1 - v_m) \cdot (\partial^2 w/\partial x \partial y) \cdot \psi . \tag{9.20}$$

The factor ψ (<1) stands for the fact that the corner leads are insufficient to force the module to follow the card curvature over its whole area. Since the saddle curvature is unchanged by Eq. (9.1) and equal to

$$\partial^2 w/\partial x \partial y = 4w_P/L^2 , \tag{9.21}$$

we can then consider the Q versus P relationship unchanged for identical modules of a module cluster. As an additional simplification resulting from the preceding analysis, the saddle curvature is also a constant with respect to the location x, y on the card; thus the module can be placed anywhere along the card, provided symmetry is preserved. This has been numerically proven by finite element calculations [2].

The present theory tends to ignore module location dependency in computing stiffness increase Δk_T; at the root of this is the fundamental assumption of a linear twist geometry by concentrated forces applied at the card corners. The latter contention is progressively further violated as a larger number of multiple modules are attached to a card, as long as uniformity in

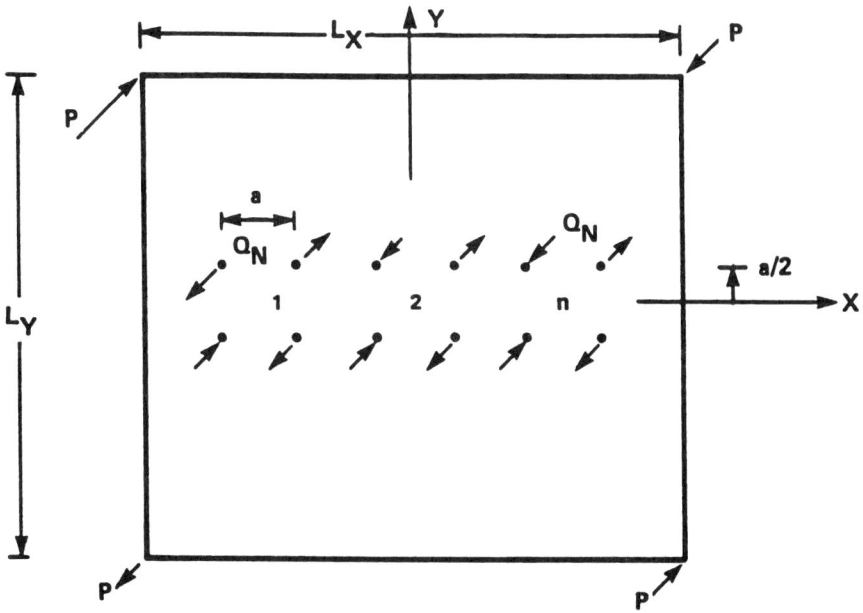

(a)

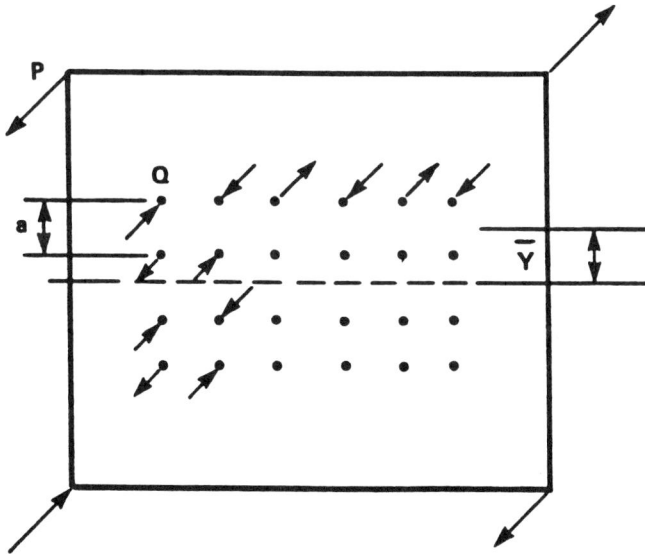

(b)

FIGURE 9.6. Modeling of lead forces in module clusters along the x-axis of the card. (a) n-modules of square size a; (b) two rows of modules straddling the x-axis.

card stiffness declines; this also leads to a decline of accuracy of the procedure. The analytical theory of the next chapter will shed light on the goodness of this assumption.

The following example will show the simplicity of the calculations required for module clusters.

Example 2: Module Clusters

Evaluate the lead forces in three configurations as shown in Fig. 9.7, for the card and module combination of Example 1. Fixed-fixed J-leads of $K = 543$ N/mm (3100 lb/in.) spring constant are considered. The twist angle is $\phi = 2.25$ deg, and we assume linear elasticity.

a) The increase of torsional stiffness is first calculated for a single 25-mm (1-in.) PLCC module. We have, from Eq. (9.12):

$$k_{T0} = 2.80\,D_c = (2.80)(2674) = 7487 \text{ N.mm/rad}$$

and from Table 9.3,

$$k_{T1} = 3.02\,D_c = (3.02)(2674) = 8075 \text{ N.mm/rad}$$

so that

$$\Delta k_T = k_{T1} - k_{T0} = 0.22\,D_c = (0.22)(2674) = 588 \text{ N.mm/rad}$$

For $\phi = 2.25$ deg, there result

$$T = k_T\,\phi = (8075)(2.25\pi/180) = 317.1 \text{ N.mm}$$

$$P = T/L = 317.1/125 = 2.54 \text{ N}.$$

From Table 9.1:

$$Q = 1.816\,P = (1.816)(2.54) = 4.607 \text{ N}$$

and from Eq. (9.20) we get $\psi = 0.131Q = 0.60$.

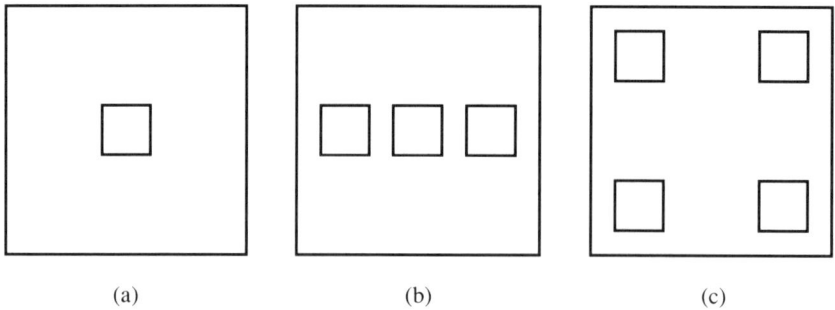

FIGURE 9.7. Three configurations of a 125-mm-square card with 25-mm modules attached. (Data from Example 1.)

b) Assuming the same form [Eq. (9.1)] for the deflected card surface, each of the three modules will add the same torsional stiffness increment:

$$k_T = k_{T0} + 3 \, \Delta k_T = 7487 + (3)(588) = 9251 \text{ N.mm/rad} .$$

Now
$$T = k_T \phi = (9251)(2.25\pi/180) = 363.1 \text{ N.mm}$$

$$P = 363.1/125 = 2.905 \text{ N} ,$$

and
$$Q = 1.816 \, P = (1.816)(2.905) = 5.28 \text{ N} ,$$

yielding a value $\psi = 0.69$.

c) The location of the modules is immaterial for the approximate theory, and for four modules we get

$$k_T = k_{T0} + 4 \cdot \Delta k_T = 7487 + (4)(588) = 9839 \text{ N.mm/rad} .$$

Then
$$T = 9839 \, (2.25\pi/180) = 386.2 \text{ N.mm}$$

and
$$P = 386.2/125 = 3.089 \text{ N} ,$$

Finally
$$Q = (1.816) \, (3.089) = 5.610 \text{ N} ,$$

which has a ψ factor of 0.73.

The results of Example 2 show that the lead force Q would rise only moderately each time we add a module, as a response to the added torsional stiffness.

It is remarked that since incremental torsional stiffnesses due to individual modules of various size and rigidity can be computed by Eq. (9.17) independently of each other, we may follow the procedure of Example 2 when several types of module are attached to the circuit card; the only required condition is the symmetry of their distribution.

5. Finite Element Check of the Approximate Theory

The three configurations (Figs. 9.7a–9.7c) of Example 2 were checked by finite element (ANSYS) calculation. The respective lead forces for the three cases are shown in Figs. 9.8a–9.8c.

In order to emphasize similarities with the approximate theory scheme, the finite element scheme included a corner lead (Fig. 9.9a) in addition to evenly spaced lead springs; this is unlike the real situation of a PLCC module which would instead have a leadless corner (Fig. 9.9b).

In addition to the corner lead force and possibly two near-corner lead forces from both sides, all other leads yield negligible values, thus we lump these five into one "equivalent" corner force Q and compare this with the

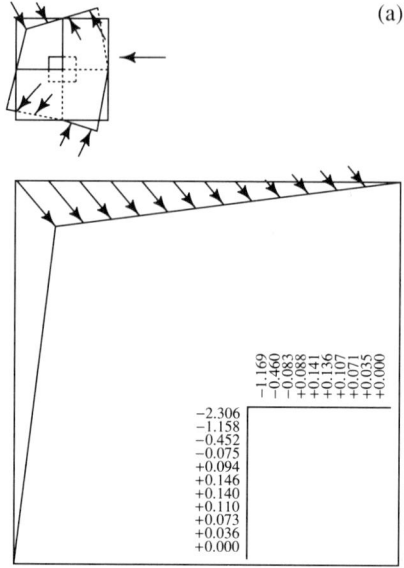

(a)

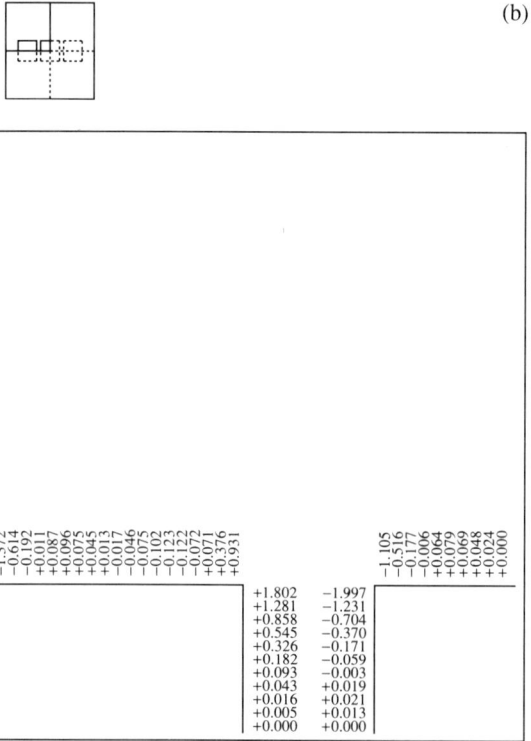

(b)

(c)

-1.373 -0.587 -0.147 -0.055 +0.125 +0.127 +0.099 +0.062 +0.023 +0.015 -0.055 -0.099 -0.144 -0.184 -0.201 -0.155 +0.020 +0.427 +1.206

-2.646	+2.495
-1.271	+1.359
-0.441	+0.637
-0.008	+0.232
+0.178	+0.033
+0.226	-0.046
+0.207	-0.063
+0.163	-0.053
+0.113	-0.033
+0.064	-0.010
+0.020	+0.013
-0.024	+0.038
-0.068	+0.066
-0.113	+0.095
-0.151	+0.119
-0.164	+0.117
-0.117	+0.054
+0.054	-0.134
+0.445	-0.542
+1.187	-1.299
+2.406	-2.525

+1.283 +0.577 +0.186 -0.001 -0.072 -0.082 -0.067 -0.043 -0.018 +0.007 +0.033 +0.060 +0.089 +0.110 +0.106 +0.041 -0.148 -0.557 -1.310

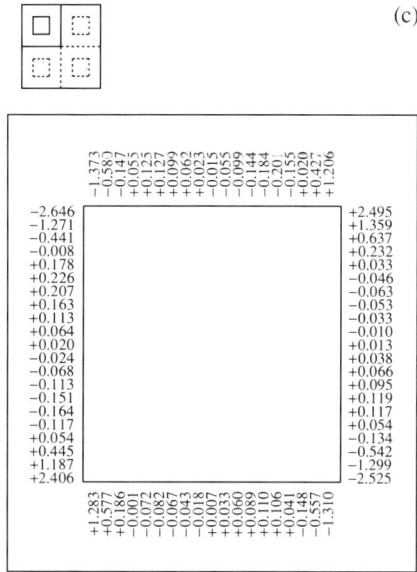

FIGURE 9.8. Continued.

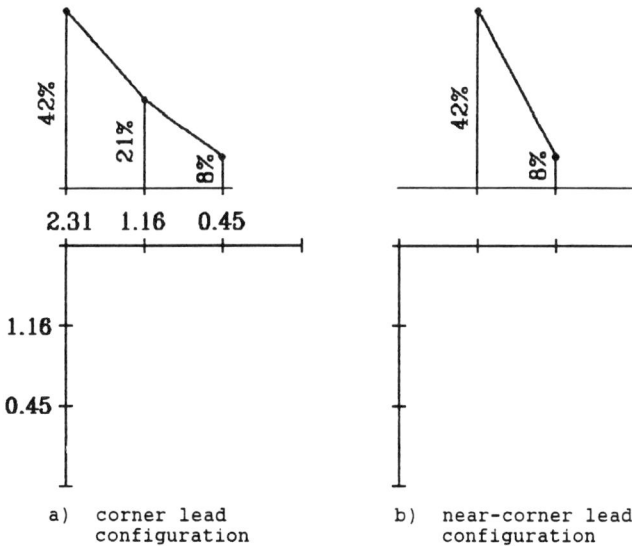

a) corner lead configuration

42%   21%   8%

2.31   1.16   0.45

1.16

0.45

b) near-corner lead configuration

42%   8%

FIGURE 9.9. Corner lead versus near-corner lead configurations; lead forces evaluated by finite elements.

FIGURE 9.8. Finite element–calculated lead forces (N) due to a $\phi = 2.25$-degree twist of a 125-mm-square card with three 25-mm-square module configurations as defined in Example 2. (a) One central module, see Fig. 9.10b; (b) three modules along the x-axis, see Fig. 9.10c; (c) four corner modules, see Fig. 9.10d.

approximate theory. The comparison is shown here:

Module	Approximate theory	Five corner leads, F.E.
1-module	4.54 N	5.23 N
3-module	5.21 N	6.04 N (5.55 outside module)
4-module	5.55 N	6.30 N

Other even higher-density module populations (see Fig. 9.10) were tested and also computed by finite elements. All the cases except the highest populated card with 16 modules gave agreement similar to this with the approximate theory. For the 16-module card a 47% higher Q was obtained by the approximate method than by finite elements.

At this point we are in the position to judge the Q-value computed by the approximate theory. First of all, since in reality we have two near-corner leads, the Q-value computed should fairly be split into two. This is confirmed by a typical finite element computation such as in Fig. 9.9, of the case of Fig. 9.10b. Allotting the corner force value half to the second position on one side and half to the second position on the other side of the corner, the fall-off

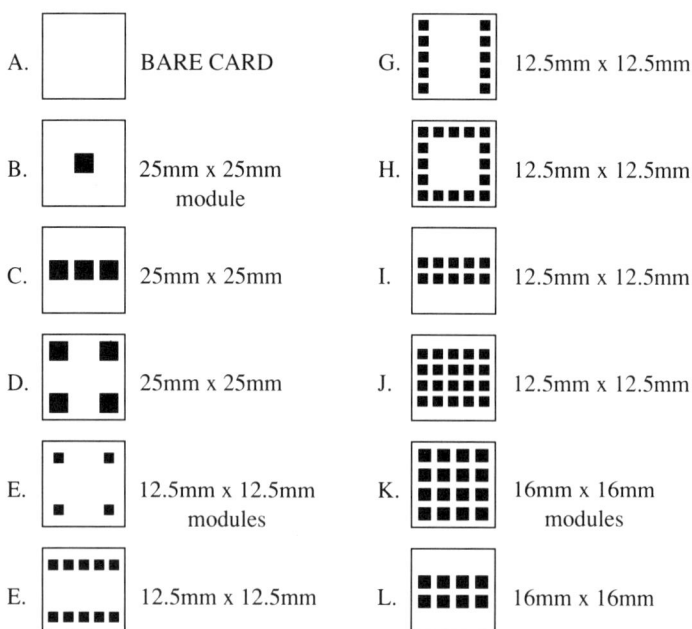

A.	BARE CARD	G.	12.5mm x 12.5mm
B.	25mm x 25mm module	H.	12.5mm x 12.5mm
C.	25mm x 25mm	I.	12.5mm x 12.5mm
D.	25mm x 25mm	J.	12.5mm x 12.5mm
E.	12.5mm x 12.5mm modules	K.	16mm x 16mm modules
E.	12.5mm x 12.5mm	L.	16mm x 16mm

FIGURE 9.10. Various configurations of modules mounted to $L = 125$-mm-square cards, torque tested and analyzed.

of the computed lead forces from 42% in the second position to 8% in the third position justifies a 50%–50% distribution to both near-corner leads.

Table 9.4 shows a comparison of the finite element torsional stiffness results, with respect to both the approximate engineering theory and experimental measurements (to be further discussed in Chapter 10), for module cluster configurations depicted in Fig. 9.10. Table 9.5 shows the

TABLE 9.4. Torsional stiffnesses by the engineering theory by finite elements and experimental methods.

Card configuration	Stiffness (Eng. theory)		Stiffness (FEM)		Stiffness (experimental)	
	N-mm/rad	(in-lbf./rad)	N-mm/rad	(in-lbf./rad.)	N-mm/rad	(in-lbf./rad)
A.	15,652	(138.5)	14,451	(127.9)	13,706	(121.3)
B.	16,355	(144.7)	17,310	(153.2)	16,135	(142.8)
C.	17,761	(157.1)	n.a.		n.a.	
D.	18,464	(163.4)	29,727	(263.1)	30,643	(271.2)
E.	17,500	(154.8)	n.a.		n.a.	
F.	19,348	(171.2)	n.a.		n.a.	
G.	19,348	(171.2)	n.a.		n.a.	
H.	23,044	(203.9)	n.a.		n.a.	
I.	19,348	(171.2)	22,835	(202.1)	24,892	(220.3)
J.	24,892	(220.2)	34,439	(304.8)	40,586	(359.2)
K.	25,764	(227.9)	74,901	(662.9)	81,172	(718.4)
L.	20,708	(183.2)	32,925	(291.4)	37,083	(328.2)

Card thickness: $h_c = 1.60$ mm (0.063 in.). Configurations listed in Fig. 9.10.

TABLE 9.5. Lead force results for $2.25°$ twisting of module populated cards shown in Fig. 9.10, by the analytical and F.E. approaches; card thickness $h_c = 1.27$ mm (0.050 in.) and module rigidity $D_m = 8619$ N.mm (76.25 in.lb).

Card config. (Fig. 9.10)	Q_{FEM} (corner lead only) N (lbf)	Q_{FEM} (5 lead sum) N (lbf)	Q_{approx} theory N (lbf)
A.	n.a.	n.a.	n.a.
B.	10.23 (2.30)	24.59 (5.53)	20.20 (4.54)
C.	11.43 (2.57)	26.86 (6.04)	23.20 (5.21)
D.	11.74 (2.64)	28.02 (6.30)	24.71 (5.55)
E.	8.318 (1.87)	21.79 (4.90)	27.86 (6.26)
F.	9.608 (2.16)	28.02 (6.30)	32.41 (7.28)
G.	9.830 (2.21)	26.69 (6.00)	32.41 (7.28)
H.	9.030 (2.03)	29.50 (6.63)	41.50 (9.33)
I.	8.986 (2.02)	24.46 (5.50)	32.41 (7.28)
J.	9.875 (2.22)	26.24 (5.90)	46.07 (10.35)
K.	11.74 (2.64)	29.35 (6.60)	43.56 (9.79)
L.	—	—	32.05 (7.20)

corresponding lead force calculations, by the approximate theory and by finite elements, for the dozen module configurations of Fig. 9.10.

6. Conclusions

The steps of the approximate engineering theory to be followed for a specific case are as follows:

1. Calculate the card and module rigidities.
2. For a rectangular card, calculate the equivalent square size by Eq. (9.14).
3. Calculate the influence coefficients w_{QP}, w_{QQ}, and W_{QQ} for card and module corresponding to their size. This involves the dimensionless influence function, $\Lambda(r, r')$ (Fig. 9.4, Table 9.1). (A simple analytical evaluation of the influence function will be described in Chapter 10.)
4. Calculate $Q(P)$ from Eq. (9.9) for one module.
5. Calculate the torsional stiffness increment k_T of a module from Eq. (9.17). Add all module stiffness increments for a module cluster to the torsional stiffness of the card.
6. Calculate the lead force as from Eq. (9.13) for any module of a given size. All modules of the same size have the same lead force, regardless of location on the card surface. (The variable $\Lambda(1; a/L)$ distinguishes the size in Eq. (9.13).)

7. Exercises and Questions

1. Enumerate similarities and differences between the circuit card flexing operations of bending and torsion.
2. Show that the influence function $\Lambda(x; \xi)$ satisfies Maxwell's law of reciprocal deflections.
3. Prove the relationships (9.14 and 9.15) for rectangular cards.
4. Consider lead-force calculation by the approximate method, for the corner lead of the one-module structure of Fig. 9.8a. If a value of $Q = 1$ N is obtained by Eq. (9.13), what is a realistic value for the actual maximum force of a PLCC lead located nearest the corner?

References

1. Engel, P.A., and Vogelmann, J.T. (1992), "Approximate Structural Analysis of Circuit Card Systems Subjected to Torsion," *ASME J. Elec. Packag.*, **114**(2), 203–210.
2. Engel, P.A. (1986), "Torque Stress Analysis for Printed Circuit Cards Carrying Peripherally Leaded Modules," ASME Winter Annual Meeting, Anaheim, Calif., Paper No. 86-WA-EEP-2.
3. Timoshenko, S.P., and Woinowsky-Krieger, S. (1959), *Theory of Plates and Shells*, 2d ed., McGraw-Hill, New York.

Chapter 10

Analytical Theory and Experimental Work in Compliant Leaded Systems Subjected to Twisting

In the previous chapter we discussed an approximate engineering approach founded on the premise that corner leads alone carry stress between module and card. The validity of this idealization deteriorates with decreasing lead spacing and lead stiffness and with increasing module stiffness.

In this chapter we will start with an analytical solution [1] for a single central module. This solution is quite general and valid for all module, lead, and card parameters; it may then be used with the torsional stiffness relations to handle module clusters. The procedure is extended to include the stiffening role of membrane stresses arising in large displacements. The stiffness increase was evidenced in experiments. Fatigue due to torsional cycling is discussed last.

1. Analytical Theory

The antisymmetrical loading of a square card with square central module under torsion was simplified in Fig. 9.2 to consist of corner forces P and Q on card and module, respectively. For the plate displacements of both card and module, an antisymmetrical influence function $\Lambda(\rho; \rho')$ was introduced. This function was nondimensional [see Eq. (9.4)] and its arguments were nondimensionalized coordinates with respect to the size of the plate to which they referred (i.e., $\rho = r/L^*$).

The influence function has been computed [1] by a truncated polynomial series using the Ritz method; a two-term solution proved extremely close to finite element results. In the ensuing treatment this Ritz solution is treated as an analytical plate solution expression. The plate coordinates and the transverse displacements will be used in their nondimensional forms, designated by lower-case symbols (x, y, and w), while the physical coordinates and displacement will be denoted by the respective upper-case symbols, thus X, Y, and W. To denote the physical card and module displacements, the subscripts c and m, respectively, will be added (W_c and W_m).

179

Let us now consider a displacement influence function along the diagonal of a square plate (the card) of size L, similar to Fig. 9.4. When the unit point load $F = 1$ is at the diagonal distance $r' = (X'^2 + Y'^2)^{1/2}$, $(X' = Y')$, the displacement can be calculated at the diagonal plate coordinates $r = (X^2 + Y^2)^{1/2}$, $(X = Y)$. Nondimensionalizing the coordinates: $x = X/(L/2)$, $y = Y/(L/2)$, $\xi = X'/(L/2)$, $\eta = Y'/(L/2)$, the nondimensional card displacement is related to the physical card displacement:

$$W_c = (L/2)^2 \cdot F \cdot w_c / D_c . \tag{10.1}$$

Similarly, the nondimensional and physical module displacements are related as

$$W_m = (a/2)^2 \cdot F \cdot w_m / D_m . \tag{10.2}$$

The nondimensional displacements w_c and w_m, or in general, w, can be written as a two-term expression according to the Ritz formulation:

$$w(v; x, y; \xi) = c_1 \, xy + c_2 \, xy(x + y) \tag{10.3}$$

where v, the Poisson ratio, is also involved in the two Ritz coefficients:

$$c_1 = \frac{\xi^2}{2(1 - v)(6 + v)} [(6 + v) + 12(1 - v)(1 - \xi)] \tag{10.4}$$

$$c_2 = -\frac{3\xi^2}{6 + v}(1 - \xi) . \tag{10.5}$$

This use of the w function thus neatly replaces the influence function Λ. In the use of Eq. (10.3), whenever the coordinates separated by a comma are identical (e.g., $x = y$), only one quantity will be noted, i.e., $w(v; x; \xi)$. The accuracy of the two-term Ritz solution with respect to finite element calculations is demonstrated by the table of Fig. 9.4 (see the values in parentheses in Table 9.1).

The nondimensional module displacement can be denoted by $w_m(v_m; x, y; \xi)$, where $x = X/(a/2)$, $y = Y/(a/2)$, and $\xi = X'/(a/2)$.

Writing the card displacement at a single corner lead $(F = Q)$, we add the contributions of P and F:

$$W_c = [(L/2)^2/D_c] \cdot \{P \cdot w(v_c; a/L; 1) - F \cdot w(v_c; a/L; a/L)\}$$

while the module displacement has only the contribution of F:

$$W_m = [(a/2)^2/D_m] \cdot \{F \cdot w(v_m; 1; 1)\} .$$

Stipulating compatibility of the displacements of module, card, and leads, $F = K(W_c - W_m)$, the corner-force solution can now be obtained, similar to Eq. (9.9):

$$f = F/P = \frac{w(v_c; 1; a_r)}{1/\mu_c + w(v_c; a_r; a_r) + \left(\dfrac{a_r}{2D_r}\right)^2 w(v_m; 1; 1)} \tag{10.6}$$

where we introduced the notations

$$a_r = a/L \,, \tag{10.7}$$

$$\mu_c = K(L/2)^2/D_c \tag{10.8}$$

$$\mu_m = K(a/2)^2/D_m \tag{10.9}$$

$$\mu_r = \mu_m/\mu_c = a_r^2/D_r \tag{10.10}$$

$$D_r = D_m/D_c \,. \tag{10.11}$$

(Other variables of the comprehensive theory being treated are listed in Table 10.1.)

The F/P (or Q/P) ratio of Example 9.1 is recalculated now; we have

$$a_r = 25/125 = 0.2; \qquad \mu_c = 26 \times (125/2)^2/2674 = 37.98,$$

$$D_r = 8680/2674 = 3.246, \qquad \mu_r = 0.2^2/3.246 = 0.01232.$$

Then

$$w(v_c; a_r; 1) = w(0.3; 0.2; 1) = 0.028571$$

$$w(v_c; a_r; a_r) = 0.021181$$

$$w(v_m; 1; 1) = w(0.4; 1; 1) = 0.83333 \,,$$

yielding by Eq. (10.6): $f = 0.738$.

This value compares with 0.700 of Example 9.1, based on the 5×5 influence matrix computed by finite elements. Figure 10.1 shows a plot of f values versus D_r (i.e., module stiffness) for a family of the μ_c (lead stiffness) parameter. The module/card size ratio parameter a_r was kept constant at 0.2. It is clear that $f = F/P$ can easily be larger than 1 as D_r and μ_c are increased.

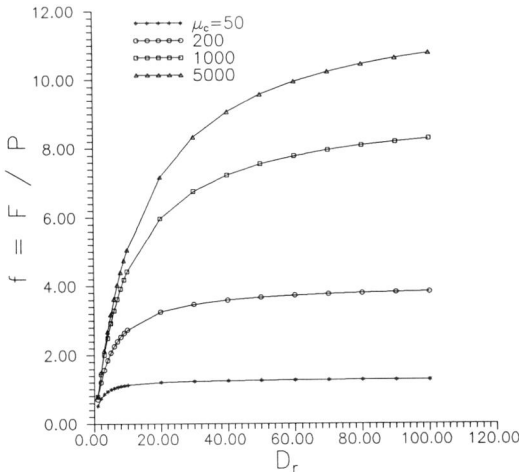

FIGURE 10.1. Force ratio $f = F/P$ for single corner lead versus module/card rigidity ratio $D_r = D_m/D_c$ and lead stiffness parameter $\mu_c = KL^2/4D_c$; $a_r = 0.2$.

We have so far treated a force applied to the diagonal of a square plate. This treatment can be extended, or rather generalized, to include two equal neighboring forces in each quadrant, with their locations symmetrical about the diagonal (Fig. 10.2a); the two forces are thus located at (ξ, η) and (η, ξ), respectively. The motivation for this extended analysis is its application to leads located around the module. The pair of forces $(\xi, \eta$ and $\eta, \xi)$ causes a displacement still optimally computed by a two-term solution in the form of Eq. (10.3), but the corresponding Ritz coefficients are, in this case:

$$c_1 = \frac{\xi\eta}{(1 - v)(6 + v)} [(6 + v) + 6(1 - v)(2 - \xi - \eta)] \tag{10.12}$$

$$c_2 = -\frac{3\xi\eta}{6 + v} [2 - \xi - \eta] . \tag{10.13}$$

The nondimensional displacements are now typically a function of five distinct parameters, v, (x, y), and (ξ, η), represented as $w(v; x, y; \xi, \eta)$. For symmetry, of course $w(v; x, y; \xi, \eta) = w(v; \xi, \eta; x, y)$ and $w(v; x, y; \xi, \eta) = w(v; y, x; \xi, \eta)$. Conversion to the corresponding physical displacements is done again by Eqs. (10.1) and (10.2).

This double-force solution is very useful, because all antisymmetric lead force configurations can be built up of pairs of equal forces located symmetrically with respect to the diagonal. The $N = n/2$ lead forces F_i spaced at distances s along half the module side (Fig. 10.2b) can then be determined by the compatibility of N displacements, expressed by simultaneous algebraic equations. For a unit load $P = 1$ applied at the card corners, we get the system of N equations:

$$\sum a_{ij} F_j = b_i \quad (i = 1, 2, \ldots N) \tag{10.14}$$

where we define

$$a_{ij} = \mu_c \cdot w(v_c; x_{ci}, a_r; x_{cj}, a_r) + \mu_m w(v_m; x_{mi}, 1; x_{mj}, 1) + \delta_{ij} , \tag{10.15}$$

δ_{ij} being the Kronecker delta, equal to one if $i = j$, and zero if $i \neq j$; also

$$b_i = \mu_c w(v_c; x_{ci}, a_r; 1) \tag{10.16}$$

and

$$x_{cj} = s(j - 1)/(L/2); \ x_{mj} = s(j - 1)/(a/2); \ i, j = 1, 2, \ldots N . \tag{10.17}$$

Figure 10.3a shows the lead force distributions for 25-mm PLCC modules centrally attached to a 125-mm-square FR-4 card. There are $N = 10$ leads spaced $s = 1.25$ mm along half the module side. The first solution, using $K = 26$ N/mm lead stiffness ($\mu_c = 37.98$, corresponding to fixed-free leads) from Example 9.1 has an almost linear distribution. The second solution, with $K = 543$ N/mm ($\mu_c = 793.1$, fixed-fixed leads), exhibits more of a sharp corner force, similar to the finite element result of Fig. 9.8a. A third lead, with an even stiffer spring ($K = 3000$ N/mm), exhibits a strong peak at the corner

FIGURE 10.2. Force applications, symmetrical with respect to the diagonal of a square plate; they are part of an antisymmetrical loading pattern. (a) Two forces (one pair); and (b) N pairs of forces, spaced along the module outline.

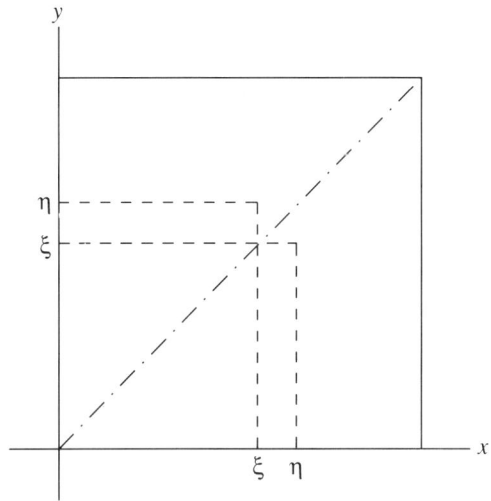

(a) Two forces
(one pair)

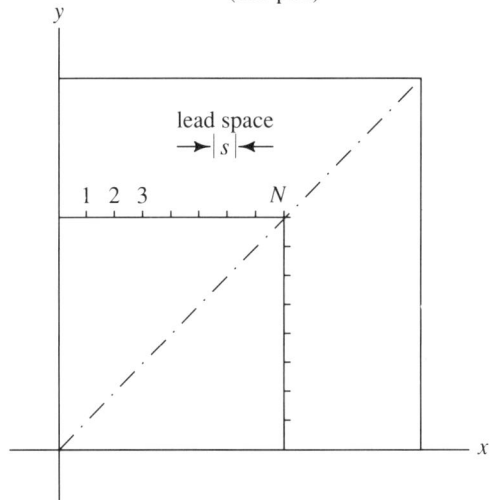

(b) N pairs of forces,
spaced along the
module outline

and sizable negative lead forces in the central region of the module edge. The corresponding graphs with double the lead spacing ($N = 5$) are shown in Fig. 10.3b, demonstrating an increased tendency for corner peaking and larger lead force values, with increasing K. Note that by the analytical method we do not have to compute leads at the corner of the module, where leads do not, in reality, occur.

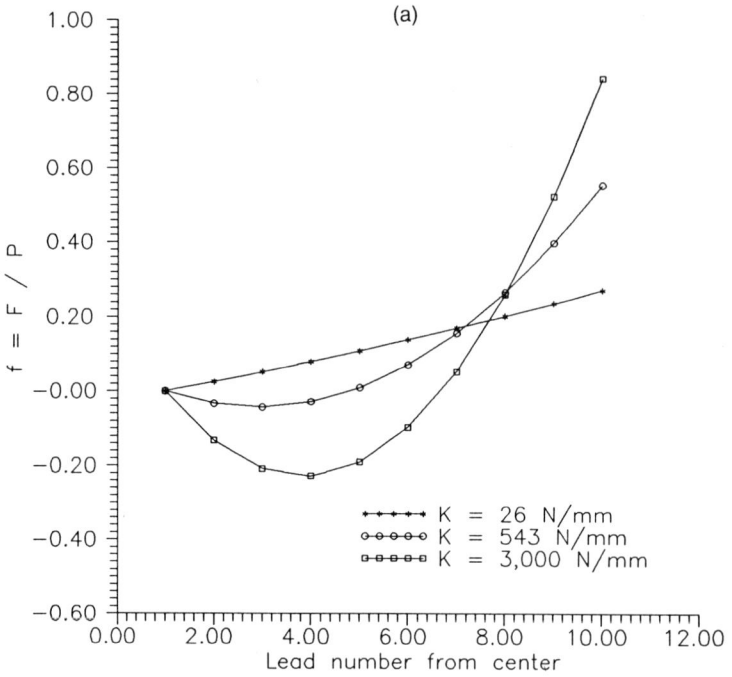

(a)

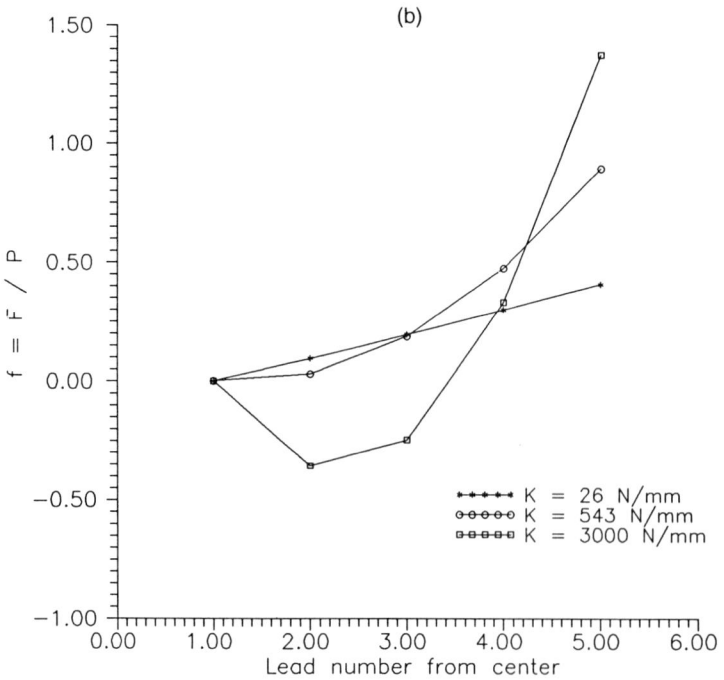

(b)

The effects of varying the module/card size ratio a_r and the module/card rigidity ratio D_r on the F_{max}/P ratio are shown in Fig. 10.4, for a multilead configuration. The card rigidity $D_c = 2674$ N.mm and card size $L = 125$ mm are fixed, as is the lead stiffness $K = 26$ N/mm, so that $\mu_c = 37.98$. $N = 10$ leads over the half width of the module were taken. We note the increase and then subsequent decrease of f_{max} with a_r, as the latter is varied from 0.1 to 0.6. The maximum occurs at about $a_r = 0.4$.

We recognize that if the module were infinitely rigid, then because of antisymmetry the rigid-body module deflections and the elastic ones also could be neglected ($W_m \approx 0$), the lead force resulting in $F(X_i) = K \cdot W_c(X_i)$. In case of a linear lead force distribution $F(\xi) = F_{max} \cdot \xi$ ($\xi = X/(a/2)$, and $F = F_{max}$ when $\xi = 1$), and the unknown F_{max} can be obtained easily as follows. Equation (10.3) is integrated as a continuous distribution along the module edge to yield $W_c(a)$ and $W_m(a)$; the compatibility of these displacements with W_{cP} (caused at $X = a$ by the external card force P) and the

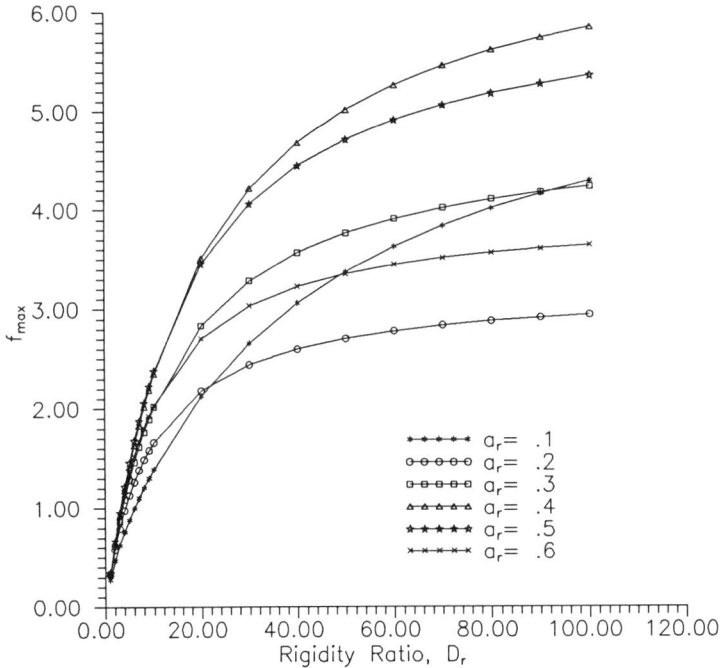

FIGURE 10.4. Maximum lead force versus module/card rigidity ratio D_r and size ratio a_r; $N = 10$.

FIGURE 10.3. Nondimensional lead force distributions f/P for a 25-mm PLCC module centrally attached to a 125-mm FR-4 card; lead stiffnesses K: 1, 26 N/mm; 2, 543 N/mm; 3, 3000 N/mm. (a) $N = 10$; (b) $N = 5$.

corner lead displacement F_{max}/K yield F_{max} from the equation

$$F_{max} =$$

$$\frac{P}{2(1-v_c)N\left[a_r^2 \dfrac{36-22v_c-45(1-v_c)a_r+21(1-v_c)a_r^2}{6(1-v_c)(6+v_c)} + \dfrac{1}{N\mu_c a_r^2} + \dfrac{12+v_m}{6D_r(1-v_m)(6+v_m)}\right]} \quad . \tag{10.18}$$

2. Torsional Stiffness

Torsional stiffness is easily expressed in the preceding analytical notation. Comparison can be made with Chapter 9, Section 2. For corner card forces P and corner leads only, we have

$$k_T = T/\phi = PL^2/2W_P , \tag{10.19}$$

which yields

$$k_T = \frac{2D_c}{w(v_c; 1; 1) - f \cdot w(v_c; 1; a_r)} \tag{10.20}$$

whence, the bare card stiffness $k_{TO} = D_c/2w(v_c; 1; 1)$ [compare with Eq. (9.12)] is augmented by the incremental stiffness due to a single central module:

$$\Delta k_T \cong \frac{2fD_c w(v_c; 1; a_r)}{w^2(v_c; 1; 1)} . \tag{10.21}$$

When n equally spaced leads are considered along each side of the square module, the card corner displacement becomes

$$w_P = [P \cdot w(v_c; 1; 1) - \sum F_i w(v_c; x_{ci}, a_r; 1, 1)]L^2/D_c \tag{10.22}$$

so that the torsional stiffness is

$$k_T = \frac{2D_c}{w(v_c; 1; 1) - \sum\limits_{i=1}^{N} f_i w(v_c; 1, 1; x_{ci}, a_r)} \tag{10.23}$$

from which the incremental stiffness $\Delta k_T = k_T - k_{TO}$ is

$$\Delta k_T = \frac{2D_c \sum\limits_{i=1}^{N} f_i w(v_c; 1, 1; x_{ci}, a_r)}{w^2(v_c; 1; 1)} . \tag{10.24}$$

It is important to note that if the card forces P, constituting the torque T, are not applied at the corner point $r' = L^*$, but rather along the diagonal at some point $r' = \xi L^*$ ($\xi < 1$), then a substantial increase of the torsional stiffness occurs. The value of b_i in Eqs. (10.14) and (10.16) is now calculated from $b_i = \mu_c \cdot w(v; x_{ci}, a_r; \xi)$. The torsional stiffness (T/ϕ) must likewise be redefined. Figure 10.5 shows the variation of torsional stiffness k_T in terms of the point-force location ξ along the diagonals of a square elastic plate; v is varied the full range from 0 to 0.5. Applying the definition of the torsional stiffness as stated in Eq. (10.19), we get

$$k_T = PL^2\xi^2/2w_P = 2\xi^2 D/w(v; \xi; \xi) . \tag{10.25}$$

In Chapter 9, an approximate torsional stiffness increase by the addition of a module cluster of m modules was calculated as $m \cdot \Delta k_T$. This approximation can again be made, and accordingly each module of the cluster will have the same lead force distribution F_i as would a single central module. Now, however, the analytical theory can simply explain the limitations of the preceding approximation.

By Eq. (10.3), the displacement, slope, and curvature of the card can be evaluated for a transverse force applied at ξ, η. If there is only a corner force P considered acting on the card, then $\xi = \eta = 1$ and $c_2 = 0$, while $c_1 = 1/1 - v$. Thus the bilinear displacement [Eq. (9.1)] holds, and the saddle

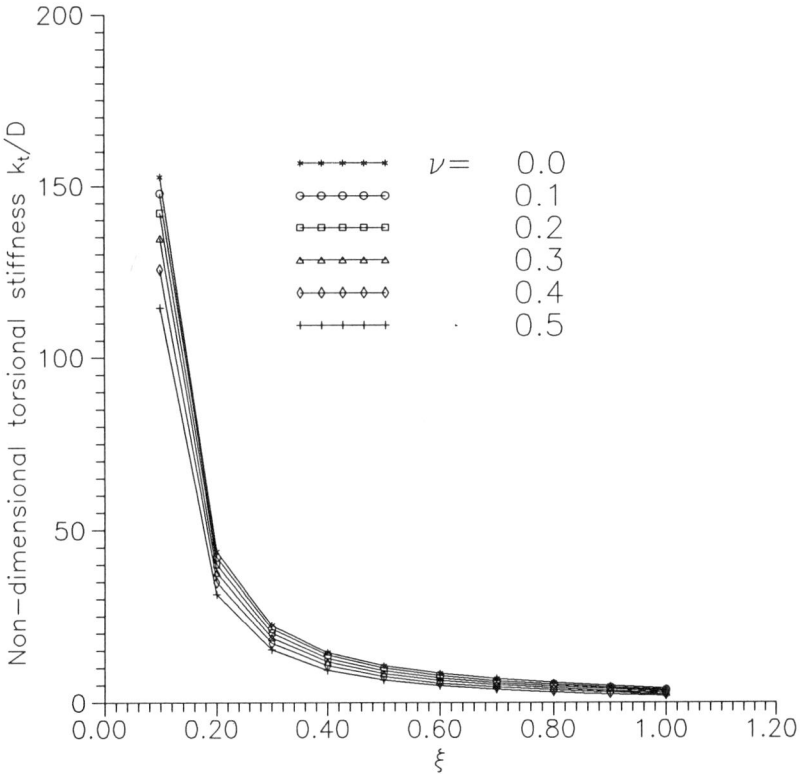

FIGURE 10.5. Torsional stiffness of a square plate for a torque $T = 2\xi PL$; the point loads P are applied along the diagonals.

curvature is constant throughout the card; we have seen that the latter causes a tendency for lead forces to be independent of the module location along the card. When only a small number of modules is attached, the effect of F_i forces altering the constant saddle curvature is also moderate. However, numerous stiff modules, especially if they are not evenly distributed over the card, may appreciably upset the bilinearity of the elastic card displacements, and saddle curvatures will increasingly be larger toward the center than in the outlying region. Lead forces are thus expected to be somewhat larger in the center of a card than on the periphery, for a linear twist applied along the card edge.

In Fig. 10.6 the torsional stiffnesses of three systems of leads (Example 9.1 with $K = 26$ N/mm, Example 9.2 with $K = 543$ N/mm, and $K = 3000$ N/mm) were plotted for a number of leads ranging from $N = 1$ to $N = 20$. $N = 0$, meaning the card alone, yielded $k_{TO} = 7487$ N.mm. External forces P were applied at the card corners. We see that the number of leads in a real system ($N > 5$) has relatively little effect on the torsional assembly stiffness.

The torsional stiffness of multiple modules may be calculated analogously to the principles described under flexure in Chapter 8. A double-sided module may be treated as two modules whose displacements are identical. Thus the incremental torsional stiffness Δk_T is doubled. Stacked modules are distinguished from the corresponding (one-story) simple modules only

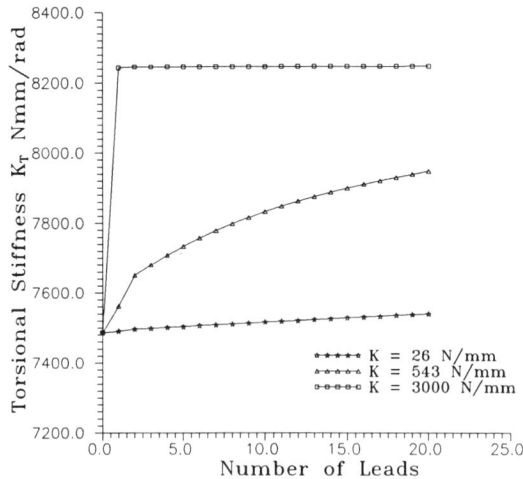

FIGURE 10.6. Torsional stiffness of PLCC ($a = 25$ mm) attached to FR-4 card ($L = 125$ mm) with three lead stiffnesses, and for lead numbers N varying up to 20. Corner application of P is assumed.

to the extent of their increased rigidity; the latter is translated into a torsional stiffness increment in the same proportion as the increased flexural rigidity.

3. Experimental Study

Two experimental apparatuses will be described here: the first imposes a linearly varying displacement $W_c(x) = cx$ at the top card edge. The second exerts point forces at the four corners of a circuit card. Both enforce an antisymmetrical displacement pattern.

The linear edge-displacement torque apparatus [2] was built accommodating populated circuit cards of the same size treated in Examples 9.1 and 9.2. Fig. 10.7a depicts the apparatus and Fig. 10.7b shows the schematic of its operation. Four equally long edge bars hold the card along their centers; the bars are linked to their neighbors at their ends and pivot about steel shafts at

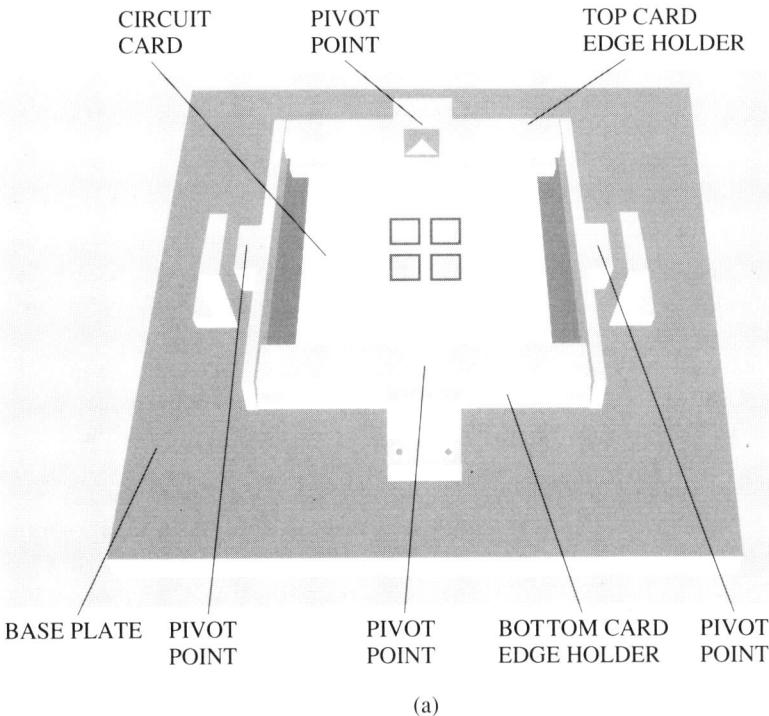

(a)

FIGURE 10.7. Edge-loading circuit card twisting apparatus (from Engel and Vogelmann [2]). (a) Components of apparatus; (b) schematic of operation.

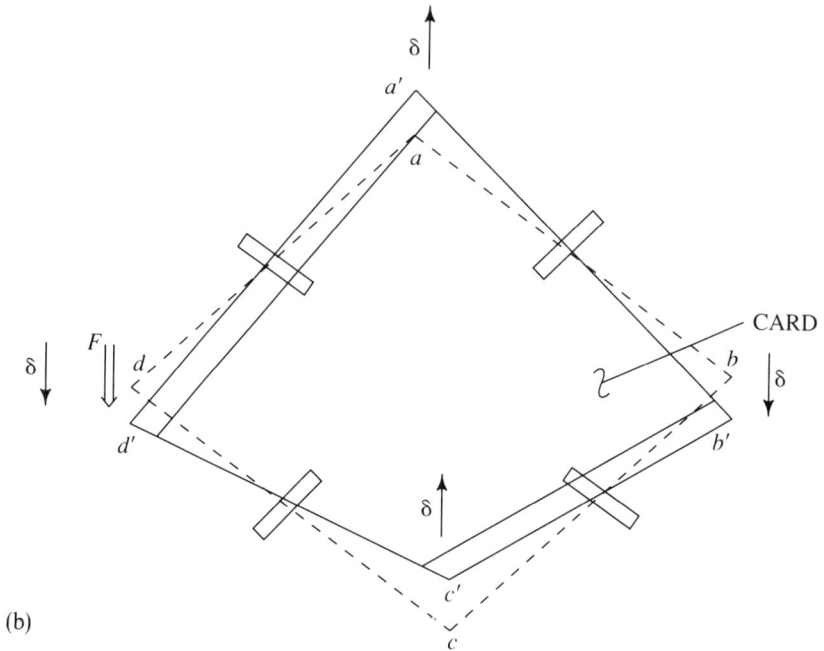

(b)

FIGURE 10.7. Continued.

midspan. The head of a load tester deflects one corner with a force F, causing a twist. By equating the external work $F \cdot \delta/2$ input into the system with the internal work $4 \cdot P \cdot \delta/2$ lifting the four corners, it is shown that the point force to which each card corner is subjected is $P = F/4$.

This apparatus was used with several module configurations of Fig. 9.10. Table 9.4 shows the experimental torsional stiffness versus those calculated by finite elements for some of these; all the stiffnesses represent the linear small-deflection range.

Another torsional apparatus [3] (Fig. 10.8) has also been used at the author's laboratory. This apparatus, connected to the MTS tester, exerts point loads according to the scheme of Fig. 10.8b along the diagonals, inward of the corners of the circuit card. The load is transmitted through posts at the corners of the card. A diagonal set of two posts supports the card from below, and the two posts of the adjacent diagonal descend from above. Each post contains a screw that has a stainless-steel ball bearing mounted to the top.

The screws would be used in two separate modes of operation: (1) in stiffness testing and (2) in fatigue testing. The ball bearing would act as a point load in the stiffness tests. In another mode of operation, for fatigue testing, the screws could be tightened down on the card, enabling a fully

TABLE 10.1. Nomenclature of the analytical theory of coupled plates under torsion.

Variable	Definition
a	= module dimension
a_r	= a/L, size ratio
c_1, c_2	= Ritz constants
D	= plate rigidity
D_c	= card rigidity
D_m	= module rigidity
D_r	= D_m/D_c, rigidity ratio
f	= F/P, force ratio
F, F_i	= lead force
K	= lead stiffness
P	= external load
x, y	= dimensionless coordinates
X, Y	= dimensional coordinates
x_{cj}	= dimensionless coordinate of the jth lead on the card
x_{mj}	= dimensionless coordinate of the jth lead on the module
w	= $DW/(a^2 F)$, nondimensional displacement
W	= displacement
ξ, η	= point load location
v	= Poisson ratio
v_c, v_m	= Poisson ratios for card and module, respectively
μ	= plate constant
μ_c	= $KL^2/4D_c$ dimensionless lead/card plate constant
μ_m	= $Ka^2/4D_m$ dimensionless lead/module constant
μ_r	= $\mu_m/\mu_c = a_r^2/D_r$

reversed loading condition. Thus the screws could be fully tightened for a "fixed" support simulation, or left loose for a "pinned" support.

The posts could be moved diagonally in and out to accommodate various size and shape cards and various load application distances.

4. Large Displacements

For transverse displacements exceeding a third of the card thickness, membrane forces become appreciable, with a progressive stiffening effect. Figure 10.9a shows the P versus w_P curve obtained for an 8.89-cm-square FR-4 circuit card of $h = 1.52$ mm (0.060 in.) thickness; the point-loading torsional apparatus of Fig. 10.8 was used. The "large displacement" region is marked by an upturn of the experimental P versus w_P curve (which may also be considered proportional to a corresponding T versus ϕ curve), hitherto a straight line. The increase in torque T for module-populated cards will continue to cause increased lead forces, as it did in the case of linear torsional deformations discussed until now; we shall evaluate this nonlinear effect quantitatively.

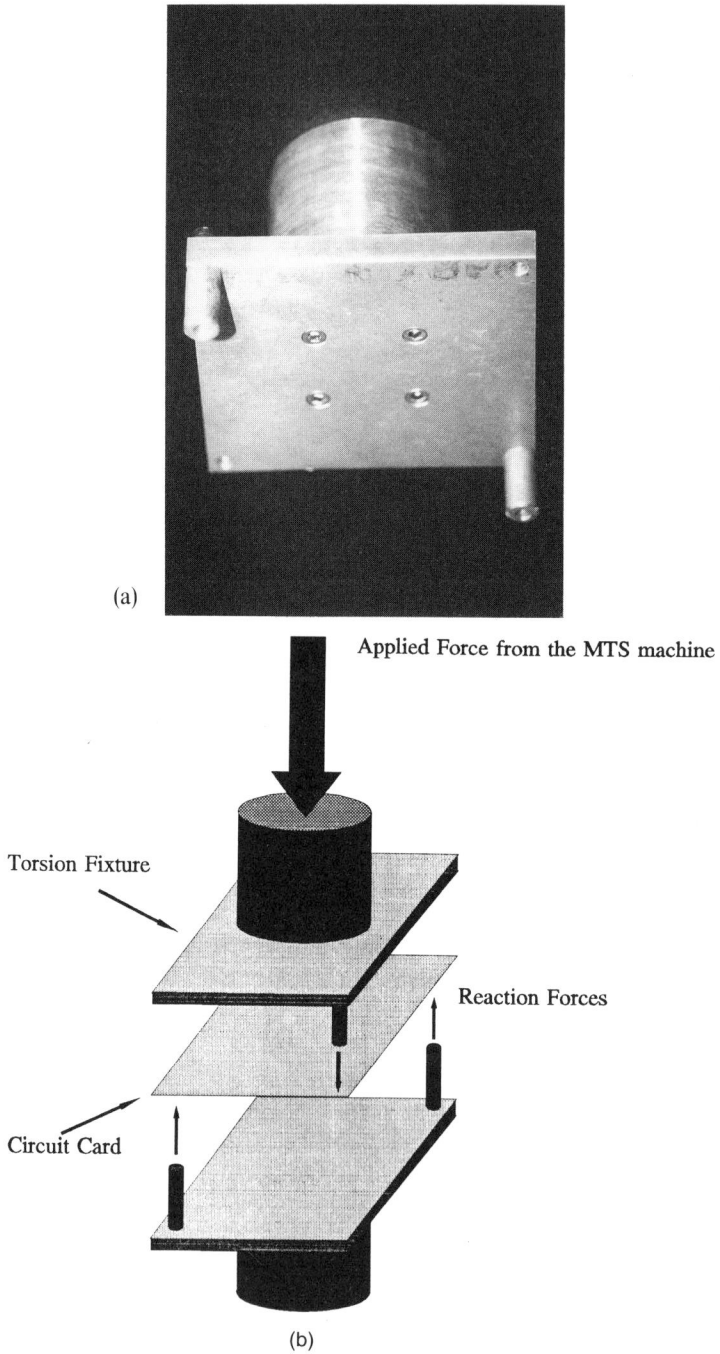

(a)

Applied Force from the MTS machine

Torsion Fixture

Reaction Forces

Circuit Card

(b)

FIGURE 10.8. Point-loading torsional testing apparatus (from Engel and Miller [3]). (a) Torsional fixture for the MTS machine; (b) schematic of operation.

(a)

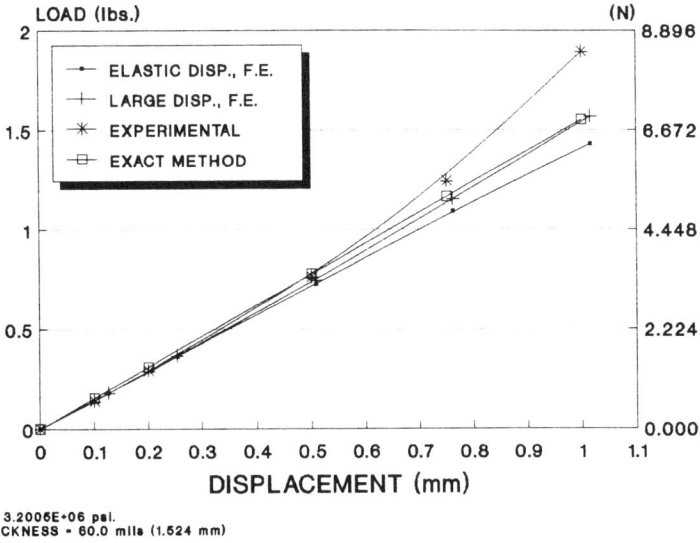

E = 3.2006E+06 psi.
THICKNESS = 60.0 mils (1.524 mm)

(b)

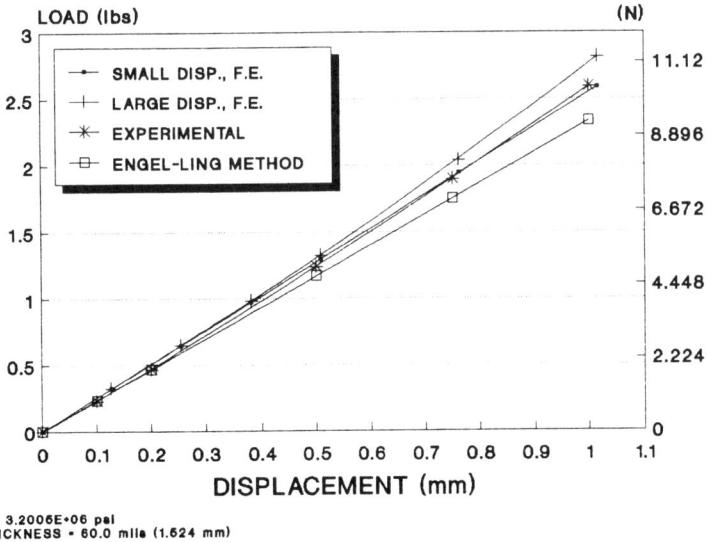

E = 3.2006E+06 psi
THICKNESS = 60.0 mils (1.524 mm)

FIGURE 10.9. Torsional stiffness test (from Engel and Miller [3]). (a) Torque of a plain circuit card; (b) torque of a module assembly. The actual plots are for P versus w_p.

Figure 10.9b shows a comparison of torsional load-deformation curves obtained by several methods for square circuit cards of size $L = 8.89$ cm and thickness $h = 1.52$ mm, described earlier; to the card an $a = 25$-mm ceramic module of 2.54 mm (0.1 in.) thickness was attached by gull wing leads at a spacing of $s = 0.673$ mm (0.0265 in.) ($N = 18$). The gull wing leads had a fixed-free spring constant of $K = 7.94$ N/mm (45.33 lb/in.). The load P was applied by the point-load tester of Fig. 10.8, at a distance 6.25 mm from the card corner along the diagonal. Note that this represents a 22% torsional stiffness increase with respect to the same card loaded at the corners, according to Fig. 10.5.

The P versus w_P relations were plotted in Fig. 10.9 for both the bare card and a single central module attachment, using four criteria: (1) Finite element numerical computation using large displacements, (2) finite element numerical computation using small displacements, (3) analytical methods, and (4) Experimental measurements. For the bare card, the simple torsion formula of Eq. (9.3) or the equivalent Eq. (9.12), served as analytical method. For the module-attached card, the computation procedure featured earlier in this chapter (to be referred as the Engel-Ling method) was used as the analysis procedure.

It is evident that at small displacements each of the four methods gives nearly the same answer. At large displacements, however, some discrepancies are increasingly showing. We conclude that experimental measurement of the nonlinear stiffness increase is desirable.

We shall now show that the nonlinear stiffness increase can be obtained by measurement of the torsional load-deflection curve of the bare card. For this purpose, the P versus w_P curves of the bare card (Fig. 10.9a) and its module-attached version (Fig. 10.9b) are compared.

The experimental data indicate that the large-displacement load versus displacement relation for the bare card can be written

$$P \approx A \cdot w_P + B \cdot w_P^2 , \tag{10.26}$$

where A and B are constants obtainable by a least square fit of the data. For the same card with the described central 25-mm module attachment, the relationship is similar:

$$P \approx C \cdot w_P + D \cdot w_P^2 , \tag{10.27}$$

with C and D being another set of constants. Now the linear-range constant C can be obtained as a scaled value of the stiffness from Eq. (10.19):

$$C = 2k_T/L^2 , \tag{10.28}$$

while D turns out to be the curvature B of the bare-card curve:

$$D = B . \tag{10.29}$$

For the data of Fig. 10.9, the maximum (corner) lead force ratio was computed from the linear analytical theory as $f = F_{max}/P = 0.1$. The stiffness relations were measured as $A = 6.5$ N/mm, $B = 2$ N/mm^2, $C = 10$ N/mm, and, confirming Eq. (10.29), $D = 2$ N/mm^2.

The significance of this is that the nonlinear deformations affected only the card, not the module and leads. Therefore, the lead forces F_i should be computed from the card force P (and not from the card displacement w_P), with recourse to the previous formula

$$F = f \cdot P .\qquad(10.30)$$

The f values are computed by the elastic analysis [Eq. (10.14)]. In the case of a $w_P = 2.5$-mm load application displacement, the maximum lead force would be computed as

$$F_{max} = f \, [(10)(2.5) + (2)(2.5)^2] = 3.75 \text{ N } (0.843 \text{ lbs}) ,$$

a much greater value than the linearly extrapolated equivalent force at $w_P = 2.5$ mm, which is read off from Fig. 10.9b as $f(25) = 2.5$ N.

The engineering treatment of the large displacement region for torsionally loaded module-populated circuit cards thus involves an experimentally, analytically (see Section 5), or numerically (by finite elements) obtained T versus ϕ curve for the card. Once this has been obtained, the large-displacement solution for the stiffness of the module-populated card can follow by Eq. (10.27), and the lead forces can be obtained for any P by Eq. (10.30).

5. Approximate Large Displacement Analysis of a Square Card

We have seen that a square card subjected to twist may be significantly stiffened by in-plane or membrane forces ($N_x = h\sigma_x$, $N_y = h\sigma_y$, $N_{xy} = h\tau_{xy}$) if the displacements become larger than $h/3$. This effect may be considered approximately, by extending the Ritz method of Section 1 to include the strain energy due to in-plane displacements u and v. The strain components resulting from large flexural displacements are written by Ref. [5] of Chap. 1:

$$\varepsilon_x = \frac{\partial u}{\partial X} + \frac{1}{2}\left(\frac{\partial w}{\partial X}\right)^2 \qquad(10.31)$$

$$\varepsilon_y = \frac{\partial v}{\partial Y} + \frac{1}{2}\left(\frac{\partial w}{\partial Y}\right)^2 \qquad(10.32)$$

$$\gamma_{xy} = \frac{\partial u}{\partial Y} + \frac{\partial v}{\partial X} + \frac{\partial w}{\partial X}\frac{\partial w}{\partial Y} . \qquad(10.33)$$

An approximate analysis procedure with all variables considered in the physical (dimensional) sense, will be introduced through the case of a corner-loaded square plate. An "ansatz" or guess, for the X, Y, Z displacement components u, v, w is needed first. Similar to Eq. (10.3), we set

$$w = c_1 X Y . \qquad(10.34)$$

For small displacements $c_1 = P/2(1 - v)D$, but this value is expected to be increased by membrane stiffness. The in-plane displacements are sought in a similar form; the simplest relation acceptable on physical grounds is

$$u = v = k_1 X Y ,\qquad (10.35)$$

for, while $u(X, Y)$ and $v(X, Y)$ both vanish at $X = 0$ and $Y = 0$, their magnitudes are expected to grow monotonically toward the card edges. Here k_1 is a constant to be determined, in addition to c_1. Now, substituting into Eqs. (10.31)–(10.33), we get for the strains

$$\varepsilon_x = \tfrac{1}{2} c_1^2 Y^2 + k_1 Y \qquad (10.36)$$

$$\varepsilon_y = \tfrac{1}{2} c_1^2 X^2 + k_1 X \qquad (10.37)$$

$$\gamma_{xy} = c_1^2 X Y + k_1 (X + Y) \qquad (10.38)$$

and equating these to the internal forces by Hooke's law:

$$\varepsilon_x = \frac{1}{hE} (N_x - v N_y) \qquad (10.39)$$

$$\varepsilon_y = \frac{1}{hE} (N_y - v N_x) \qquad (10.40)$$

$$\gamma_{xy} = \frac{N_{xy}}{hG} . \qquad (10.41)$$

Equating the two sets of expressions, we can attempt to calculate c_1 and k_1. Force-boundary conditions for the twisted card would require

$$X = \pm L/2 : \qquad N_x = N_{xy} = 0 \qquad (10.42)$$

$$Y = \pm L/2 : \qquad N_y = N_{xy} = 0 . \qquad (10.43)$$

Clearly, these cannot be satisfied by any choice of c_1 and k_1; however, we shall try to satisfy them in the average sense over the half-length $L/2$, setting

$$\frac{1}{L/2} \left[\int_0^{L/2} N_x \, dY \right]_{x = L/2} = 0 \qquad (10.44)$$

$$\frac{1}{L/2} \left[\int_0^{L/2} N_y \, dX \right]_{y = L/2} = 0 \qquad (10.45)$$

$$\frac{1}{L/2} \left[\int_0^{L/2} N_{xy} \, dY \right]_{X = L/2} = \frac{1}{L/2} \left[\int_0^{L/2} N_{xy} \, dX \right]_{Y = L/2} = 0 . \qquad (10.46)$$

Two choices are available: that of satisfying either Eqs. (10.44) and (10.45) or Eq. (10.46). The former choice of average normal stresses yields the relation

$$k_1 = - \frac{1 + 3v}{6(1 + 2v)} L c_1^2 \qquad (10.47)$$

whereas the latter choice, satisfying the average shears, leads to

$$k_1 = -\frac{L}{6} c_1^2 . \tag{10.48}$$

Note that in either case the average membrane forces are satisfied along the total side length L. The general form of the k_1 versus c_1 relationship is then

$$k_1 = -\psi L c_1^2 . \tag{10.49}$$

where ψ is a constant, a function of v.

We shall adopt Eq. (10.47), which renders for the constant of proportionality

$$\psi = \frac{1 + 3v}{6(1 + 2v)} . \tag{10.50}$$

Clearly, the two equations (10.47) and (10.48) coalesce if $v = 0$, yielding $\psi = 1/6$. The unknown c_1 will now be determined using the Ritz method. Separating the strain energy into two parts; U_1 due to bending and U_2 due to membrane action, Eq. (1.73) gives for the former

$$U_1 = (1 - v) DL^2 c_1^2 . \tag{10.51}$$

The in-plane internal forces result in

$$U_2 = \frac{1}{2} \int_{-L/2}^{L/2} \int_{-L/2}^{L/2} (N_x \varepsilon_x + N_y \varepsilon_y + N_{xy} \gamma_{xy}) dX \, dY , \tag{10.52}$$

which is evaluated from Eqs. (10.39)–(10.41), using Eq. (10.47). The potential energy of the external corner loads twisting the plate is

$$\Omega = -4Pw\left(\frac{L}{2}, \frac{L}{2}\right) . \tag{10.53}$$

Now, the total potential energy V is composed of the strain energy, $U = U_1 + U_2$, added to Ω, thus $V = U + \Omega$, so that the equilibrium configuration is determined by $\partial V/\partial c_1 = 0$. The outcome, a cubic equation in c_1, is

$$C_1 = \frac{P}{2(1 - v) D} - \frac{L^4}{120(1 - v) h^2}$$
$$\times [28 - 150\psi + (225 - 15v)\psi^2] c_1^3 . \tag{10.54}$$

For the data of Fig. 10.8, c_1 was thus evaluated. Figure 10.10 shows the respective large displacement curves resulting from the Ritz-type analytical method and the finite element method. Discrepancies could be further minimized if a better set of assumptions than those of Eqs. (10.35) were made for u and v. Such improvements could of course be obtained through computing the shapes of $u(x)$, $v(x)$, etc., by finite elements, if desired.

TORSIONAL STIFFNESS OF SQUARE CARD
CORNER FORCE VS CORNER DISPLACEMENT

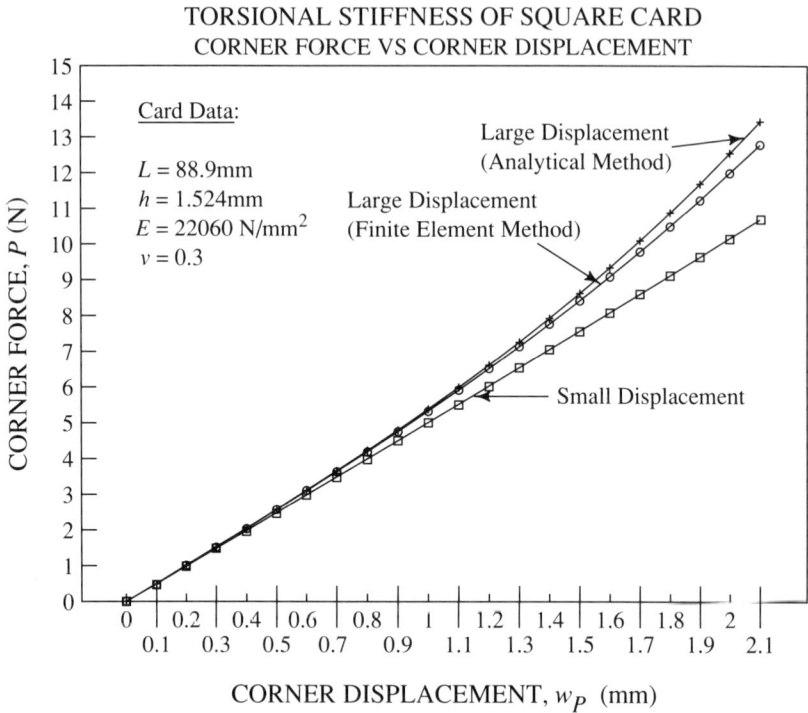

FIGURE 10.10. Torsional stiffness of a square card computed for large displacements.

6. Torsional Fatigue

The apparatus exhibited in Fig. 10.8 was used for fatigue testing. The MTS machine was used in a low-cycle displacement-controlled mode in which the circuit card was subjected to a fully reversed load at the constant frequency of 1/30 Hz. Tests were mostly run in an automatic mode. Failure was defined as a partial or total separation of the lead from the rest of the structure. In order to determine the time at which failure occurred, test runs were filmed, with the camera focused on two sides of the module and the attached leads. The camera was attached to a VCR and a monitor. A monitor was used to allow for periodic evaluation of the testing while it was performed. After completion of the test, signaled by a drastic load drop, the tape could be viewed; this permitted an easier and more accurate cycle count than visual inspection.

   Both gull wing leaded and J-leaded modules were torsionally fatigue-tested in various test runs. Failures predominantly occurred at corner leads. Gull wing leads were found to fail as a result of flexural fatigue at the bend, while J-leads fatigued at the solder joints; this was in total agreement with the experience gained in the flexural fatigue runs discussed in Chapter 8. An example of gull wing failures is shown in Fig. 10.11a: the lead left of the

(a)

(b)

FIGURE 10.11. Typical torsional fatigue failures of compliant leads. (a) Gull wing lead failure; (b) J-lead failure.

module corner is severely bent, and the one on the right of the corner has broken near its interface with the module. The photo in Fig. 10.11b shows failed J-leads at the corner, partially separated at their solder joints.

The following is the description of a test series on square epoxy-glass cards of 8.89-cm (3.5-in.) square size and 1.5-mm (0.060-in.) thickness. Loads were applied in the fatigue tester of Fig. 10.8, at points located at a load line 5 mm (0.2 in.) from the edge of the card. A central ceramic square module $a = 25$ mm (1 in.) was surface-mounted by gull wing leads in the middle, at

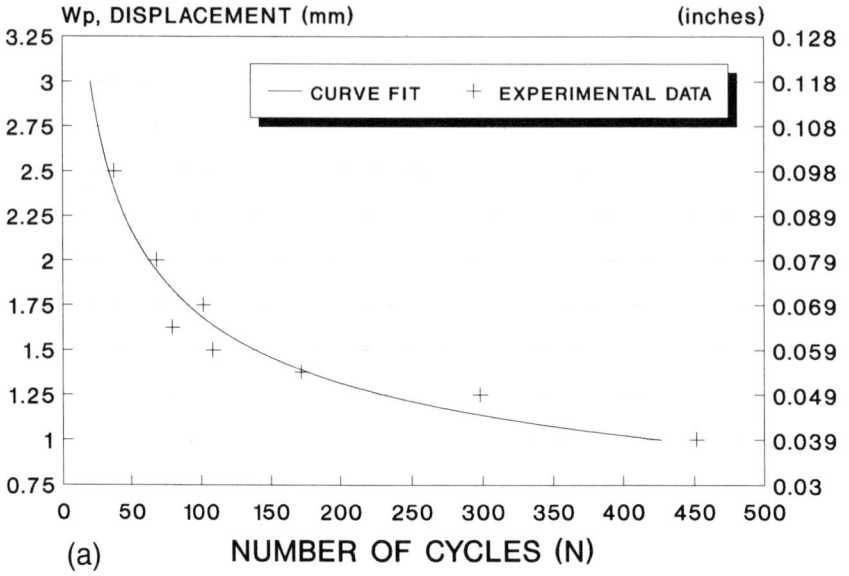

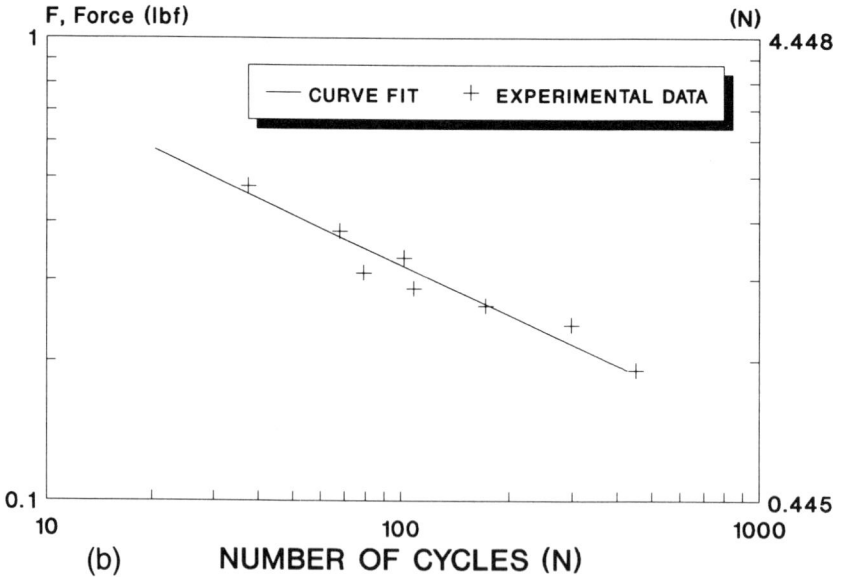

FIGURE 10.12. Displacement controlled fatigue results for gull wing leads. (a) w_P versus N (number of cycles); (b) F_{max} versus N.

a spacing $s = 1.25$ mm (0.05 in.). The spring constant for a lead was computed in a fixed-free configuration as 7.96 N/mm (45.4 lb/in.). The plate constants for card and module were measured by three-point bending tests, as $D_c = 7.16$ kN. mm (63.3 lb-in.) and $D_m = 986.5$ kN. mm (8728 lb-in.), respectively.

The low-cycle fatigue curve obtained for this test series is shown in Fig. 10.12a, in terms of a w_P versus N (the number of cycles) plot. Using the analytical theory to convert w_P into maximum lead force F, the results were also rendered as a log-log plot in Fig. 10.12b.

7. Conclusions

The analytical Engel-Ling theory was shown to produce reliable data, commensurate with linear finite element results, for square module/card systems with arbitrary lead spacing. Stiffness results were also confirmed by experiments in the linear load range; this supplies further evidence of the soundness of considering leads as rows of z-directional linear springs during flexing of these systems. The lead force distribution for closely spaced, not very stiff leads was found to be nearly linear and not concentrated at the corners.

Large torsional displacements were analyzed by superimposing experimental, nonlinear, torsional stiffness data of the bare card onto the linear stiffness increase represented by the addition of a module. Approximate large displacement analysis was shown through the simple example involving a bare card. Torsional fatigue results of gull wing leaded and J-leaded modules were discussed.

8. Exercises and Questions

1. Outline a mathematical derivation of c_1, c_2 in Eqs. (10.4) and (10.5) by the Ritz method.
2. Show the numerical closeness of Eq. (10.6) to the values tabulated in Table 9.1.
3. Discuss the role of each variable in Table 10.1 influencing the torsional stiffness and maximum lead force of a square module/lead/card system.
4. What technological needs would justify the application of the edge-displacement torsional tester shown in Fig. 10.7? Or of the point-loading torsional apparatus of Fig. 10.8?

References

1. Engel, P.A., and Ling, Y. (1992), "Torsion of Elastically Coupled Plates with Applications to Electronic Packaging," *Proc. ASME/JSME Electronics Packaging Conf.*, Milpitas, Calif. pp. 575–581.
2. Engel, P.A., and Vogelmann, J.T. (1992), "Approximate Structural Analysis of Circuit Card Systems Subjected to Torsion," *ASME J. Elec. Packag.*, **114**(2), 203–210.
3. Engel, P.A., and Miller, T.M. (1992), "Torsional Stiffness and Fatigue Testing of Surface Mounted Compliant Leaded Systems," *Proc. 42nd ECTC Conf.*, San Diego, Calif., pp. 557–562.

Chapter 11

Thermal Stresses in Compliant Leaded Systems

1. Motivation for Analysis

The stresses of handling (flexing), shock, and vibration can be reduced by engineering a safer environment for the circuit board system. Thermal stress, however, is induced in the operational mode, owing to the differential expansion of interconnected components. Even before functional stress cycles can test the fatigue strength of the leads, severe thermal loads may arise during the cooling accompanying manufacturing, i.e., the soldering (reflow) process.

In surface-mount technology, great advantages lie in both allowing denser pad spacing (more I/O) and increased packaging density. An additional but substantive benefit is the greater flexibility of leads to withstand mismatch stresses. The mismatch Δu for nonuniform heating or cooling of module and card was given by Eq. (1.51) and for uniform heating by Eq. (1.52). We note that a constant ΔT_1 and ΔT_2 characterizing Eq. (1.51) is a mere abstraction in "power cycling," when nonuniform temperatures are defined by transients. For SMT modules, the distance r to the corner pin is greatest, determining the maximum mismatch in a square module.

According to Eq. (1.51), the greatest manufacturing stress is created if the material with the greater α cools faster. This will be the case if a ceramic module ($\alpha_1 = 6.5$ ppm/$^\circ$C) is attached to an FR-4 circuit card ($\alpha_2 = 15$ ppm/$^\circ$C), provided the module cools slower in the wake of the reflow process; this means, from $T_0 = 183\,^\circ$C, the melting point of Sn-Pb 60–40 solder, to room temperature, 20 $^\circ$C. The transient cooling process continues until the latter is reached (Fig. 11.1). In Fig. 11.2, Δu for various module materials (α_1) versus an FR-4 card is plotted, assuming $\Delta T_2 = 160\,^\circ$C (family A of curves). It is worth noting that while the manufacturing stress for glass ceramic substrate modules is lowest for the given transient temperatures (family A), this same material results in high uniform temperature mismatch; see family B.

Whereas circuit card flexing involves transverse displacement of the total structure, in thermal loading it usually suffices to consider the in-plane

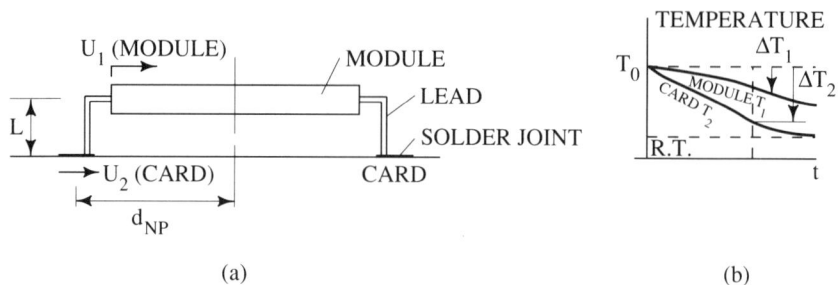

FIGURE 11.1. Differential thermal expansion of module and card arising during solidification of solder joint. (a) Schematic of module and leads; (b) cooling of module and card.

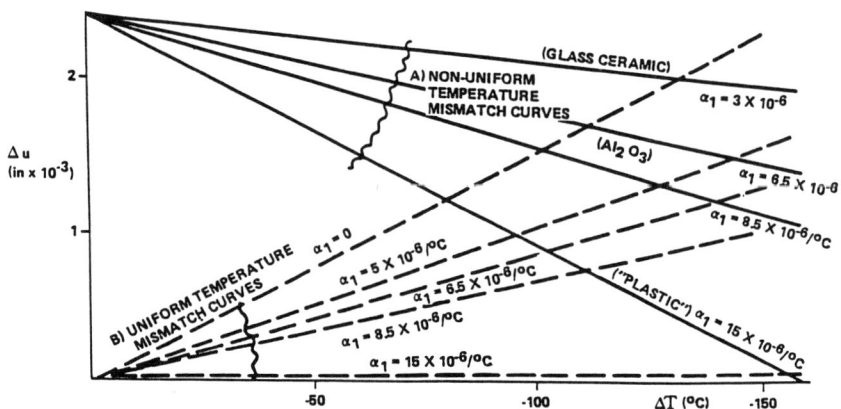

FIGURE 11.2. Mismatch of modules of different materials cooling with respect to FR-4 card ($d_{NP} = 25$ mm or 1 in.). (a) nonuniform temperatures: $\Delta T_2 = -160\,°C$; (b) uniform temperature mismatch, $\Delta T_2 = \Delta T_1$ varied.

displacements just in the vicinity of the module (see also Chapter 6 for pin-in-hole construction). This thermal displacement is, in many respects, equivalent to a mechanically induced stretch; Solomon [1] tested SMT joints bridging two cards cyclically subjected to a pull (Fig. 11.3). Mechanical stretching does not, of course, account for thermal changes in material behavior, or for strain dependence on distance from a neutral point.

In cyclic thermal loading both the solder joint and the leads may be highly stressed, but the failure mechanism commonly of concern is cyclic solder fatigue. Thus thermal stress investigations are all motivated by fatigue prediction [1–14]. Thermal fatigue evaluation, on the other hand, serves the creation of a statistical statement of functionality, i.e., the establishment of reliability.

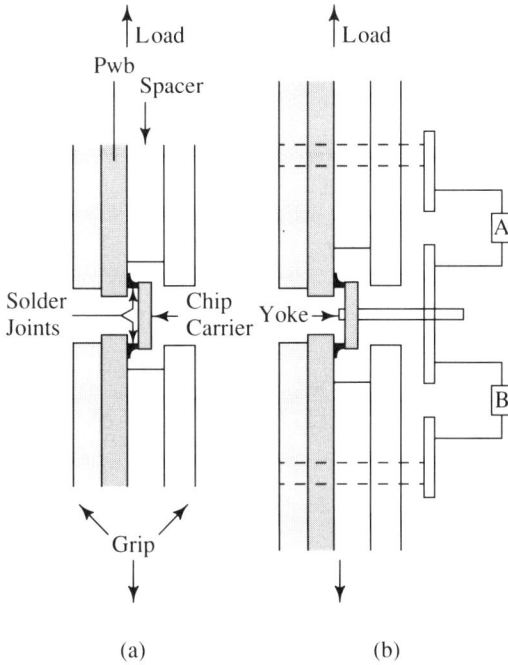

FIGURE 11.3. Schematic showing thermal stress test. (a) Without the yoke and extensometer; (b) with yoke and extensometer (from Solomon [1], © 1989 IEEE).

In order to appraise the thermal stresses on the leads and solder joint, several methods have been advanced, such as analytical schemes and conceptual approaches, numerical (finite element) computations, and experimental measurement techniques. These will be explored next.

2. Analytical Lead Stress Computation

2.1. Plane Frame Analysis

A plane frame analysis of a sloping lead fixed at the foot to a solder pad and joining an inflexible module (α_1) is shown in Fig. 11.4. The differential displacement for uniform ΔT is

$$\Delta u = (\alpha_1 r_1 - \alpha_2 [r_1 + L \cos \theta]) \Delta T \approx (\alpha_1 - \alpha_2) r_1 \Delta T . \qquad (11.1)$$

The bending moment M, shear V, and axial force F acting on the lead attached to the solder pad are written for fixed-fixed boundary conditions as

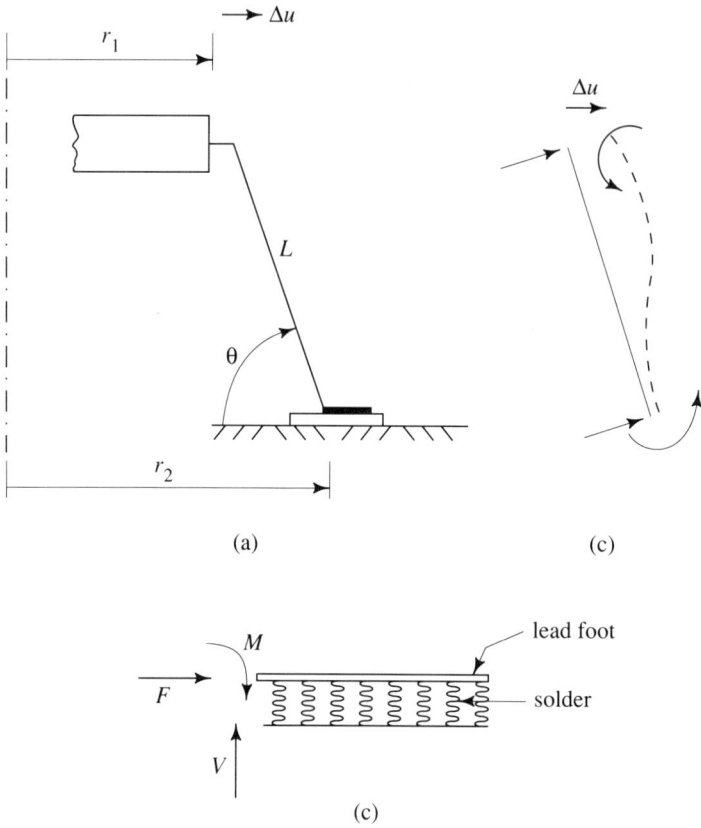

FIGURE 11.4. Schematic of surface mount subjected to thermal displacement. (a) Module, lead, and solder joint; (b) standoff part of lead; (c) lead foot, a beam on elastic foundation.

$$M = \frac{6EI \sin \theta \cdot \Delta u}{L^2} \tag{11.2}$$

$$V = \frac{EI \sin 2\theta}{L^3} \left(6 - \frac{AL^2}{2I} \right) \Delta u \tag{11.3}$$

$$F = \frac{2EI \sin^2 \theta}{L^3} \left(6 + \frac{AL^2}{2I} \cot^2 \theta \right) \Delta u . \tag{11.4}$$

These may be treated as end loads on a beam segment supported on an elastic foundation; see Section 2 of Chapter 1. Note that for the same lead length L, M is reduced but V and F may increase as the angle θ deviates from a rectangle.

2.2. Three-Dimensional Frame Analysis

Kitano et al. [8] considered the three-dimensional thermal deformations of an SMT module with the lead foot immersed in a triangular solder fillet h_s deep and h_l wide on both sides Fig. 11.5a. The module, lead, and card expanded owing to a uniform temperature rise ΔT; first the yz and then the xz-plane will be considered to relate the thermal displacements to elastic forces.

In the yz-plane (Fig. 11.5c), the vertical lead segment l_z has the bending moment variation in terms of the reaction moment M_y,

$$M_1(y) = F_y \cdot x_1 - M_y \tag{11.5}$$

whereas the horizontal segment l_y has

$$M_2(y) = F_y(l_z - h_s) - M_y . \tag{11.6}$$

The boundary conditions on the two beam segments

$$x_1 = 0: \frac{dw_1}{dx_1} = \phi_y \tag{11.7}$$

$$w_1 = -\phi_y \cdot h_s \tag{11.8}$$

$$\left. \begin{aligned} x_1 = l_z - h_s \\ x_2 = 0 \end{aligned} \right\} : \quad \begin{aligned} \frac{dw_1}{dx_1} = \frac{dw_2}{dx_2} \\ w_1 = \delta_y \end{aligned} \tag{11.9} \tag{11.10}$$

$$x_2 = l_y : \frac{dw_2}{dx_2} = 0 . \tag{11.11}$$

By integration of the beam equation $d^2w/dx^2 = M/E_L I_y$, we get two expressions for the displacements δ_y and ϕ_y, in terms of the internal force quantities F_y and M_y:

$$\phi_y = \left[F_y(l_z - h_s)\left(l_y + \frac{l_z - h_s}{2}\right) - M_y(l_y + l_z - h_s) \right] / E_L I_y \tag{11.12}$$

and

$$\delta_y = \phi_y l_z - [F_y(l_z - h_s)^3/6 - M_y(l_z - h_s)^2/2]/E_L I_y . \tag{11.13}$$

Two additional equations are obtained by stipulating constant shear stress τ_y against the lead foot; thus

$$M_y = hbh_s\tau_y , \tag{11.14}$$

and approximating the thermal expansion:

$$\delta_y = [(l_p + l_y)\alpha_c - l_p\alpha_m - l_y\alpha_L]\Delta T . \tag{11.15}$$

Thus F_y may be eliminated from the two equations (11.12) and (11.13), and we can solve for ϕ_y. The shear strain in the solder, in the yz-plane, is finally

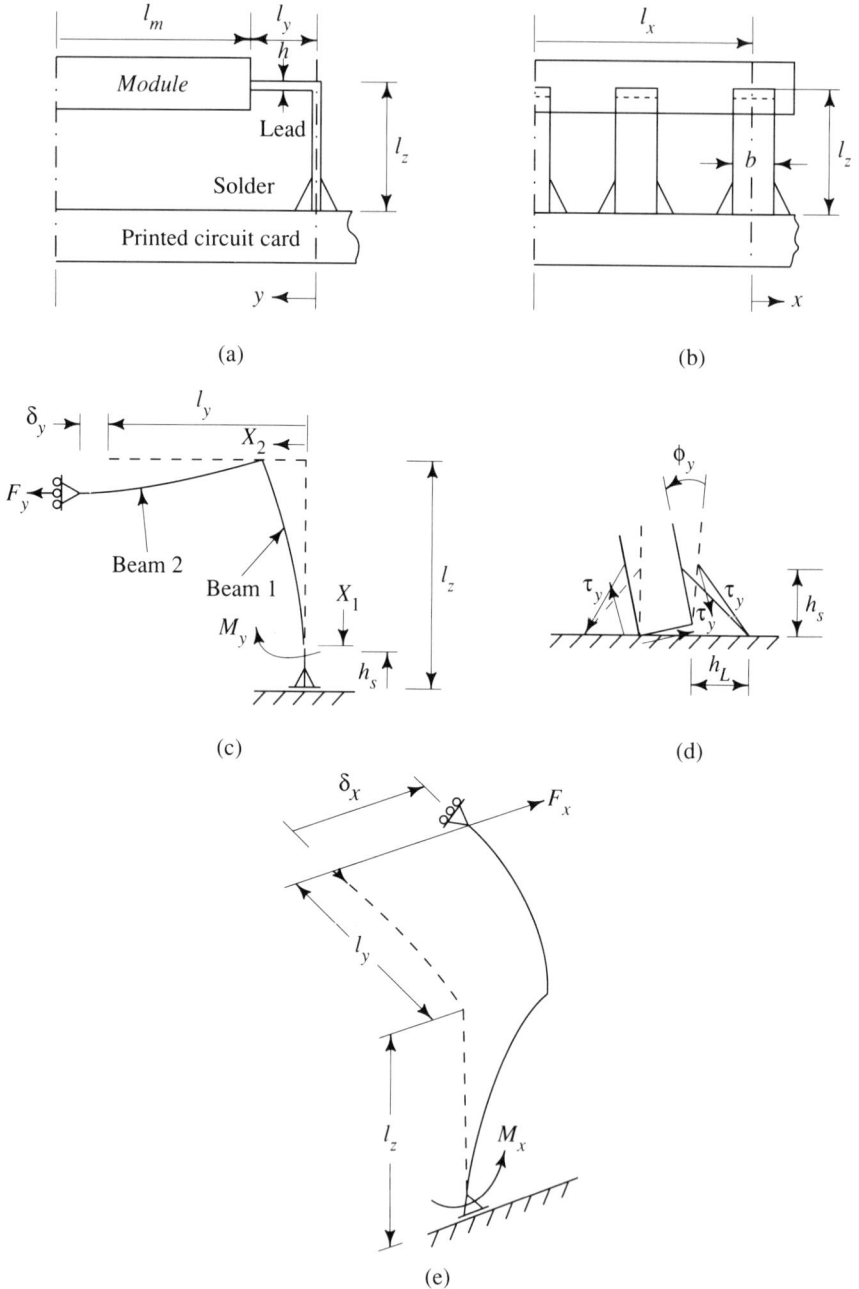

FIGURE 11.5. Schematic for three-dimensional thermal analysis. (a) Dimensions in the yz-plane; (b) dimensions in the xz-plane; (c) deformation in the yz-plane; (d) Soldered lead tilt in the yz-plane; (e) deformations in the xz-plane (from Kitano et al. [8]).

obtained by the approximation:

$$\gamma_y = \phi_y + \phi_y \frac{h}{2h_L} = \left(1 + \frac{h}{2h_L}\right)\phi_y . \tag{11.16}$$

Modeling the xz-plane thermal expansion, the treatment (Fig. 11.5e) is similar: we have to add a twist, however, to the bending effect. In the bottom part of the lead, the transverse elastic line has the curvature due to bending moment $M(x)$:

$$\frac{d^2w_1}{dx_1^2} = \frac{M(x)}{E_L I_x} \tag{11.17}$$

and the twist, due to torque $T(x)$:

$$d\phi/dx = \frac{T(x)}{G_L J} . \tag{11.18}$$

The boundary conditions now are

$$x_1 = 0: \frac{dw_1}{dx_1} = -\phi_x \tag{11.19}$$

$$w_1 = -\phi_x \cdot h_s \tag{11.20}$$

$$\left.\begin{array}{c} x_1 = l_z - h_s \\ \\ x_2 = 0 \end{array}\right\} : \begin{array}{c} \dfrac{dw_1}{dx_1} = \phi_2 \quad (11.21) \\ \\ \dfrac{dw_2}{dx_2} = \phi_1 \quad (11.22) \end{array}$$

$$x_2 = l_y: w_2 = \delta_y \tag{11.23}$$

$$\frac{d\phi_2}{dx_2} = 0 \tag{11.24}$$

By equilibrium considerations, the torque along the two beam segments can be written:

$$T_2 = -(M_y - F_y l_z) \tag{11.25}$$

$$T_1 = F_y l_x . \tag{11.26}$$

The solution is

$$\phi_x = l_y[F_x(l_z - h_s) - M_x]/G_L J + [F_x(l_z - h_s)^2/2 - M_x(l_z - h_s)]/E_L I_x \tag{11.27}$$

$$\delta_x = \phi_x l_z + [-F_x\{(l_z - h_s)^3 + l_y^3\}/6 + M_x(l_z - h_s)^2/2 + M_c l_y^2/2]/E_L I_x \tag{11.28}$$

where

$$M_c = \frac{F_x l_y^2/2E_L I_x + F_x l_y(l_z - h_s)/G_L J}{l_y/E_L I_x + (l_z - h_s)/G_L J} . \tag{11.29}$$

A constant shear stress on the lead foot is related to the moment M_x:

$$M_x = h b h_s \tau_x . \tag{11.30}$$

The thermal expansion is written

$$\delta_x = l_x(\alpha_m - \alpha_c) \Delta T , \tag{11.31}$$

and in a manner similar to Eq. (11.16) the shear strain in the solder is approximated

$$\gamma_x = \left(1 + \frac{b}{2h_L} \right) \phi_x . \tag{11.32}$$

Finally, the total shear strain on which the solder fatigue computation is to be based is vectorially summed from Eqs. (11.16) and (11.32):

$$\gamma = \sqrt{\gamma_x^2 + \gamma_y^2} . \tag{11.33}$$

The testing equipment, designed to induce this strain mechanically, is sketched in Fig. 11.6. Kitano et al. verified the stress analysis results by performing various stiffness tests. The stress analysis was then incorporated in a solder joint fatigue study.

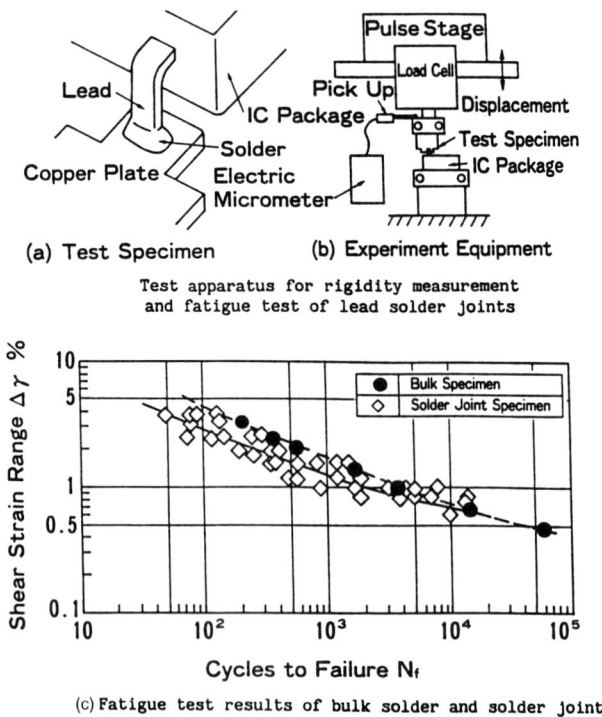

(a) Test Specimen   (b) Experiment Equipment

Test apparatus for rigidity measurement
and fatigue test of lead solder joints

(c) Fatigue test results of bulk solder and solder joint

FIGURE 11.6. Test apparatus for rigidity measurement and fatigue test of leaded solder joints (from Kitano et al. [8]).

3. Finite Element Thermal Stress Analysis

In a thermal stress study of surface mounts, Lau and his co-workers [e.g., 11–15] performed a series of finite element analyses for a great variety of component leads and solder joint configurations. Reference [11] is on SOIC assemblies (Fig. 11.7) attached by gull wing leads to 1.75-mm (0.062-in.) thick FR-4 circuit cards. The leads were 0.6 mm (0.024 in.) wide and 0.0125 mm (0.005 in.) thick. Sn-Pb 63–37 eutectic solder was used. A temperature cycle $\Delta T = 180\,^\circ\mathrm{C}$ (from -55° to $125\,^\circ\mathrm{C}$) was considered. Three-dimensional solid elements with 20 nodal points (only 8 are shown in the figures for greater clarity) were used in conjunction with the ABAQUS program [16]. This analysis was a linearly elastic one, indicating likely places of incipient cracking.

Figure 11.8a shows von Mises stress contours on the plastic package; the maximum stress is at the outer interface from where the lead emerges. Figure 11.8b is a stress contour plot for the lead, which is heavily stressed in the inner surfaces of the bends, but especially near the package. Recall the fatigue failures in gull wing leads (Fig. 3.18); these may have been originated by manufacturing thermal stress.

The solder joint was modeled in Fig. 11.9, exhibiting the highest stresses near the outer toe of the joint and at the inside upper tip of the solder fillet. In Ref. [15] various solder geometries were also studied, such as joints without a fillet at the tip of the lead, solder on top of the lead, and solder on the side of the lead. Very little effect on solder stress due to fillet shape was found, but lack of a solder fillet at the tip of the lead raised the stress. Standoff height of the SOIC was an important factor in combatting high stress. A void was also modeled inside the solder by considering six various sizes of a circular cavity of radius R: from 0.00312 to 0.05 mm (0.000125 to 0.002 in.). Due to stress concentration effects, solder stresses were found to increase with the size of the void.

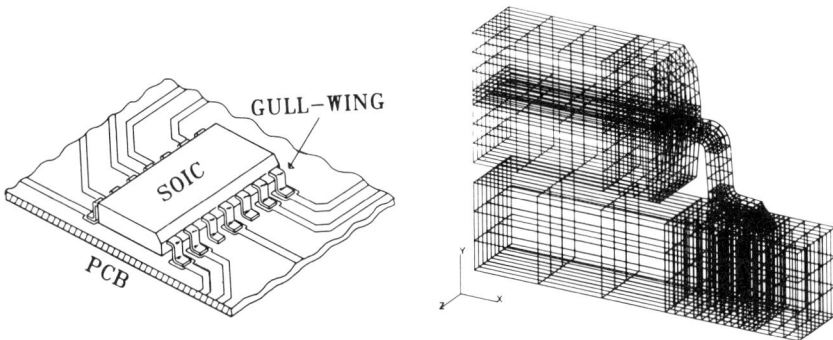

FIGURE 11.7. Three-dimensional Finite element model for a SOIC surface-mounted assembly (from Lau et al. [11], © 1987 IEEE).

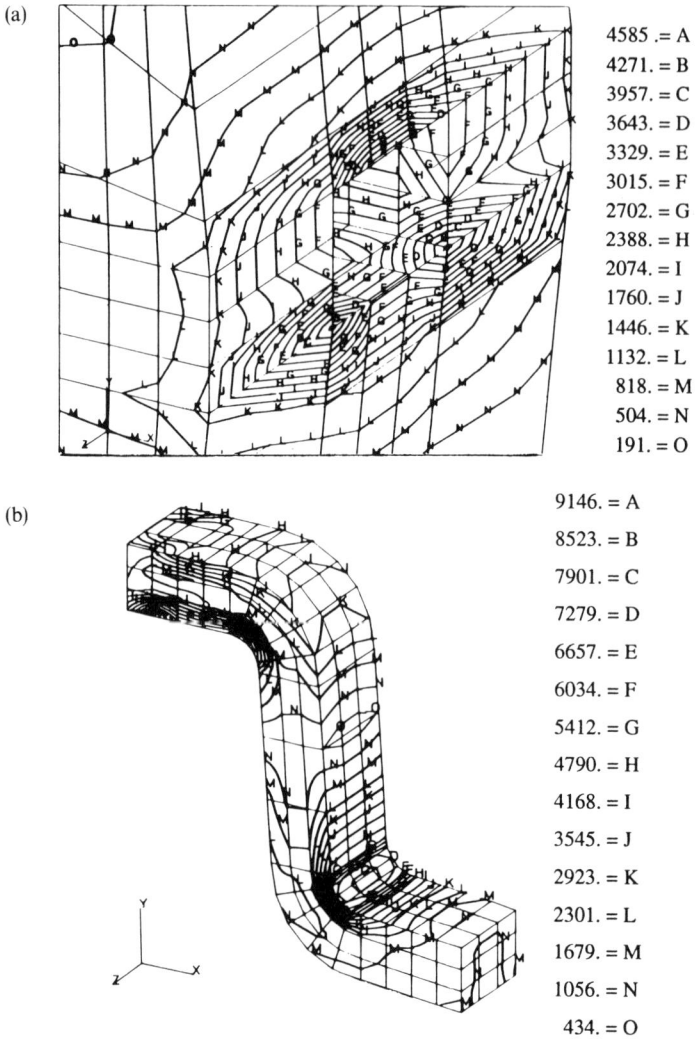

FIGURE 11.8. von Mises stress acting on the gull wing leaded package; stress contours (psi) (a) stresses on package at lead exit (b) stresses on lead (from Lau et al. [11], © 1987 IEEE).

In Ref. [13], Lau modeled J-leads surface mounting PLCC modules to FR-4 cards using 1296 solid elements; 864 for the J-lead and 432 for the solder joint. Since thermal loads are equivalent to mechanical displacements, Lau computed a 12×12 stiffness matrix; this corresponded to six degrees of freedom (three displacements and three rotations) of a point on the top of the lead, and six at the bottom, on the interface with solder (Fig. 11.10). This and Lau's stiffness matrix for gull wings are reproduced in Tables 11.1 and 11.2, respectively.

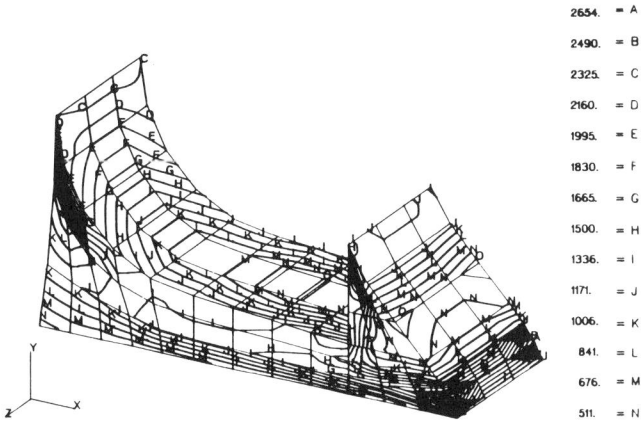

2654.	= A
2490.	= B
2325.	= C
2160.	= D
1995.	= E
1830.	= F
1665.	= G
1500.	= H
1336.	= I
1171.	= J
1006.	= K
841.	= L
676.	= M
511.	= N

FIGURE 11.9. von Mises stress acting on the solder joint; stress contours (psi) (from Lau et al. [11], © 1987 IEEE).

It was generally found that the maximum von Mises solder stress occurred near the tip of the outer fillet (Fig. 11.11), from where cracks could propagate down the interface with the lead (see Fig. 3.14).

Pan [17] studied accumulated deformations from several load cycles in surface solder joints by finite elements and compared them with experimental observations. He used an elastic-plastic model and included a steady-state creep, described by the law $d\varepsilon/dt = A(\sinh B\sigma)^n \cdot d^m \cdot \exp(-Q/RT)$. The macroscopic deformations were well shown by the model. Two failure modes due to thermal cycling, a tensile and a shear mode, were indicated by the study.

4. Stress Reduction in Compliant Leaded Solder Joints

Hall's analysis [9] of stress reduction in the solder posts of leadless ceramic chip carriers (Chapter 4) has been followed by Clech and Augis [2], who showed that the same concepts were valid for leaded surface mounts. Figure 11.12 illustrates that the differential thermal joint displacement Δu is comprised of two effects: the elastic lead deflection $X_L = V/K_L$ and the plastic solder shear $X_s = h_s\gamma$. Here the lead deflection is computed as the shear force V divided by the longitudinal stiffness K_L of the lead, which has all of its other degrees of freedom at the foot fixed. At a peripheral joint of the module this displacement would be in the diagonal direction r. Equating the thermal mismatch to $X_L + X_s$, e.g., for a uniform temperature rise ΔT:

$$h_s\gamma + V/K_L = r \cdot \Delta\alpha \cdot \Delta T . \qquad (11.34)$$

The shear stress τ is calculated as the force V divided by the effective solder area A, which is the product of the width of the lead and the minimum solder

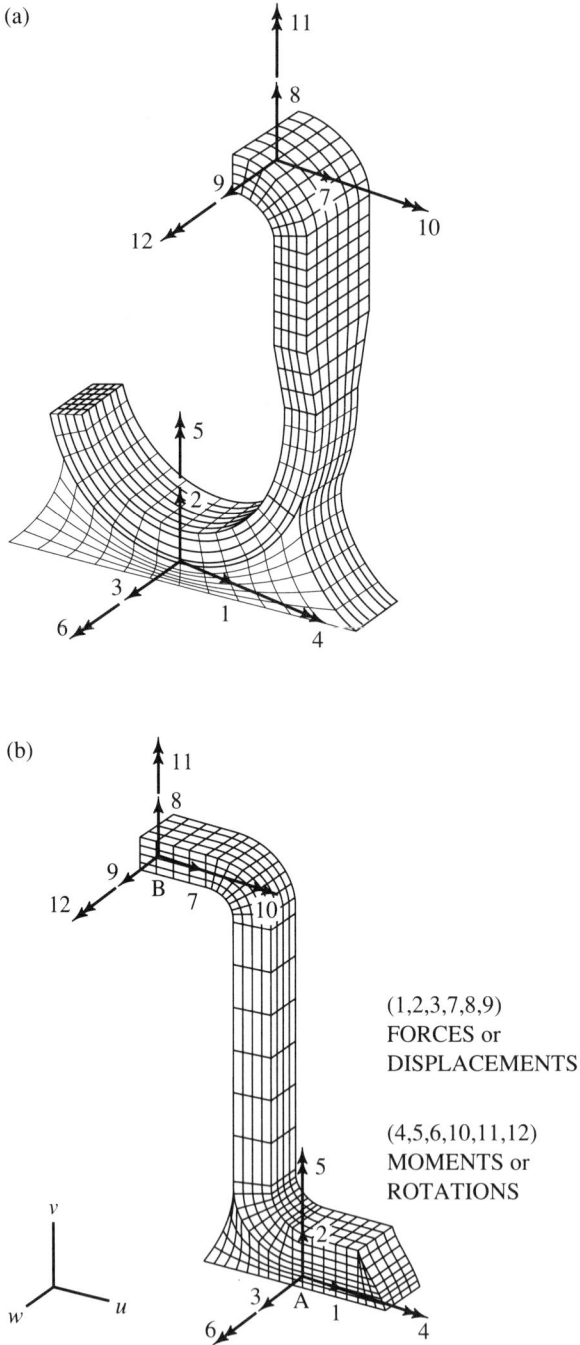

FIGURE 11.10. Finite element model and coordinate system for surface mount leads. (a) J-lead; (b) gull wing (from Lau [13]).

TABLE 11.1. Stiffness matrix (12×12) for J-lead in Fig. 11.10a.

$$[K] = \begin{bmatrix}
298 & 105 & 0 & 0 & 0 & -13 & -298 & -105 & 0 & 0 & 0 & -14 \\
105 & 8973 & 0 & 0 & 0 & 233 & -105 & -8973 & 0 & 0 & 0 & -109 \\
0 & 0 & 665 & 26 & -21 & 0 & 0 & 0 & -665 & 39 & 11 & 0 \\
0 & 0 & 26 & 1.7 & -0.8 & 0 & 0 & 0 & -26 & 0.9 & 0.5 & 0 \\
0 & 0 & -21 & -0.8 & 0.9 & 0 & 0 & 0 & 21 & -1.2 & -0.6 & 0 \\
-13 & 233 & 0 & 0 & 0 & 7 & 13 & -233 & 0 & 0 & 0 & -2.3 \\
-298 & -105 & 0 & 0 & 0 & 13 & 298 & 105 & 0 & 0 & 0 & 14 \\
-105 & -8973 & 0 & 0 & 0 & -233 & 105 & 8973 & 0 & 0 & 0 & 109 \\
0 & 0 & -665 & -26 & 21 & 0 & 0 & 0 & 665 & -39 & -11 & 0 \\
0 & 0 & 39 & 0.9 & -1.2 & 0 & 0 & 0 & -39 & 3 & 0.6 & 0 \\
0 & 0 & 11 & 0.5 & -0.6 & 0 & 0 & 0 & -11 & 0.6 & 0.4 & 0 \\
-14 & -109 & 0 & 0 & 0 & -2.3 & 14 & 109 & 0 & 0 & 0 & 2.1
\end{bmatrix}$$

TABLE 11.2. Stiffness matrix (12×12) for gull wing in Fig. 11.10b.

$$[K] = \begin{bmatrix}
165 & -325 & 0 & 0 & 0 & -3 & -165 & 325 & 0 & 0 & 0 & 2 \\
-325 & 1648 & 0 & 0 & 0 & -6 & 325 & -1648 & 0 & 0 & 0 & 21 \\
0 & 0 & 138 & 6.1 & 1.3 & 0 & 0 & 0 & -138 & 2.3 & 2.6 & 0 \\
0 & 0 & 6.1 & 0.3 & 0.1 & 0 & 0 & 0 & -6.1 & 0 & 0.1 & 0 \\
0 & 0 & 1.3 & 0.1 & 0.1 & 0 & 0 & 0 & -1.3 & 0 & 0 & 0 \\
-3 & -6 & 0 & 0 & 0 & 0.2 & 3 & 6 & 0 & 0 & 0 & 0.1 \\
-165 & 325 & 0 & 0 & 0 & 3 & 165 & -325 & 0 & 0 & 0 & -2.1 \\
325 & -1648 & 0 & 0 & 0 & 6 & -325 & 1648 & 0 & 0 & 0 & 21 \\
0 & 0 & -138 & -6.1 & -1.3 & 0 & 0 & 0 & 138 & -2.3 & -2.6 & 0 \\
0 & 0 & 2.3 & 0 & 0 & 0 & 0 & 0 & -2.3 & 0.1 & 0.1 & 0 \\
0 & 0 & 2.6 & 0.1 & 0 & 0 & 0 & 0 & -2.6 & 0.1 & 0.1 & 0 \\
2.0 & 21 & 0 & 0 & 0 & 0.1 & -2.1 & 21 & 0 & 0 & 0 & 0.3
\end{bmatrix}$$

joint length. Using $\tau = V/A$ and substituting into Eq. (11.34), we get

$$\gamma + \tau/k' = r \cdot \varDelta\alpha \cdot \varDelta T/h_s \qquad (11.35)$$

where the reduced stiffness k' of the leaded assembly is defined as

$$k' = K_L h_s/A . \qquad (11.36)$$

This k' is the isothermal line of the τ versus γ diagram; note that Eq. (11.35) is akin to Eq. (4.17) (Fig. 11.13). The k' of leaded surface mount joints, however, is two to three orders of magnitude smaller than the equivalent kH/A of the solder posts of LCCC devices. Some comparative numbers [18] for k' are 5.1 MPa (734 1b/in².) for J-leads of a 68-I/O PLCC assembly; for an 84-I/O leadless chip carrier on FR-4, $k' = 626$ MPa (90,800 1b/in.²) was obtained.

Solder strain can again be thought of as the sum of an elastic part γ_e and a plastic part γ_p. For the elastic part of the solder strain we have $\gamma_e = \tau/G(T_s)$, with T_s being the average solder joint temperature, determining the shear modulus G. The plastic strain, meanwhile, may be computed from a creep power law

$$d\gamma_p/dt = c(T_s)\tau^n \qquad (11.37)$$

(a)

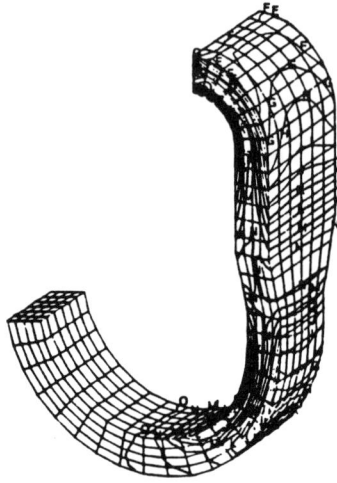

4504 = A
4193 = B
3883 = C
3572 = D
3261 = E
2951 = F
2640 = G
2329 = H
2019 = I
1708 = J
1398 = K
1087 = L
776 = M
466 = N
155 = O

(b)

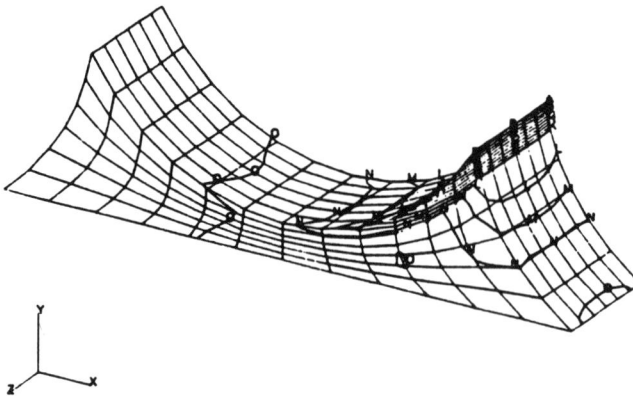

2083 = A
1940 = B
1797 = C
1654 = D
1510 = E
1367 = F
1224 = G
1081 = H
938 = I
795 = J
651 = K
508 = L
365 = M
222 = N
79 = O

FIGURE 11.11. Finite element analysis of J-leaded surface mount. (a) von Mises stress contours for unit displacement in 1-direction; (b) von Mises stress contours in the solder joint for displacement in the 1-direction (from Lau [13]).

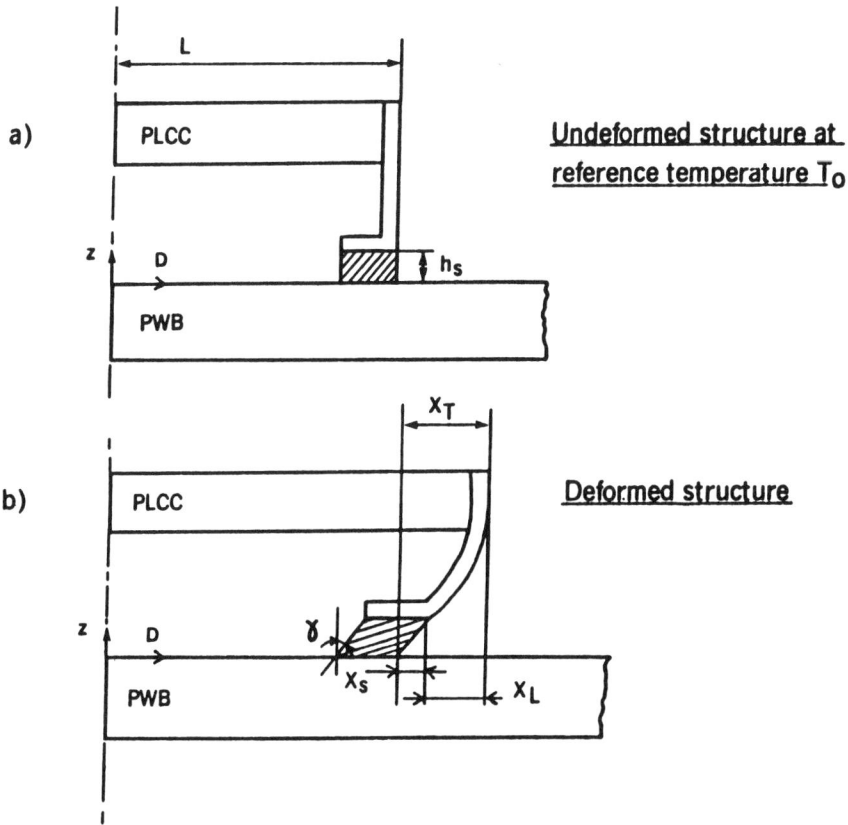

FIGURE 11.12. The compliant lead deflection accommodating thermal mismatch.

where the coefficient $c(T_s)$ is an Arrhenius function and n is a creep coefficient.

These mathematical procedures allow computation of the hysteresis loop, i.e., the cyclic energy dissipation for a solder joint thermally loaded into the plastic range. The energy dissipation is further related to fatigue damage [19]. As we shall next see, the area of the hysteresis loop has crucial significance in computing thermal fatigue for the joint.

5. Thermal Fatigue in Solder Joints of Compliant Leads

Generally, the cause of solder joint failure is cyclic fatigue from repeated thermal cycles under mismatch conditions. The work of Engelmaier [20] provides a computational predictive method: it is based on evaluating the cumulative cyclic fatigue damage in the solder. As described in the previous section, solder tends to respond to applied strains by time-dependent plastic

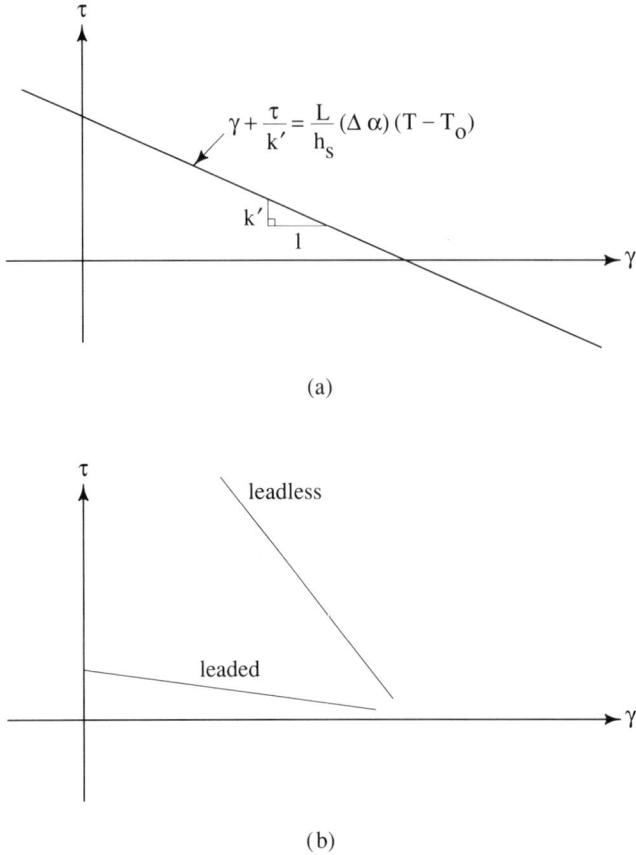

$$\gamma + \frac{\tau}{k'} = \frac{L}{h_s} (\Delta \alpha)(T - T_0)$$

(a)

(b)

FIGURE 11.13. Isothermal stress reduction. (a) General stress reduction line; (b) comparison of leaded and leadless structures (from Clech and Augis [18]).

deformation; this stress-reduction process consists of stress relaxation and creep. The plastic strain component γ_p is known to produce far greater fatigue damage than does the elastic strain; γ_p also increases with temperature and applied stress level.

The cyclic fatigue damage is proportional to the area of the hysteresis loop of the stress-strain diagram. The latter can, of course, be reduced by good compliant design such as CTE matching. Stylized hysteresis curves of both leadless and compliant leaded solder joints are shown in Fig. 11.14. The leadless system is highly stressed, that is, up to the yield point τ_y of solder. A cycle of the leadless system is much bigger than the cycle corresponding to the low-stressed compliant lead; after a much smaller amplitude of stress, however, the leaded system develops a strain γ the same as the leadless unit, corresponding to the mismatch displacement. This is reflected also in the vastly different reduced stiffness constants k' depicted in Fig. 11.13.

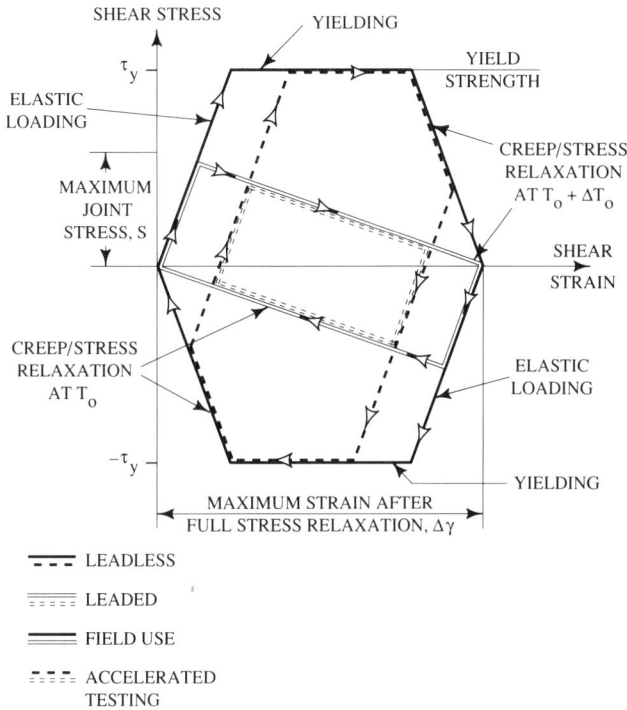

FIGURE 11.14. Cumulative fatigue damage and cyclic strain energy (from Engelmaier [20]).

The mean cyclic shear fatigue life N_f^* is proportional to the inverse of the viscoplastic strain energy density per cycle, ΔW, through Morrow's equation [21]

$$N_f^* = \frac{1}{2}\left[\frac{\Delta W}{W_f'}\right]^{1/c} \tag{11.38}$$

where W_f' is a material constant and c is between -0.5 and -0.7 for metals. This equation is the origin of the Coffin-Manson fatigue equation [22]:

$$N_f^* = \frac{1}{2}\left[\frac{\gamma_p}{2\varepsilon_f'}\right]^{1/c} \tag{11.39}$$

where γ_p is the temperature-dependent plastic strain and $2\varepsilon_f'$ is a material constant. Engelmaier deduced a method [23] by which, instead of having to determine the time- and temperature-dependent γ_p, the maximum available displacement shear γ is used in Eq. (11.39); meanwhile, he gave the following definition of coefficients for Sn-Pb 60-40 or 63-37 solders:

$$2\varepsilon_f' = 0.65 , \tag{11.40}$$

and

$$c = -0.442 - 6 \times 10^{-4} T_{sj} + 1.74 \times 10^{-2} \ln\left(1 + \frac{360}{t_D}\right) \quad (11.41)$$

to be used in conjunction with Eq. (11.39); here the mean cyclic solder joint temperature T_{sj} and the half-cycle dwell time t_D (in minutes) are involved. The total cyclic strain range γ to be used in Eq. (11.39) includes a "form factor" F^* and the thermal expansion mismatch, which for uniform, nonuniform, and power-cycle-type temperature expansions is symbolically written as $\Delta u = \Delta(\alpha \Delta T)$ we get, analogously to Eq. (1.53):

$$\gamma = F^*(r/h)\, \Delta(\alpha \Delta T) . \quad (11.42)$$

If the component is compliant leaded, a more complex procedure is needed because the cyclic viscoplastic strain energy is more difficult to determine here. The reason may lie in the fact that [20] while leadless solder joints (with large k') can totally relax even after a low-temperature cycle, the creep in leaded joints (with small k') is greatly retarded and incomplete even at long dwell times and high temperatures.

The hysteresis loop for leaded joints is marked by the stress-reduction line at the temperature extremes (Fig. 11.14). The maximum applied stress is $\tau = K_L r\, \Delta(\alpha \Delta T)/A$, which determines the stress reduction rate during dwell at a constant temperature. Short of having the extent of stress degradation, a proportionality can be written for the cyclic hysteresis energy,

$$\Delta W \propto \frac{K_L [r\Delta(\alpha \Delta T)]^2}{Ah} . \quad (11.43)$$

A plot of ΔW versus N_f^*, measured for several systems, produced Fig. 11.15, thus showing that c, the slope of the logarithmic plot, is again close to -0.5, as for leadless Sn-Pb 60-40 or 63-37 solder joints.

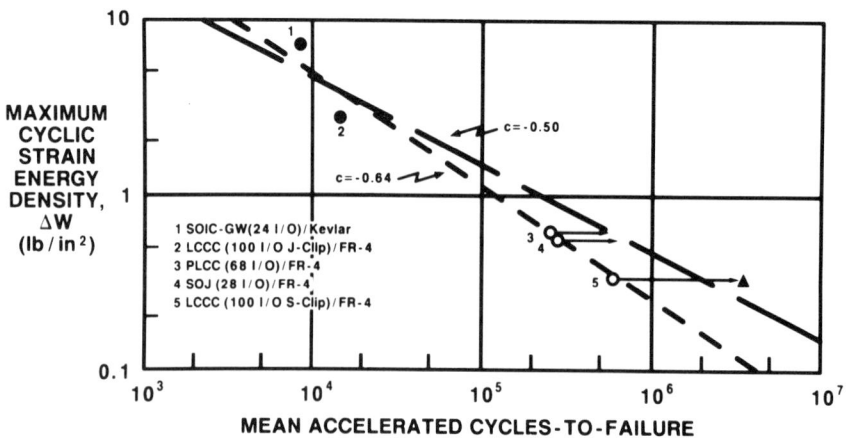

FIGURE 11.15. Plot of test results from accelerated functional and thermal cycling tests for five different leaded surface mount component assemblies (from Engelmaier [20]).

Figure 11.14 shows the difference in hysteresis loops (causing viscoplastic solder damage), whether for accelerated test (*t*) cycles or operational use (*u*) cycles. The accelerated cycles have insufficient dwell time for stress relaxation, while the use cycles allow nearly complete relaxation. For life prediction of

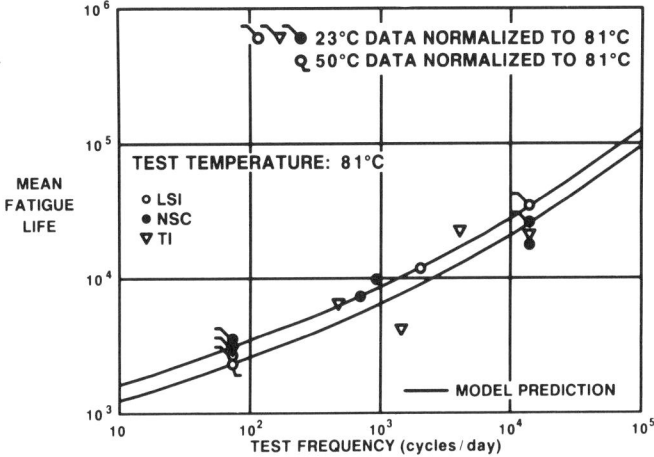

FIGURE 11.16. Plot of test results from accelerated isothermal mechanical cycling tests of PLCCs employing three test temperatures and near-square-wave cycle shapes (from Engelmaier [20]).

TABLE 11.3. Surface mount figure of merit equations. Based on Ref. [26].

1. Leadless

$$FM = 0.288 \frac{h}{L\,\Delta\alpha\,\Delta T\sqrt{N}} \left\{ \frac{\ln[1 - F(N)]}{\ln(0.99)} \right\}^{1/2\beta}$$

2. Leaded

$$FM = 56.25 \frac{Ah}{k(L\,\Delta\alpha\,\Delta T)^2\sqrt{N}} \left\{ \frac{\ln[1 - F(N)]}{\ln(0.99)} \right\}^{1/2\beta}$$

Quantities Involved	PLCC 84 lead
where h = solder joint height (in.)	0.02
A = solder joint area (in.^{2})	0.0016
L = diagonal length of module (in.)	1.5
k = lead compliance (lb/in.)	3000
$\Delta\alpha$ = thermal expansion mismatch (in./in. °C)	1.5×10^{-6}
ΔT = field temperature increase (°C)	30
N = on/off cycles	2500
$F(N)$ = allowable fraction fails	0.0001
β = Weibull shape factor	2
[~ 2 leaded; ~ 4 leadless]	0.831

u specimens, a method involving the "acceleration transform" can be used [23]. It uses accelerated test results (t) produced at levels not far from use conditions. We get

$$N_f^*(u) = \frac{1}{2}\left[(2N_f^*(t))^{c_t} \frac{\Delta W(u)}{\Delta W_t} \right]^{1/c_u} ,$$ (11.44)

which is derived by equating the denominators W' in two expressions of Eq. (11.39):

$$N_f^*(t) = \frac{1}{2}\left[\frac{\Delta W(t)}{W'} \right]^{1/c_t} , \qquad N_f^*(u) = \frac{1}{2}\left[\frac{\Delta W(u)}{W'} \right]^{1/c_u} .$$

This procedure was used successfully in several test series (Fig. 11.16).

Note that in order to compare the relative damage expected, the subscripts u and t can be assigned to two alternative test conditions, computing $N_f^*(t)/N_f^*(u)$, the inverse ratio of damage.

TABLE 11.4. Design parameters for sample applications of the surface mount figures of merit. From Engelmaier [20].

Component	L_D mm (in.)	h mm (in.)	K N/mm (lb/in.)	A mm^2 (10^{-6} sq in.)	$\Delta\alpha$ (ppm/°C)
Leadless SM Attachments on CTE-Tailored PWBs					
84 I/O CCC (pillars) [28]	10.3 (0.407)	0.508 (0.020)	–	–	2.0
84 I/O CCC (balls) [29]	10.3 (0.407)	0.229 (0.009)	–	–	2.0
68 I/O CCC (castellated)	15.7 (0.620)	0.0762 (0.003)	–	–	2.0
Leadless SM Attachments on FR-4 PWBs					
RC 1206/MELF	15.2 (0.060)	0.127 (0.005)	–	–	11.5
RC 1820/MELF	2.29 (0.090)	0.127 (0.005)	–	–	11.5
Leaded SM Attachments on FR-4 PWBs					
28 I/O SOJ	9.96 (0.392)	0.127 (0.005)	25.4 (145)	0.522 (810)	3·0
68 I/O PLCC-J	17.1 (0.674)	0.127 (0.005)	9.64 (55)	0.548 (850)	3.0
24 I/O SOIC-gull wing	8.56 (0.337)	0.127 (0.005)	36.3 (207)	0.465 (720)	3.0
100 I/O CCC (S-clips) [24]	23.3 (0.919)	0.127 (0.005)	1.31 (7.5)	0.522 (810)	11.5
100 I/O CCC (J-clips) [24]	23.3 (0.919)	0.127 (0.005)	10.2 (58)	0.516 (800)	11.5

6. Thermal Cycling Reliability

The reliability of a solder joint, i.e., the probability of its survival for a given cycle number N at the functional load level, is an important consideration for product manufacturers. Engelmaier [24] found that the wear-out type of failure mechanism of thermally cycled solder joints had a Weibull probability distribution [25] around the mean cyclic life N_f^*. Leadless solder mounts tend to have a Weibull slope $\beta = 4$, and leaded ones $\beta = 2$. For a design with n components, the design mean fatigue life $N_f^*(D)$ can then be written as follows, for the number N of operating cycles:

$$N_f^*(D) = N \left[\frac{-n \ln 2}{\ln[1 - F(N)]} \right]^{1/\beta}, \qquad (11.45)$$

where $F(N)$ is the cumulative failure probability after N cycles. It is noted that Eq. (11.45) is valid for a module (component), not any single solder joint; for example, any single square module has four corners and thus eight corner leads, any one possibly failing after a certain number of stress cycles.

Engelmaier and co-workers [26] developed reliability figures of merit (FM) combining the fatigue mechanism of Eq. (11.39) with the probability distribution consideration of Eq. (11.45) (see Table 11.3). The FM helps to simply evaluate a given design: FM > 1 means "favorable," while FM < 0.7 means "unfavorable." Between 0.7 and 1.0, a grey area exists. We note that

TABLE 11.5. Reliability figures of merit FM(rel) for the assemblies in Table 11.4 for various acceptable failure probability levels, a design life of 20 years (7300 cycles of operation) and a cyclic use environment resulting in $\Delta T_e = 35°C$. From Engelmaier [20].

Component	Acceptable Failure Probability, $F(N)$, %				
	0.0001	0.001	0.01	0.1	1
Leadless SM attachments on CTE-tailored PWBs					
84 I/O CCC (pillars) [28]	0.75	1.00	1.3	1.8	2.4
84 I/O CCC (balls) [29]	0.34	0.45	0.60	0.80	1.06
68 I/O CCC (castellated)	0.07	0.10	0.13	0.17	0.23
Leadless SM attachments on FR-4 PWBs					
RC 1206/MELF	0.22	0.29	0.39	0.52	0.70
RC 1820/MELF	0.15	0.20	0.26	0.35	0.47
Leaded SM attachments on FR-4 PWBs					
28 I/O SOJ	1.1	1.9	3.4	6.1	10.9
68 I/O PLCC-J	1.01	1.8	3.2	5.7	10.2
24 I/O SOIC-gull wing	0.91	1.6	2.9	5.1	9.1
100 I/0 CCC (S-clips) [24]	0.26	0.46	0.82	1.5	2.6
100 I/O CCC (J-clips) [24]	0.03	0.06	0.10	0.19	0.33

the scale factor outside the bracket for leadless surface mounts is dimension-less, but for leaded ones it is 0.388 MPa or 56.25 psi.

Analogously to surface mounts, figures of merit can also be attached to modules (components) [20]:

$$FM(comp, leadless) = 20\,h/r \qquad (11.46)$$

$$FM(comp, leaded) = (S.F.)\,Ah/K_L r^2 \qquad (11.47)$$

where the scale factor S.F. $= 6.9 \times 10^4$ MPa in SI (10^7 lb/in.^{2} in English units).

Table 11.4 shows some design figures and Table 11.5 shows reliability figures for popular leaded attachments [20].

The temperature cycling reliability of 28- and 32-pin TSOPs (thin small outline package) was studied by Lau et al. [27].

7. Exercises and Questions

1. In what respect is the lead response expected to be different in thermal loading versus flexing?
2. Assuming an inverted L-shaped formation for a gull wing lead, such as in Fig. 11.5, calculate the bending moment and torque variation for corner leads of a square, plastic leaded module (similar to a PLCC). Consider $\Delta T = 50\,°C$ and FR-4 card. The Amzirc lead dimensions are $l_z = 2.0$ mm, $b = 0.3$ mm, $h = 0.15$ mm, $l_y = 0.5$ mm, and $l_m = 12$ mm. The eutectic Sn-Pb solder joint has $h_s = h_L = 0.25$ mm.
3. Calculate the spring constant K_L [Eq. (11.34)] and the reduced stiffness k' [Eq. (11.36)] for the leads of Problem 2.
4. What material properties would have to be included in a finite element program destined to help calculate fatigue life by the likes of Eq. (11.44)?

References

1. Solomon, H.D. (1989), "Low Cycle Fatigue of Surface mounted Chip Car-rier/Printed Wiring Board Joints," *Proc. 39th Elec. Comp. and Tech. Conf., IEEE*, pp. 277–292.
2. Clech, J-P.M., and Augis, J.A. (1988), "Temperature Cycling, Structural Response and Attachment Reliability of Surface-Mounted Leaded Packages," *Proc. Int'l Elec. Packag. Soc. Conf.*, Dallas, Tex. pp. 305–324.
3. Taylor, J.R., and Pedder, D.J. (1982), "Joint Strength and Thermal Fatigue in Chip Carrier Assembly," *ISHM Int'l J. Hybrid Microelec.*, **5**, 209–214.
4. Sherry, W.M., Erich, J.S., Bartschat, M.K., and Prinz, F.B. (1985), "The Effect of Joint Design on the Thermal Fatigue Life of Leadless Chip Carrier Solder Joints," *IEEE Trans.*, **CHMT-8**(4), 417–426.
5. Sinnadurai, N., Cooper, K., and Woodhouse, J. (1986), "Assessing the Joints in Surface Mounted Assemblies," *Microelec. J.*, **17**(2), 21–32.
6. Kotlowitz, R.W., and Engelmaier, W. (1986), "Impact of Lead Compliance on the Solder Attachment Reliability of Leaded Surface Mounted Devices," *Proc. Int'l, Elec. Packag. Soc. Conf.*, San Diego, Calif., pp. 841–865.

7. Engel, P.A., and Lee, L-C. (1986), "Surface Solder Stress Design for Thermal Loading of Modules," *Proc. NEPCON East Conf.*, Boston, Mass., pp. 263–270.

8. Kitano, M., Kawai, S., and Simizu, I. (1986), "Thermal Fatigue Strength Estimation of Solder Joints of Surface Mount IC Packages," *Proc. NEPCON East Conf.*, Boston, Mass., pp. 373–383.

9. Hall, P.M. (1987), "Creep and Stress Relaxation in Solder Joints in Surface Mounted Chip Carriers," *IEEE Trans.*, **CHMT-10**(4), 556–565.

10. Jahsman, W.E., and Jain, P. (1990), "Comparison of Predicted and Measured Lead Stiffnesses of Surface Mount Packages," *Proc. 40th Elec. Packag. Comp. and Tech. Conf.*, **2,** 926–933.

11. Lau, J.H., Rice, D.W., and Avery, P.A. (1987), "Elastoplastic Analysis of Surface-Mount Solder Joints," *IEEE Trans.*, **CHMT-10** (3), 346-357.

12. Lau, J.H., and Harkins, C.G. (1988), "Thermal Stress Analysis of SOIC Packages and Interconnections," *IEEE Trans.* **CHMT-11**, 380–389.

13. Lau, J.H. (1987), "Stiffness of PLCC 'J' Lead and Its Effect on Solder Joint Reliability," ASME Winter Annual Meeting, Paper No. 87/WAM-EEP-3.

14. Lau, J.H., Harkins, C.G., Rice, D.W., Kral, J., and Wells, B. (1987), "Experimental Analysis of SMT Solder Joints Under Mechanical Fatigue," *Proc. IEEE 37th Elec. Components Conf.*, pp. 589–597.

15. Lau, J.H. (1989), "Thermal Stress Analysis of SMT PQFP Packages and Interconnections," *ASME J. Elec. Packag.*, **111**(1), 2–8.

16. ABAQUS Finite Element Program, HKS Inc., Providence, R.I.

17. Pan, T-Y. (1991), "Thermal Cycling Induced Plastic Deformation in Solder Joints – Part I: Accumulated Deformation in Surface Mount Joints," *ASME J. Elec. Packag.*, **113**(1), 8–15.

18. Clech, J-P.M., and Augis, J.A. (1987), "Engineering Analysis of Thermal Cycling Accelerated Tests for Surface Mount Attachment Reliability Evaluation, "*Proc. 7th Int'l Elec. Packag. Conf.*, Boston, Mass., Vol. 1, pp. 385–411.

19. Stone, D., Wilson, H., Subrahmanyan, R., and Li, C-Y. (1986), "Mechanisms of Damage Accumulation in Solders During Thermal Fatigue," *Proc. IEEE Elec. Comp. Conf.*, pp. 630–635.

20. Engelmaier, W. (1989), "Surface Mount Solder Joint Long-Term Reliability: Design, Testing, Prediction," *Soldering and Surf. Mount Technol.* No. 1, pp. 14–22.

21. Morrow, J.D. (1964), "Cyclic Plastic Strain Energy and Fatigue of Metals," *ASTM STP*, **378**, 45–87.

22. Manson, S.S. (1966), *Thermal Stress and Low Cycle Fatigue*, McGraw-Hill, New York.

23. Engelmaier, W. (1985), "Functional Cycling and Surface Mounting Attachment Reliability," *Circuit World*, **11**(3), 61–72.

24. Engelmaier, W. (1988), "Surface Mount Attachment Reliability of Clip Leaded Ceramic Chip Carriers on FR-4 Circuit Boards," *IEPS J.* **9**(4), 3–11.

25. Lloyd, D.K., and Lipow, M. (1962), *Reliability: Management, Methods, and Mathematics*, Prentice-Hall, Englewood Cliffs, N.J.

26. Clech, J-P.M., Engelmaier, W., Kotlowitz, R.W., and Augis, J.A. (1989), "Surface Mount Solder Attachment Reliability Figures of Merit – Design for Reliability Tools," *Proc. SMART V. Conf.*, New Orleans, La.

27. Lau, J.H., et al, (1992), "Solder Joint Reliability of a Thin Small Outline Package (TSOP)," *Proc. 42nd ECTC*, San Diego, Calif., pp. 519–532.

Chapter 12

Dynamic Response of Circuit Card Systems

1. General Considerations

In general, modules are supported by stiff leads and possess a higher K/M ratio than the cards and boards to which they are attached. In the module/lead/card system, the card is the most flexible element; the fundamental natural frequency f_1 for circuit cards is generally between 20 and 50 Hz, and even for multimodule stiff cards or stiffened boards, it seldom exceeds 200 Hz. With the stiffnesses and dimensions of modules and leads discussed in Chapters 3, 7, and 8 (consider $K = 100$ N/mm and the mass of a 25-mm module $m = 2$ g supported on 68 gull wing leads), the lowest frequency of the lead-supported module is 9.3 kHz. The fundamental frequency f_1 of the leads alone is, taking a lumped mass of 2 mg, about 36 kHz.

Consequently, circuit card vibration results in quasistatic bending of the leads. Dynamic analysis of the card, based on known excitation and boundary conditions, will yield its flexural mode shapes; this will suffice in computing vibration response of the leads and solder joints. Circuit card flexural modes, on the other hand, are closely related to the analysis concepts discussed in the previous chapters. The calculation of lead response, for known lead-deformation shapes as those of frame structures, was treated by Steinberg [1].

The free vibration of populated circuit cards will be affected by both the mass and stiffness added by modules (plus connectors and cabling). Boundary conditions, the way the card is supported or connected to the higher-hierarchy structure, play a decisive role. For example, fixed-edge conditions are virtually impossible to achieve; edge connectors may provide an elastic support. Forced vibration levels for design are usually provided by specifications. Transient vibration or shock response is equally important. Circuit cards are most often tested as parts of a larger assembly; their attachment to boards, boxes, and frames, and the dynamics of electrical contacts in general, will be discussed in Chapter 14.

2. Dynamic Load Levels and Measurement

Vibration input levels for the design of electronics equipment vary widely from industry to industry, in the military branches, etc. Increased recognition of the significance of random vibration is widely seen throughout the application spectrum.

IBM has collected data of the environment its machines have seen [2, 3]. Table 12.1 distinguishes random vibration levels for Class V1 (floor-mounted) and Class V2 (table-mounted) equipment types. In Table 12.1, SS means sample size, and RTGP means real-time g waveform peak values. It was found that horizontal vibration was typically half the vertical vibration. Measured vibration levels for tabletop machines were generally lower than those for floor-mounted ones.

The shock environment seen by IBM equipment is shown in Table 12.2. Floor-mounted and tabletop-mounted equipment are designated S1 and S2, respectively. Indications were that machines weighing over 200 kg tended to attenuate floor-originated shock inputs, while machines lighter than 100 kg often transmitted significant levels through the structure.

Contrasting these relatively mild commercial levels, avionics specifications for the U.S. Air Force demand survival of a rigorous environment. Table 12.3 is an example [4] for the loads to be endured in 15 years of service by a transformer mounted on an avionics circuit board. It is precisely this milieu that spurs achievement of high reliability in printed circuit cards and boards.

TABLE 12.1. Vibration data analysis summary. From Frey [3].

Class	SS	Avg. g rms	Max. g rms	Avg. RTGP
VI < = 600 Kg	184	0.0133	0.1510	0.0490
VI > 600 Kg	50	0.0101	0.0348	0.0373
VI 60 Hz	79	0.0074	0.0402	
V2	55	0.0105	0.0491	0.0546

TABLE 12.2. S1 and S2 shock data summary. From Frey [3].

Class	SS	Average g			Average PW			Y-axis max g	
		x	y	z	x	y	z	g	ms
S1	48	0.34	0.47	0.26	1.8	6.6	1.6	3.0	@ 0.6
S2	25	1.56	3.61	2.28	3.1	2.1	2.1	20.5	@ 2.3

1. y is vertical axis. x and z are horizontal.

2. PW is pulse width in milliseconds.

3. In the case of complex shock pulses with ringing, the g and PW values are for the first half-sine cycle of the waveform.

TABLE 12.3. Operating requirements for the example of a transformer mounted on a PCB over 15 years. From Steinberg [4].

Condition	Definition	Number to failure	Unit
A	ESS random vibration screen PCB response 11.2 G rms 3 axes,	1.0	hour
B	Captive flight vibration PCB response 6.1 G rms,	2160	hour
C	Free-flight vibration PCB response 15.9 G rms,	1.0	hour
D	Ground transportation vibration PCB response 3.8 G rms,	840	hour

Standard measurement techniques [5] for vibration levels in circuit card systems rely on attachment of local sensors. Strain gauges measure the time-varying strain at a strategic location. They are light and relatively small but have several drawbacks. Their attachment to low-modulus materials and thermal applications must be handled with special care.

Small lightweight accelerometers are available; their frequency range (over 10 KHz range) is high enough for circuit cards and boards. For module attachment, accelerometers must be selected with care for the higher frequencies required.

Experimental modal analysis [6] is a technique that reconstructs vibration mode shapes from accelerometer measurements strategically located throughout the structure under consideration. Graphic results include FRFs (frequency response functions).

The laser Doppler vibrometer [7] is a surface velocity-measurement device that operates on the basis of the Doppler effect: a beam of light reflected from a moving surface is also changed in frequency. While the fractional change can be very small (as low as 10^{-8}), it can be accurately measured using optical interferometry in conjunction with electronic frequency measurement. The wide velocity range accommodated by the LDV stretches between 10^{-9} and 10^1 m/s. It also allows accurate measurement of smaller displacement amplitudes at higher frequencies and larger displacements at lower frequencies than some other optical techniques such as electronic speckle pattern interferometry (ESPI).

3. Vibration Analysis in Circuit Card Systems

It is difficult to vibration-analyze plates unless they are rectangular or polar symmetrical, and their material isotropic, homogeneous, and linearly elastic. Certainly, the addition of modules, amounting to a reinforcement, will add complexity to the calculations, which may only be surmountable by finite element methods coupled with substructuring. The flexural analysis methods described in Chapters 7–10 can be utilized in analyzing such systems.

Pitarresi and his co-workers [8, 9] adapted the smeared property technique to finite element analysis of populated circuit cards. By this method, the area of a module cluster is replaced by an equivalent orthotropic plate that has the same thickness as the card but a uniform specific mass $\rho = m/L_x L_y h$ and uniform orthotropic stiffness properties E_x, E_y, v_{xy}, and G_{xy}. Such an equivalent-area segment can be inserted in the finite element analysis of the whole card, producing improved free vibration input data (Ref. [1] of Chapter 7).

The orthotropic constants have elastostatic functions and thus can be obtained by static loading, experimentally, analytically, or by a hybrid of the two methods. E_x and E_y can be measured by three-point bending tests (Chapter 3), thus $E_x = K L_x^3 / 4 L_y h^3$ will yield one of the orthotropic moduli. Alternatively, these can be analytically calculated by the methods of Chapters 7 and 8, as for coupled beams or plates. The shear modulus can be determined by a torsion test (Chapter 3) of the module cluster area, and $G_{xy} = 3 K_T L_x / L_y h^3$ could be averaged in two directions. It would alternatively be calculated by the methods of Chapters 9 and 10, for module clusters under torsion. For v_{xy} a reasonable assumption may suffice (i.e., that of the lead frame, etc.); alternatively an iterative procedure can be used (Ref. [4] of Chapter 7).

The goodness of the smeared property technique was tested on several cases, involving not only the first but up to five natural modes of free vibration. Experimental and analytical vibration response correlations were calculated by a goodness-of-fit procedure called MAC (modal assurance criterion), expressed as [6, 10]:

$$\text{MAC}(\{\phi_e\}_j, \{\phi_f\}_k) = \frac{|\{\phi_e\}_j^T \{\phi_f\}_k|^2}{\{\phi_e\}_j^T \{\phi_e\}_j \{\phi_f\}_k^T \{\phi_f\}_k} \tag{12.1}$$

where $\{\phi_e\}_j$ = experimental mode shape vector j, $j = 1, \ldots n_e$; $\{\phi_f\}_k$ = finite element mode shape vector k, $k = 1, \ldots n_f$; n_e = number of experimental mode shape vectors; and n_f = number of finite element mode shape vectors.

A value of 1 would mean absolute correlation, and 0 would mean no correlation at all. The ith mode would have the circular frequency ω_i and mode shape vector ϕ_i associated with it.

For experimental modal analysis, the module-populated card of Fig. 12.1 was suspended, creating a free boundary condition, and impacted with a light hammer. Modal parameters were obtained by measuring frequency response functions (FRF) at strategic points throughout the structure. Finite element analysis of the same structure involved the smeared properties (Table 12.4). In Table 12.4 the G_{xy} values of the shear modulus were computed by orthotropic plate theory, while the G'_{xy} values were obtained by torsional testing. Finite element analysis agreed well with the latter method.

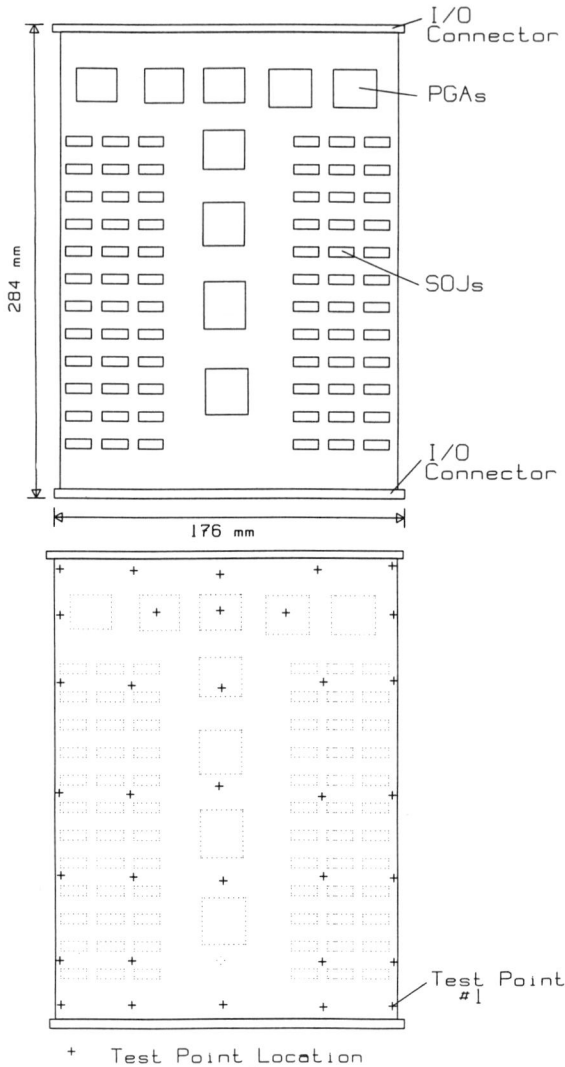

FIGURE 12.1. Circuit card (176 × 284 mm) set to free vibration by an initial impact. Note accelerometers at marked locations and the I/O connector at the ends. Square PGA modules and SOJs are attached (from Pitarresi et al. [8]).

The correlation between experiment and FE analysis of the heavily populated structure is shown in Table 12.5; up to the first 11 modes were used. The dependence of support locations was also studied [11] for their effect on optimization and analytical predictability. The support locations, by definition, would be optimized if they yielded the highest possible natural frequency.

Another application of modal analysis to circuit cards was given by Wong et al. [12]. If the bare circuit card has M, K, and C (mass-, stiffness-, and

TABLE 12.4. Smeared modeling region (SMR) material properties. From Pitarresi et al. [8].

Smeared modeling region (SMR)	E_x (GPa)	E_y (GPa)	G'_{xy} (GPa)	G_{xy} (GPa)	ρ (kg/m$^3 \times 10^{-3}$)
PGA-1	52.3	51.0	54.0	23.0	8.50
PGA-2	52.3	51.0	54.0	23.0	7.28
SOJ-1	28.4	20.9	26.1	10.6	6.17
SOJ-2	26.4	22.3	26.1	10.6	6.19
Unpopulated PWB	19.3	19.3	(2)	8.62	2.46
Connector-1	14.6	(1)	(2)	6.52	3.27
Connector-2	14.6	(1)	(2)	6.52	1.70

TABLE 12.5. Correlation of large FE model with experiment. From Pitarresi et al. [8].

Mode No.	FEM freq (Hz)	EXP freq (Hz)	MAC
1	32.7	33.8	0.98
2	54.5	62.1	0.99
3	94.9	93.0	0.96
4	123.0	126.0	0.97
5	140.0	152.0	0.91
6	193.0	183.0	0.88
7	210.0	201.0	0.82
8	218.0	246.0	0.80
9	301.0	352.0	0.85
10	308.0	291.0	0.75
11	324.0	336.0	0.64

damping-) matrices, then addition of module(s) would change those matrices to M^*, K^*, and C^*. The difference between the bare and populated structures is reflected in the modification matrices, $\Delta M = M^* - M$, $\Delta K = K^* - K$, $\Delta C = C^* - C$. Having measured, by the modal analysis technique, both sets of matrices, the modification matrices can be obtained. It is now in the realm of computational equipment, see, e.g., [13], to establish a single mass modification scheme that would produce a frequency response function (Figs. 12.2 and 12.3) equal to that of the modified structure. Wong et al. found that a simple lumped-mass modification would achieve the same effect as the addition of a module. (In cases where this was not so, addition of a rotary moment of inertia could help.) This procedure of "lumping" represents an alternative procedure to "smearing" the properties.

Lau and Keely [14] investigated lead vibrations in several surface-mounted systems: in SOICs (with gull wing leads), PLCCs (with J-leads), and plastic quad flatpacks (PQFPs with gull wings). For experimental response

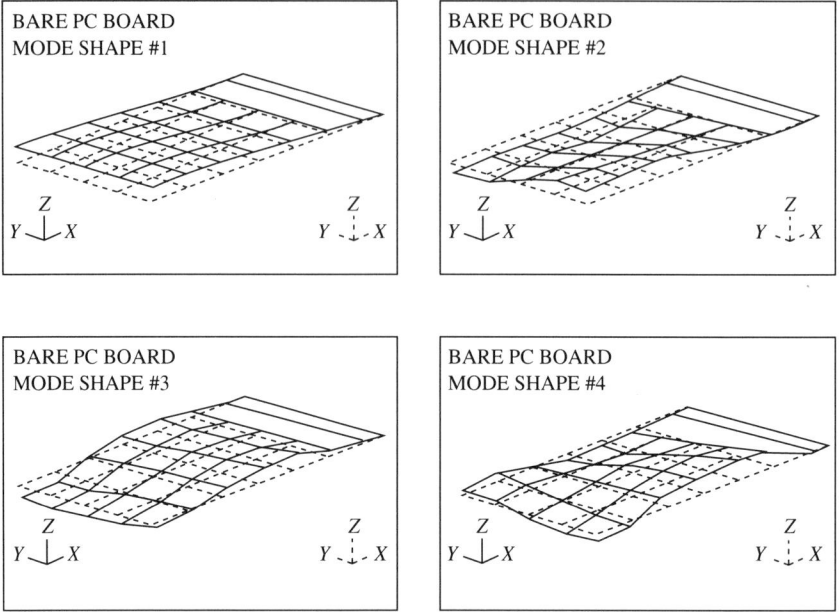

FIGURE 12.2. Circuit board mode shapes (experimental) (from Wong et al. [12]).

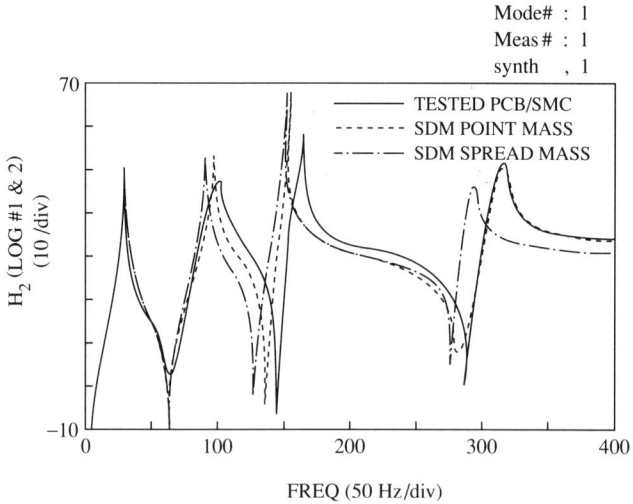

FIGURE 12.3. Effect of point mass and spread mass modifications on predicted frequency response functions (from Wong et al. [12]).

measurements of the leads, they subjected the surface-mounted structure to a high-velocity air stream; they then focused a laser Doppler vibrometer on the upper bends of the vibrating leads, where light was best reflected. This way the natural frequencies of leads could be determined and compared to the finite element-computed frequencies of the same leads for the finite element schemes of leads; see Figs. 11.10a and 11.10b.

The leads were checked both in the soldered condition and with some observational solder joints severed. This resulted in two kinds of lead structures: fixed-fixed and fixed-free (cantilevered) leads. Calculations showed that the cantilevered structures had natural frequencies 5 to 10 times lower than the fixed-fixed structures with full solder bond. For example, a typical cantilevered structure would have $f_1 = 30\,\text{kHz}$ and the corresponding soldered one $f_1 = 150\,\text{kHz}$. There was good agreement between experimental and finite element analyses. The vibration spectra for four kinds of leads are shown in Fig. 12.4. The effect of the shifting of the fundamental frequency was so strong that it was suggested to be used as the basis of an inspection procedure for solder joint integrity. It is remarked that at the high frequencies of free vibration fully soldered lead joints develop lesser effects of an elastic foundation than under the static loading of three-point bending tests described in Chapter 3.

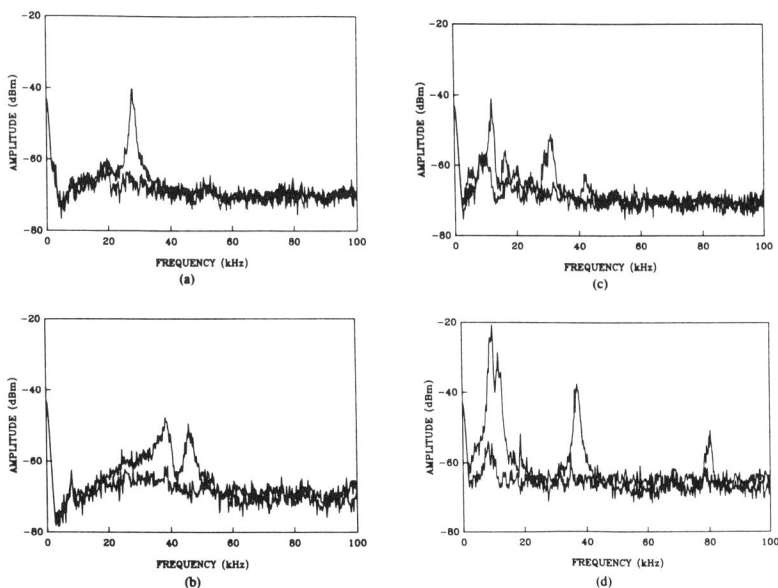

FIGURE 12.4. Vibration spectra for one soldered and one unsoldered joint for each lead type. Bad joint spectra have peaks; good joint spectra are fairly flat. (a) Wide SOIC; gull wing lead thickness, 0.24 mm; (b) narrow SOIC, gull wing lead thickness, 0.23 mm; (c) PLCC, J-lead thickness 0.255 mm; (d) PQFP, gull wing lead thickness 0.15 mm (from Lau and Keely [14] © 1989 IEEE).

4. Transient Vibration and Shock Response

The problem of a module-reinforced card can be represented in its simplest form as a beam to which a lumped parameter system is attached (Fig. 12.5); this is also called a dynamic vibration absorber [15]. Writing the dynamic beam equation

$$EI \frac{\partial^4 y}{\partial x^4} + \rho A \frac{\partial^2 y}{\partial t^2} = f_i(t)\delta(x - x_0) + f(x, t) \qquad (12.2)$$

where the excitation is

$$f_i(t) = - m\ddot{w}(t) = k[w(t) - y(x_0, t)] + c[\dot{w}(t) - \dot{y}(x_0, t)] , \qquad (12.3)$$

Keltie and Ozisik [16] obtained the solution using the Laplace transform technique [17]. In the special case of an undamped simple resonator mass m attached to a beam of mass m_b and the stiffness ratio of the attachment to that of the beam fixed at a value of 5, they calculated the ratio of frequencies f_r of the attachment to that of the beam f_b and other response properties of the system. They found that the peak acceleration level of the beam was insensitive to the attached mass, except if $m \gg m_b$. For $5 < f_r/f_b < 20$, the peak responses of the mass were amplified over those of the beam.

For circuit card systems, especially frame and box assemblies, shock-response data for mounted components are expediently viewed in terms of the damage boundary curve (DBC) or fragility curve [5, 18–21]. The latter is a curve $PQRS$ constructed in the peak-acceleration (a) versus velocity change (ΔV) coordinates for transient response of a mass m_2 attached to the supporting mass m_1; m_1 is subjected to a force P; see Fig. 12.6. Below the damage curve the product has been tested to survive without damage, whereas above it damage would result. Elements of the fragility analysis are described as follows.

Writing the equation of motion for the two masses,

$$m_1 \ddot{x}_1 + k_2(x_1 - x_2) = P - m_1 g \qquad (12.4)$$

$$m_2 \ddot{x}_2 + k_2(x_2 - x_1) = - m_2 g . \qquad (12.5)$$

FIGURE 12.5. Schematic for dynamic analysis of a beam.

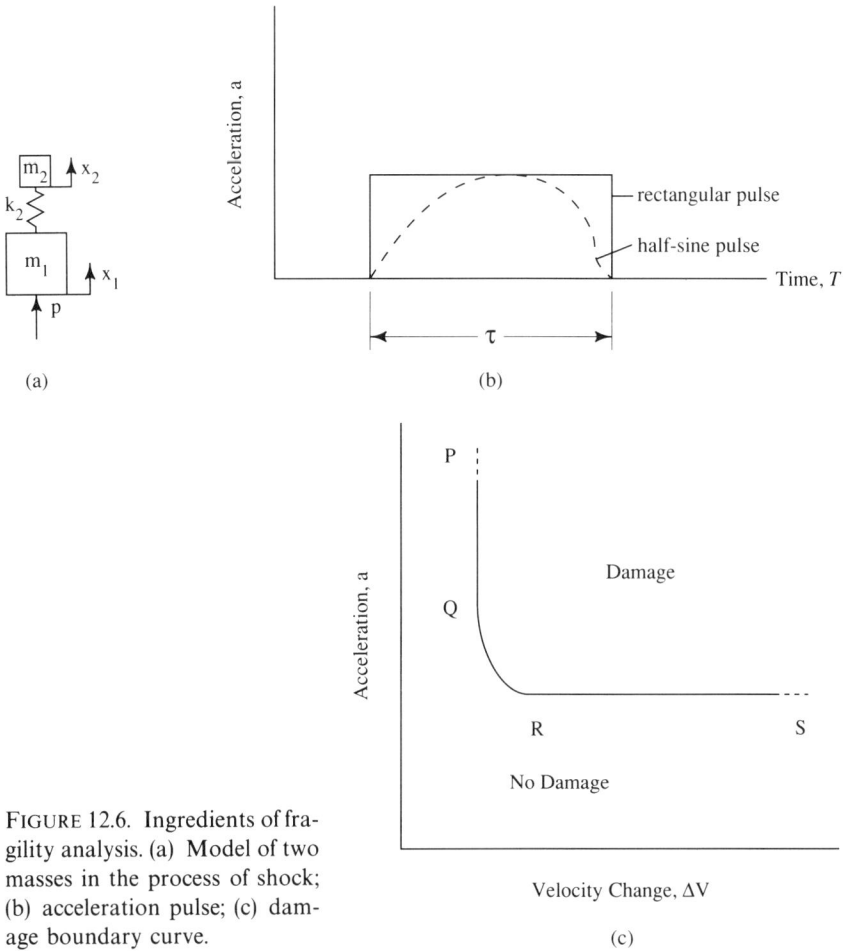

FIGURE 12.6. Ingredients of fragility analysis. (a) Model of two masses in the process of shock; (b) acceleration pulse; (c) damage boundary curve.

Now, introducing $m = m_1 + m_2$, $z = x_2 - x_1$ and $x = (m_1 x_1 + m_2 x_2)/m$, $\alpha = m_2/m_1$, and $\omega^2 = (1 + \alpha)k_2/m_2$, the following differential equations result:

$$\ddot{z} + \omega^2 z = -P/m_1 , \tag{12.6}$$

$$\ddot{x} = -g + P/m . \tag{12.7}$$

The solution for z is, in a drop test where t' is time measured from the start of the drop, and t is the time from the start of the pulse:

$$z = z_0 \cos \omega t' + \frac{m_2(a_1 + g)}{k_2}(\cos \omega t - 1) . \tag{12.8}$$

The fragile component will probably fail due to an excessive spring force $k_2 z$, proportional to the nondimensional load factor

$$n = -k_2 z/g m_2 . \tag{12.9}$$

Drop tests may be conducted applying a programmed constant force P and acceleration $a_1 = d^2x/dt^2$ of the mass center. On the horizontal portion RS of the DBC curve, failure occurs due to an excessive value of n reached before the acceleration pulse ends. It can be shown that this occurs when $(1 + 2a_1/g) < n_{max} < (3 + 2a_1/g)$.

In the vertical segment PQ of the damage boundary, failure occurs due to reaching an extreme acceleration $a_2 = \Delta V \cdot \omega$ of the component after the end of the pulse (3 milliseconds is prescribed by ASTM [19]). This region is associated with small drop heights.

In this procedure, the pulse shape plays a minor role. A major strength of the fragility curve process is that the drop test itself may be simulated by finite element response computations on a few well-chosen configurations [21].

Smearing the mass of components and disregarding their stiffness contribution, Suhir [22] analyzed the dynamic response of rectangular $(a \times b)$ circuit cards, subjected to a uniform constant acceleration applied to the support contour. He considered the boundary conditions either simply supported or fixed and provided for elongated shaped cards as well. Large deformations, with membrane stresses, led to a nonlinear problem.

Linear thin plate theory always overestimates the deflections and bending stresses and is ordinarily deemed conservative, especially when the frequency f_1 can be considered independent of the deflections. However, when deflections are large, the nonlinear frequency may be substantially increased over the corresponding linear frequencies, and the resulting accelerations are increased in proportion to f_1^2. This necessitates nonlinear design for avionic systems, where accelerations over 100 g are common. Suhir's dynamic formulation made use of the following differential equation determining the time-function $z(t)$ of the response:

$$\ddot{z} + \lambda^2 z + \alpha z^3 = q , \tag{12.10}$$

where the displacement $w(x, y, t) = w_c(t) - w_1(x, y) \cdot z(t)$; $w_c(t)$ is the contour displacement, $w_1(x, y)$ the nondimensional shape of the fundamental mode, and $q = c\ddot{w}_c$ is the acceleration with $c = \int_A w_1 \, dS/\int_A w_1^2 \, dS$. The other quantities were as follows, depending on boundary conditions:

Simple Supports	Fixed Supports
$w_1 = \cos\dfrac{\pi x}{a} \cdot \cos\dfrac{\pi y}{b}$	$\cos^2\dfrac{\pi x}{a} \cdot \cos^2\dfrac{\pi y}{b}$
$\lambda = \pi^2 \dfrac{a^2 + b^2}{a^2 b^2} \sqrt{\dfrac{D}{m}}$	$\dfrac{4\pi^2}{3a^2 b^2} \sqrt{[3(a^4 + b^4) + 2a^2 b^2]D/m}$
$\alpha = \dfrac{3\pi^4}{4} \dfrac{D}{mh^2} \dfrac{(3 - v^2)(a^4 + b^4) + 4va^2 b^2}{a^4 b^4}$	$\dfrac{\pi^4 Eh}{18ma^4 b^4}\left[\dfrac{9(a^4 + b^4 + 2va^2 b^2)}{4(1 - v^2)} + \dfrac{17}{8}(a^4 + b^4) \right.$
	$\left. + \dfrac{12a^4 b^4}{(a^2 + b^2)^2} + \dfrac{5a^4 b^4}{(4a^2 + b^2)^2} + \dfrac{5a^4 b^4}{(a^2 + 4b^2)^2} \right].$
$c = \dfrac{16}{\pi^2} = 1{\cdot}621$	$(4/3)^2 = 1.778 \tag{12.11}$

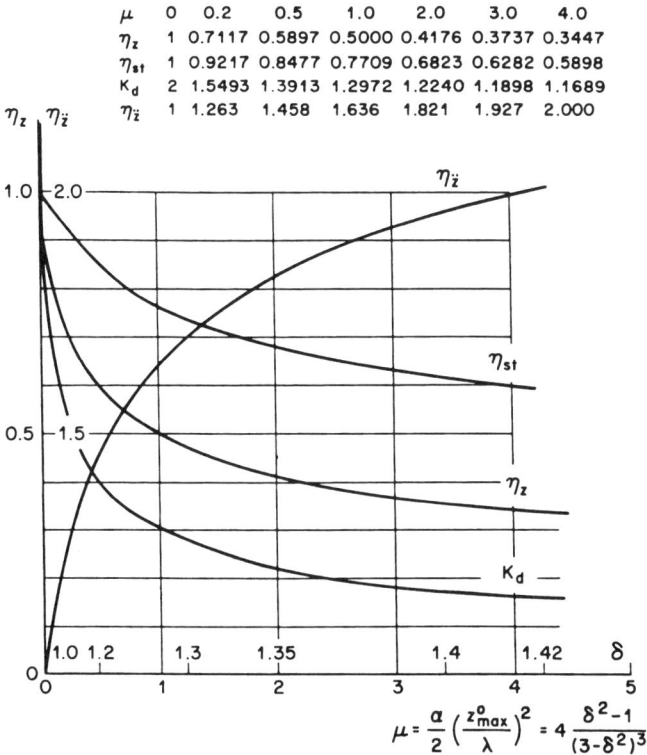

μ	0	0.2	0.5	1.0	2.0	3.0	4.0
η_z	1	0.7117	0.5897	0.5000	0.4176	0.3737	0.3447
η_{st}	1	0.9217	0.8477	0.7709	0.6823	0.6282	0.5898
K_d	2	1.5493	1.3913	1.2972	1.2240	1.1898	1.1689
$\eta_{\ddot{z}}$	1	1.263	1.458	1.636	1.821	1.927	2.000

$$\mu = \frac{a}{2}\left(\frac{z_{max}^0}{\lambda}\right)^2 = 4\,\frac{\delta^2 - 1}{(3-\delta^2)^3}$$

FIGURE 12.7. Factors reflecting the effect of nonlinearity on the dynamic (η_z) and static (η_{st}) displacements on the induced acceleration ($\eta_{\ddot{z}}$) and the dynamic factor (K_d) as functions of the dimensionless parameter μ of nonlinearity. (From Suhir [22]).

Suhir's results for the maximum dynamic and static displacements η_z and η_{st}, respectively (the latter meaning static application of q), the acceleration factor $\eta_{\ddot{z}}$ and the dynamic factor $K_d = z_{max}/z_{st}$, can be viewed in the non-dimensional plots of Fig. 12.7. The analysis permits the mapping of central areas of the board for identification of maximum negative inertia forces tending to cause tension in lead attachments.

5. Fatigue Considerations

Vibration response of circuit card system components may be considered as resulting from sinusoidal (harmonic) or random-type input. There are hybrids, of course, as some specifications superimpose a sinusoidal excitation on top of a random spectrum.

The damage done by n double-sided pulses is a crucial consideration in aerospace structures where high-cycle fatigue is the major delimiter of life.

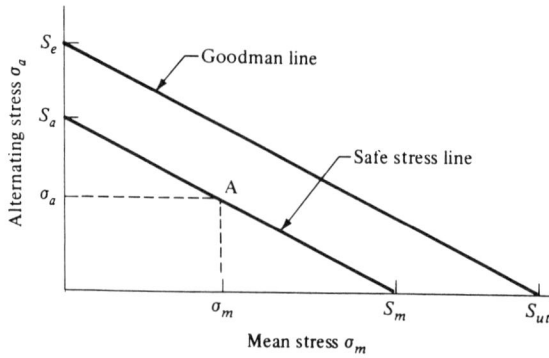

FIGURE 12.8. The "safe stress line" parallel to the Goodman line is a construction for strength prediction in repeated loading. S_e: endurance limit; S_a: $n\sigma_a$; S_m: $n\sigma_m$; n: factor of safety; S_{uf}: ultimate stress.

For structures exhibiting an endurance limit S_e, the designer endeavors to limit the stress under this threshold, thus enabling infinite product life. Standard procedures, such as the Goodman curve [23] (Fig. 12.8), can translate from symmetrical cycles to a nonzero mean stress σ_m, and the alternating stress σ_a is then expressible in terms of the ultimate stress S_u and S_e for the same life; the resulting equation is

$$\sigma_a/S_e + \sigma_m/S_u = 1 . \tag{12.12}$$

This is valid for modules, their leads, circuit lines, and connectors subjected to n cycles of vibration of the circuit card or board.

In random vibration response, there is a need to evaluate not only one but several statistical levels of loading; these are ordinarily taken as multiples of the standard deviation of acceleration response, e.g., 1σ of G_{rms}, 2σ of G_{rms}, 3σ of G_{rms}. The summing up of damage from each loading mode i is done by Miner's criterion [24]:

$$\sum(n_i/N_i) = R \quad (R = 1, \text{ or often, e.g., in avionics, } 0.7) \tag{12.13}$$

In the ith mode of damage, n_i is the number of cycles applied, with N_i being the failure limit in that damage mode alone.

Example

A sensitive component attached to a board has a natural frequency $f_1 = 100$ Hz at which the random vibration environment of the board has a power spectral density $p = 0.25$ g^2/cps normal to the mounting plane. For the specification A of Table 12.3, determine the life of the component in hours, using $R = 0.7$ in Miner's criterion [Eq. (12.13)].

The G_{rms} response is calculated [25]

$$G_{rms} = (0.5\pi p f_1 Q)^{1/2} \tag{12.14}$$

where Q, the magnification at resonance, is $(f_1)^{1/2}$. Thus

$$G_{rms} = [0.5\pi(0.25)(100)(10)]^{1/2} = 19.8 .$$

Solving the following fatigue equation [4] with $\beta = 6.4$, and based on a fatigue life of $N_b = 20 \times 10^6$ cycles:

$$N_a = N_b(G_b/G_a)^\beta \tag{12.15}$$

we get, for 1σ load level, the number of life cycles:

$$N_1 = 20 \times 10^6 (19.8/11.2)^{6.4} = 7670 \times 10^6 \text{ cycles}$$

For 2σ load:

$$N_2 = 20 \times 10^6 (19.8/2 \times 11.2)^{6.4} = 9.08 \times 10^6 \text{ cycles} ,$$

and for 3σ load:

$$N_3 = 20 \times 10^6 (19.8/3 \times 11.2)^{6.4} = 0.678 \times 10^6 \text{ cycles} .$$

In Gaussian distributions, 1σ occurs with a 68.3% frequency, 2σ with 27.1%, and 3σ with 4.33%. Therefore we get

$$n_1 = 0.683 \times 100 \times 3600 \, h = 0.2456 \times 10^6 \, h \text{ cycles}$$

$$n_2 = 0.271 \times 100 \times 3600 \, h = 0.09756 \times 10^6 \, h \text{ cycles}$$

$$n_3 = 0.0433 \times 100 \times 3600 \, h = 0.01559 \times 10^6 \, h \text{ cycles} .$$

Thus the number of hours, h, is computed from Eq. (12.13):

$$h(0.2456/7670 + 0.09756/9.08 + 0.01559/0.0678) = 0.7 .$$

A component life of $h = 2.9$ hours results. It is apparent that the 3σ load has an overwhelming contribution, and one could have simply ignored the 1σ and 2σ levels.

A combined vibrational and thermal solder joint criterion was proposed by Barker et al. [26] using contributions of thermal (th) and vibrational (v) damage in Miner's criterion:

$$R = D_{th} + D_v = (n_{th}/N_{th}) + (n_v/N_v) . \tag{12.16}$$

The thermal cycles are much less frequent than vibrational ones, so the vibration damage is expressed in terms of the ratio f_v/f_{th}, for n_{th} thermal blocks:

$$D_v = (n_{th}/N_v)(f_v/f_{th}) . \tag{12.17}$$

The total combined damage is

$$D_T = D_v + D_{th} = n_{th}([f_v/f_{th}]/N_v + 1/N_{th}) \tag{12.18}$$

so that the solder joint life, for combined vibrational and thermal cycles, is expressed by the number of thermal blocks:

$$N_f = 1/D_T . \tag{12.19}$$

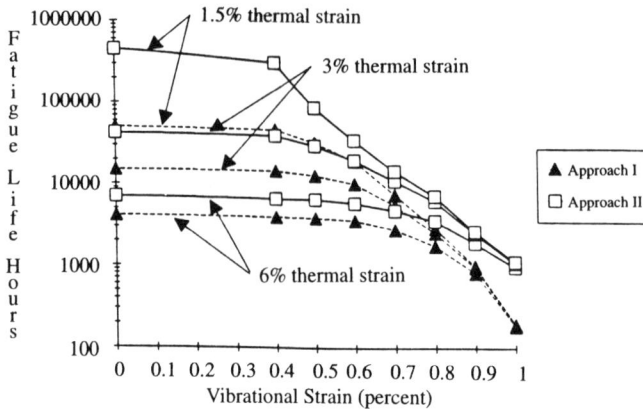

FIGURE 12.9. Vibrational effects on thermal fatigue life (from Barker et al. [26]).

Barker et al. considered two approaches: by one, thermal strain was assumed totally inelastic and the vibrational one totally elastic. By the second approach, the total (elastic and inelastic) strains were included in both thermal and vibrational fatigue prediction. Figure 12.9 shows the respective theoretical projections by both methods.

6. Conclusions

The largest dynamic responses of all applications are expected in avionic environments. Dynamic analysis should determine the excitation to a component mounted onto a circuit card or board; the lead forces, however, can be calculated by static analysis based on the deformed board shape. In the vibration response analysis we may lump or smear the mounted-on components. For shock damage criteria, the boundary damage curve (fragility analysis) can be used, especial since the curve can be generated with the aid of finite element calculations. The shock response of avionics circuit boards should be calculated based on a nonlinear fundamental vibration mode. Random vibration is of outstanding importance for potential fatigue failures.

7. Exercises and Questions

1. How would the static flexing analyses of Chapters 7–10 be utilized for dynamic loading applications?
2. Calculate the in-plane natural frequencies of an Amzirc gull wing lead, based on the flexibility or stiffness matrix given in Section 1 of Chapter 8, and mass distribution as per the data of Problem 11.2.

3. Use Suhir's shock analysis [Eq. (12.11)] to calculate the maximum acceleration of a 25 mm square PLCC module mounted to a 12.5-cm-square, 1.5 mm-thick FR-4 circuit card, the latter simply supported on a frame that is given a half-sinusoid displacement of 1 mm amplitude and 0.01 s duration.
4. What are reasonable dynamic qualifying methods for circuit card systems in electronics packaging production?

References

1. Steinberg, D.S. (1988), *Vibration Analysis for Electronic Equipment*, John Wiley, New York.
2. Skinner, D.W., and Zable, J.L. (1978), "The Business Machine Vibration Environment," *Proc. Inst. Env. Sci.*, pp. 1–8.
3. Frey, R.A. (1991), "The Vibration and Shock Environment for Commercial Computer Systems," *Proc. Inst. Env. Sci.*, Los Angeles, Calif., pp. 658–662.
4. Steinberg, D.S. (1989), "Tougher Tests for Military Electronics," *Machine Design*, May 25, pp. 105–109.
5. Harris, C.M., and Crede, C.E. (1976), *Shock and Vibration Handbook*, McGraw-Hill, New York.
6. Ewins, D.J. (1984), *Modal Testing: Theory and Practice*, John Wiley, New York.
7. Drain, L.E. (1980), *The Laser Doppler Technique*, John Wiley, New York.
8. Pitarresi, J.M., Caletka, D.V., Caldwell, R., and Smith, D.V. (1991), "The 'Smeared' Property Technique for the FE Vibration Analysis of Printed Circuit Boards," *ASME J. Elec. Packag.*, **113**(3), 250–257.
9. Pitarresi, J.M., and Primavera, A.A. (1992), "Comparison of Modeling Techniques for the Vibration Analysis of Printed Circuit Cards," *ASME J. Elec. Packag.*, **114**(4), 378–383.
10. Ewins, D., He, J., and Lieven, N. (1988), "A Review of the Error of Matrix Method (EMM) for Structural Dynamic Model Comparison," *Proc. Int'l Conf. by the European Space Agency*, Noordwijk, The Netherlands, pp. 1–8.
11. Pitarresi, J.M., and Di Edwardo, A.V. (1991), "Systematic Improvement of Support Locations for Vibrating Circuit Cards," ASME Winter Annual Meeting, Paper No. 91-WA-EEP-34.
12. Wong, T.L., Stevens, K.K., and Wang, G. (1991), "Experimental Modal Analysis and Dynamic Response Prediction of PC Boards with Surface Mount Electronic Components," *ASME J. Elec. Packag.*, **113**(3), 244–249.
13. *Structural Dynamics Modification 6.0 Operation Manual* (1984), Structural Measurement Systems Inc., San Jose, Calif.
14. Lau, J.H., and Keely, C.A. (1989), "Dynamic Characterization of Surface Mount Component Leads for Solder Joint Inspection," *IEEE Trans.*, **CHMT-12**(4), 594–602.
15. Snowdon, J.C. (1966), "Vibration of Cantilever Beams to which Dynamic Absorbers are Attached," *J. Acoust. Soc. Am.*, **39**(5), 878–886.
16. Keltie, R.F., and Ozisik, H. (1990), "Transient Structural Response of Built-Up Mechanical Systems," ASME Winter Annual Meeting, Dallas, Tex.
17. Wylie, C.R., Jr. (1960), *Advanced Engineering Mathematics*, McGraw-Hill, New York.

18. Newton, R.E. (1989), "The Damage Boundary Revisited," ASME Winter Annual Meeting, San Francisco, Calif.
19. ASTM Specs. D3332-74, "Mechanical Shock Fragility of Products," Philadelphia, Pa.
20. Kornhauser, M. (1954), "Prediction and Evaluation of Sensitivity to Transient Accelerations," *J. Appl. Mech.*, **21**(4), 371–380.
21. Caletka, D.V., Caldwell, R.N., and Vogelmann, J.T. (1990), "Damage Boundary Curves: A Computational (FEM) Approach," *ASME J. Elec. Packag.*, **112**(3), 198–203.
22. Suhir, E. (1992), "Nonlinear Dynamic Response of a Flexible Thin Plate to Constant Acceleration Applied to Its Support Contour, with Application to Printed Circuit Boards, Used in Avionic Packaging," *Int. J. Solids Struct.*, **29**(1), 41–55.
23. Shigley, J.E., and Mischke, C.R. (1989), *Mechanical Engineering Design*, 5th ed., McGraw-Hill, New York.
24. Miner, M.A. (1945), "Cumulative Damage in Fatigue," *J. Appl. Mech.*, **12**(3), A159–A164.
25. Crandall, S.H. (1958), *Random Vibration*, Technology Press, John Wiley, New York.
26. Barker, D., Vodzak, J., Dasgupta, A., and Pecht, M. (1990), "Combined Vibrational and Thermal Solder Joint Fatigue – A Generalized Strain Versus Life Approach," *ASME J. Elec. Packag.*, **112**(2), 129–134.

Chapter 13

Plated Holes in Cards and Boards

1. Introduction

Plated-through holes (PTH) provide a current path between internal and/or external conductive patterns across the depth of a multilayer interconnection card or board (MIB). A (programmable) via connects two inner-plane (IP) copper layers for signal conduction (Fig. 13.1). Boards are becoming thicker ($h > 5$ mm), and hole diameters are getting smaller ($d < 0.5$ mm) [1], as

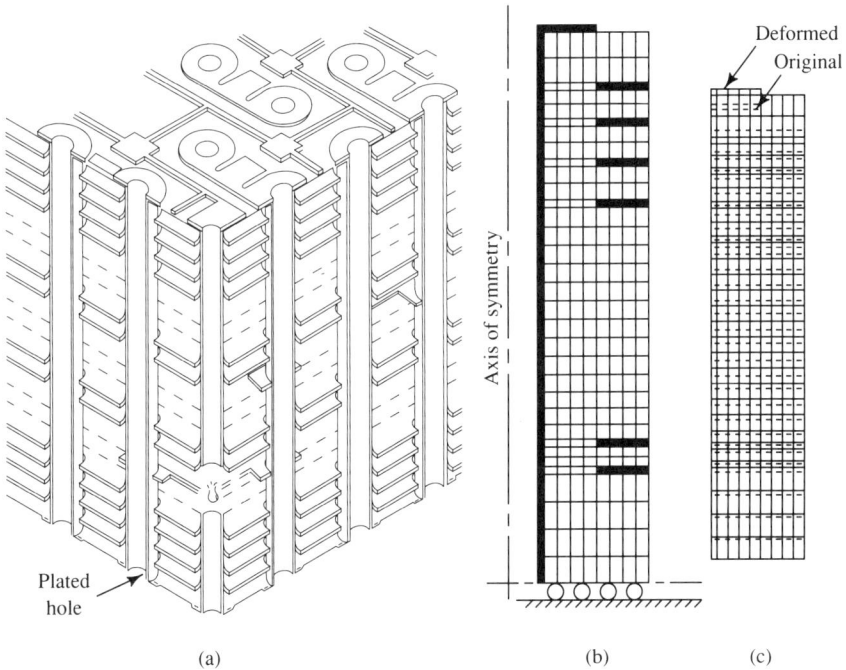

FIGURE 13.1. Circuit board and plated hole structure. (a) Pictorial view; (b) finite element model; (c) finite element plot.

a trend in industry. This growth of the aspect ratio h/d is also putting higher stresses on the structure while presenting a great challenge to the high-quality production (drilling and plating) of a PTH. Seraphim [2] describes the technology developed for the IBM 3081 circuit board.

In the MIB, consecutive cores of copper planes and reinforced polymer dielectric (e.g., epoxy glass, polyimide, Kevlar) are laminated under high pressure and temperature. While the thickness and dielectric constant must be kept small for spacial economy and for higher propagation speeds, the polymer must also be suitable for load carrying.

MIBs must have adequate dimensional stability, allowing accurate locating of the holes. The delicate multipart mechanical drilling operation [3] is subject to drill wobble and wander problems, which may highly stress the drill and leave residual stress in the board. Flank wear [4] forces frequent exchange of drill bits. The thermal aspects and measurement methods of laser drilling are discussed in Ref. [5].

Holes are lined by an approximately 25-μm (1-mil) copper sleeve or barrel, by the electroless or electroplating method. Lands or pads on both sides of the hole and at some places along the hole are provided. Lands would usually have the thickness of the barrel; signal and power planes are, as a basic or frequent dimension, 35 μm (1.4 mil = "1 ounce Cu") thick. PTHs are regularly filled by solder for a connection to a lead (pin) or soldered connector. Cyclic operation is ordinarily between 20° and 60 °C, but soldering temperatures can reach 225 °C. During the latter, a rather small temperature lag of 10 °C is estimated between PTH and board.

The prominent loading sources on PTHs are: (1) module versus card mismatch; (2) z-direction mismatch of sleeve versus laminate; and (3) mechanical connector loads, such as from compliant connectors [16]. Only the two first loadings, both thermal in origin and often associated with pin-grid arrays (PGAs), will be discussed here.

2. Thermal Stress from Module-to-Card Mismatch

The thermal mismatch between a module and the card below was considered as pin loading in Chapter 5. A ceramic module and FR-4 card would make for a relatively large $\Delta\alpha$; a $\Delta T = 100\,°C$ would result in a mismatch $\Delta u = 25\,\mu m$ (1 mil), approximately. The solder joint consists of roughly 45-degree fillets and a barrel. Fillet cracks may easily arise during manufacture and relatively low loading conditions, but solder barrel cracks (Fig. 3.12), when excessive, may constitute mechanical failure.

In an effort to appraise structural performance of the solder joint in several types of PGAs of Kovar, alloy 42, and Amzirc pins, the author performed a reliability study in 1981–82 [6, 7]. A large number of specimens that had undergone 670 accelerated test cycles in an oven at $\Delta T = 75\,°C$ were cross sectioned. In a stress analysis, solder fillets, the braze joint, pin geometry, and pin plasticity were taken into account. Evaluating the maximum solder stress

σ_s [by Eq. (5.25)] for the cracked specimens, a threshold stress σ_a for solder was then deduced: by definition, for a maximum solder pressure less than σ_a, a solder joint would be safe from barrel cracks. In addition, a stress-reduction factor f (< 1.0) was introduced [see Eq. (6.41)] to stand for the stress-reducing effect of module and card bending; f was calculated with the aid of finite element methods for various geometries such as pin, card, and module size; pin spacing; and pin-attachment boundary conditions (Fig. 13.2).

The barrel crack statistical distribution was heavy (but not exclusive) on the corner pins. A numerical value of $\sigma_a = 29.5$ MPa (4280 psi) was obtained for Sn-Pb 63-37 solder, circular pins of 0.4 and 0.5 mm (16 and 20 mil) diameter, and ceramic module sizes ranging from 24 to 37 mm. It can

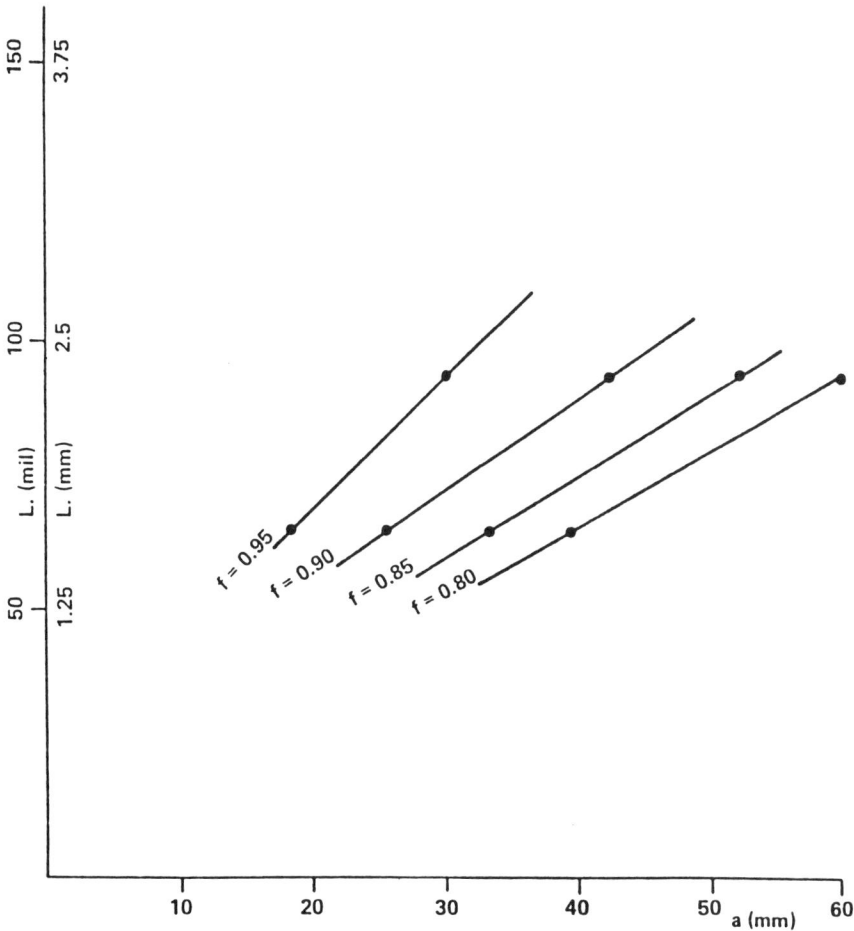

FIGURE 13.2. System reduction factors for 0.4-mm (16-mil) diameter pins of standoff L, spaced 2.5 mm (0.10 in.) in a pin-grid array (from Engel et al. [7]).

generally be said that since thermal mismatch is a displacement constraint, barrel cracks have a stress-relieving effect. By reference to a stress analysis procedure, crack growth could also be incorporated into the design approach.

3. Thermal Stress from Barrel-to-Board Mismatch

3.1. Failure Mechanisms

Epoxy glass and copper are thermally matched in the xy-plane, but in the z-direction, FR-4 has a thermal expansivity $\alpha = 50 \times 10^{-6}/°C$ or more, much higher than that of the copper barrel ($\alpha = 17 \times 10^{-6}/°C$) of the PTH. There are of course several other dielectric materials selected for board applications, Kevlar ($\alpha = 5 \times 10^{-6}/°C$) and polyimide ($\alpha = 50 \times 10^{-6}/°C$) among them; few, however, have the strength or other combined desirable properties of FR-4.

The transverse mismatch gives rise to the following major structural failure modes: (1) copper barrel tension and compression (σ_z); (2) copper delamination (τ_{rz}); and (3) copper inner-plane and land bending (σ_r).

Two manufacturing operations thermally stressing the PTH are the cooling following plating and, following this, the solder reflow operation when a pin or connector is attached. In an empty solder barrel, σ_z tends to be higher, because the barrel alone holds down the FR-4. A solidified solder mass and pin inside the barrel shares in the load, and σ_z of the PTH is proportionally lower, for the same ΔT.

In addition to the severe manufacturing thermal loads, operational thermal cycles and thermal shocks must be accounted for. Tensile stresses are induced in the barrel at heating, and compressive ones during cooling. An initial void or stress concentration area in the barrel may thin out under tensile stress and under a following compressive cycle, it may buckle.

Polymers usually display a sharp increase of thermal expansion at the glass temperature transition, T_g; for FR-4, this occurs at 110° to 125°C. Figure 13.3a shows a hot-plate measurement of PTH expansion employed by Wild [8] and others. A cross sectioned, thin PTH sample is placed on a hot plate, the latter being mounted on an X-Y micrometer stage. This technique is very useful since it eliminates temperature gradients. By this method, the expansion of the laminate and the PTH barrel could be measured separately and in assembly. Adhesion of PTH to board may also be studied through the temperature range. At the T_g of the board, strains increase at a higher rate, thus sensitivity to soldering increases.

3.2. Structural Modeling

Stress analysis efforts date back to the 1960s [e.g., 9–13]. Oiens's analysis of 1976 [10, 11] used the simple elastic model of a parallel assembly of rods (like

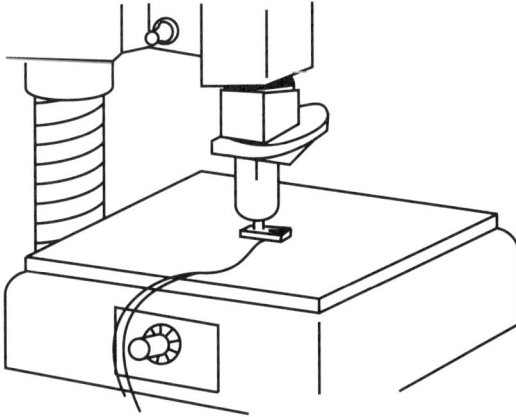

(a)

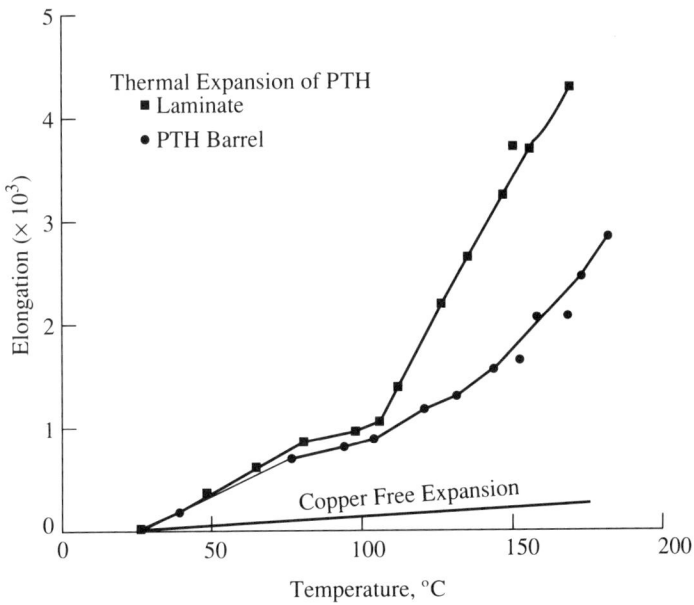

(b)

FIGURE 13.3. Thermal strain measurement in a board. (a) Hot plate experiment; (b) elongations measured for board and barrel (from Lee et al. [17], © 1984 International Business Machines Corporation; reprinted with permission).

Fig. 1.7), standing for the copper barrel and the MIB. Other "rods," such as the solder and the pin, may be added, producing the axial stress σ_z in the ith parallel member as shown in Eq. (13.1); for example, the larger the copper area, the smaller the axial stress it gets.

$$\sigma_{z,i} = \left(\frac{\sum\limits_{n}^{n} E_j A_j \alpha_j}{\sum\limits_{n} E_j A_j} - \alpha_i \right) E_i \Delta T . \tag{13.1}$$

Oien's analysis neglected not only the parallel assembly's interaction along concentric cylindrical interfaces, but also the free thermal strain $\alpha_c \cdot \Delta T$ in the barrel. Then, for a barrel modulus E_b, MIB modulus E_m, average barrel thickness t_b, pad radius r_1, and hole radius r_0, the one-dimensional model for the barrel stress yields the copper stress

$$\sigma_z = \frac{\Delta(\alpha T) E_b}{1 + \dfrac{E_b t_b r_0}{2 E_m (r_1 - r_0)(r_1 + r_0)}} . \tag{13.2}$$

This model was pursued to obtain a conceptual understanding of the interrelationship of the copper barrel and MIB as the temperature is raised in the PTH. The barrel is the higher stressed component at first; it soon reaches the yield stress, losing stiffness. The MIB next undergoes softening as it reaches the glass transition temperature T_g. During the ensuing unloading process, the copper gets stiffer again. Oien also gave a simple model for PTH land rotations.

The axially symmetrical thermal analysis of a simplified (empty barrel plus board) PTH can be performed by the following numerical method of collocation or point matching. The copper barrel, a homogeneous cylinder is defined by its length h, thickness t, outer radius a, and elastic properties E_b, v_b, α_b. The MIB is represented by a slab of thickness h with a hole of radius a, and elastic properties E_m, v_m, and α_m. The board, of course, has orthotropic behavior, and this can be taken into account; for example, the CTE in-plane (α_m) is different from that in the z-direction, $\alpha_{z,m}$. On the board–barrel interface there arise a pair of equal and opposite tractions: the radial pressure $p(z) = \sigma_r(a, z)$ and the shear $q(z) = \tau_{rz}(a, z)$. The midplane of the board is a plane of symmetry where $z = 0$, and the half-length $h/2$ may be divided into n equal segments $\Delta z = h/2n$. Now at the ith segment, discrete pressure elements p_i and shears q_i are prescribed, intended to approximate $p(z)$ and $q(z)$, respectively (Fig. 13.4).

The radial and axial displacements (u, w of the barrel and U, W for the MIB) at any location i (i.e., at halfway of lamella i) due to the tractions p_j, q_j along the barrel (u_i, w_i) and the board (U_i, W_i) can be written as

$$u_i = a_{ij} p_j + b_{ij} q_j \tag{13.3}$$

$$w_i = c_{ij} p_j + d_{ij} q_j \tag{13.4}$$

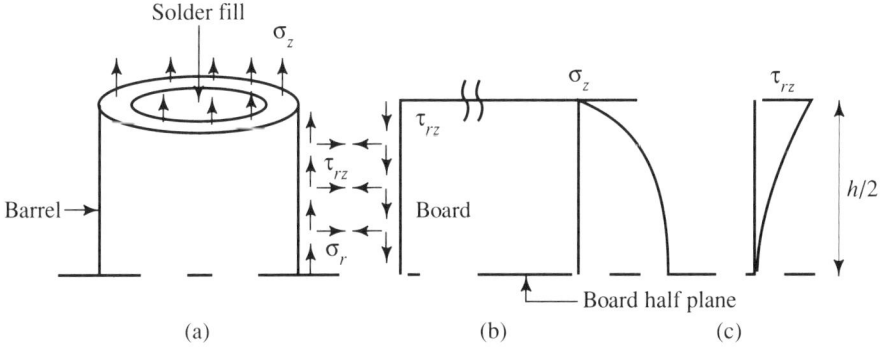

FIGURE 13.4. Basic stress components (σ_z and τ_{rz}) in a plated hole. (a) Stresses on barrel; (b) shear stress on MIB; (c) σ_z and τ_{rz} distributions.

$$U_i = A_{ij}p_j + B_{ij}q_j \tag{13.5}$$

$$W_i = C_{ij}p_j + D_{ij}q_j . \tag{13.6}$$

The free thermal displacements are

$$u_{T,i} = \alpha_b a\,\Delta T, \qquad w_i = \alpha_b \cdot (i - 1/2)\Delta z \cdot \Delta T \tag{13.7}$$

$$U_{T,i} = \alpha_m a\,\Delta T, \qquad W_i = \alpha_{z,m} \cdot (i - 1/2)\Delta z \cdot \Delta T . \tag{13.8}$$

Using linear superposition, the total displacements of all n points along the interface can be written as

$$u_i = \sum(a_{ij}p_j + b_{ij}q_j) + u_{T,i} \tag{13.9}$$

$$w_i = \sum(c_{ij}p_j + d_{ij}q_j) + w_{T,i} \tag{13.10}$$

$$U_i = \sum(A_{ij}p_j + B_{ij}q_j) + U_{T,i} \tag{13.11}$$

$$W_i = \sum(C_{ij}p_j + D_{ij}q_j) + W_{T,i} . \tag{13.12}$$

The compatibility of neighboring points from the two bodies along the interface demands

$$u_i = U_i \tag{13.13}$$

$$w_i = W_i , \tag{13.14}$$

so that $2n$ equations in the $2n$ unknown tractions p_j and q_j result. The eight sets of influence coefficients $a_{ij}, \ldots D_{ij}$ are obtainable from elastic theory; such computations have been performed by Conway and Farnham [14], who treated the axially symmetrical problem of shrink-fitting a sleeve on a cylinder.

Now the axial barrel stress can be written as an integral of the shear tractions $q(z)$; since

$$2\pi a t \cdot \sigma_z(z_i) = 2\pi a \cdot \int_{z_i}^{h/2} \tau_{rz}(z)\,dz,$$

we get for the barrel stress at the mid-plane $z = 0$

$$\sigma_z(0) = \left(\frac{h}{2t}\right)^n \sum_{j=1}^{n} q_j. \qquad (13.15)$$

The solution of this problem yields a monotonically decreasing σ_z barrel stress with a maximum at the midplane $z = 0$ (Fig. 13.4). The shear tractions τ_{rz}, on the other hand, increase away from the midplane and peak at the board surfaces $z = \pm h/2$. Poor adhesion between the barrel and the MIB backing is conducive to delamination due to τ_{rz}; this represents functional risks and greatly increased barrel stress [15].

If there is material (solder) filling out the inside of the copper barrel, then an internal solid cylinder can be added, with another set of pressures and tractions to account for by the preceding analysis. In real constructions, pads may be envisioned displaced by pressure of the bulging polymer; they may have an initial slope from the barrel deformation.

3.3. Finite Element Models

Because of the great complexity of an actual PTH structure, finite element analysis has been found to be an expeditious tool [e.g., 15–21]. In order to economically represent a PTH in polar symmetry, the FE region can be confined to a circular disk of the MIB with a radius equal to half the PTH spacing. At this outer radius the boundary conditions include zero slope $\partial w/\partial r = 0$.

Analyzing a 4.5-mm (0.18-in.) thick MIB of FR-4 and copper, Lee et al. [17] used the ANSYS program with axially symmetric elastic ring elements (Fig. 13.1b). They found a longitudinal barrel stress of 27.6 MPa (4000 psi) per $\Delta T = 10\,°C$ in the central region of an empty barrel 25 µm (1 mil) thick. This stress declined with increasing barrel thickness (Fig. 13.5). Maximum barrel stress at the midplane has been found to increase [15] when an intermittent land was added. Thus, plastic barrel stresses can be expected, and substantial copper ductility is required for manufacturing and cyclic loads.

Bhandarkar et al. [21] investigated the influence of several important design variables upon the critical stresses; they used both axisymmetrical and three-dimensional FE analyses. Figures 13.6a and 13.6b compare the σ_z distribution for both FR-4 and Kevlar boards, with two inner planes located on both sides of the midplane and a land on both top and bottom. The presence of the inner plane locally reduced the von Mises stress σ_e in the barrel for FR-4 boards, while for Kevlar boards, σ_e increased. The reason was sought in the in-plane mismatch of Kevlar versus Cu, while none existed between FR-4 and Cu. Nevertheless, the inner planes did not drastically alter the σ_z distribution with either board material used. Barrel stress increased with higher aspect ratios h/d. The barrel stress somewhat increased with increased PTH separation/hole radius ratio, but leveled out around a ratio of 8.5.

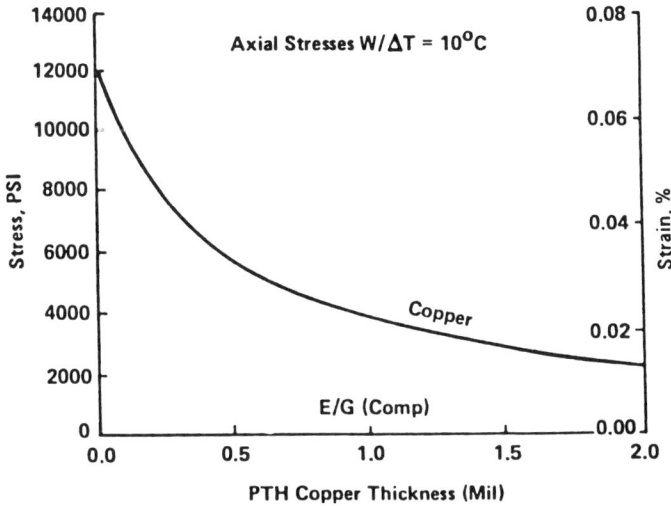

FIGURE 13.5. Effect of copper thickness on axial stress (from Chen et al. [18]).

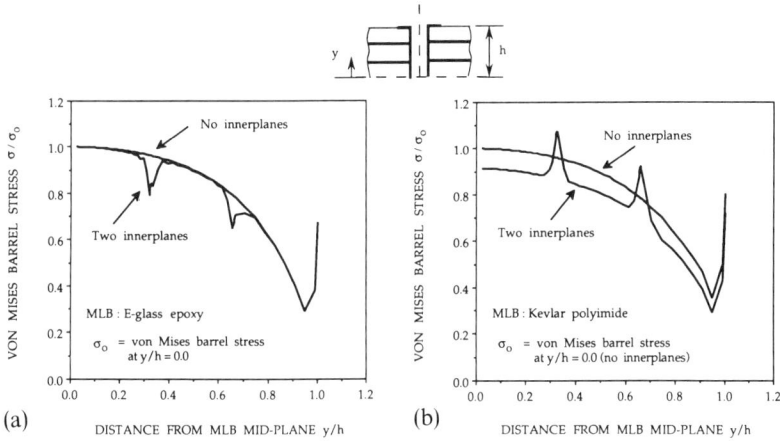

FIGURE 13.6. Longitudinal barrel stress distribution along the PTH (a) FR-4 board; (b) Kevlar polyimide board (from Bhandarkar et al. [21]).

Lau et al. [20] used a finite element program with three-dimensional orthotropic elasto-plastic elements for solving several configurations of a soldered electroplated copper PTH, Fig. 13.7. The solder, made of eutectic Sn/Pb 63–37 alloy, was modeled with a bilinear stress-strain curve, with $E = 10\,\text{GPa}$ (1.5 Mpsi), $\sigma_y = 8.3\,\text{MPa}$ (1200 psi), and strain hardening parameter $n = 0.1$. Three lands were included: one each at the ends, and one at the middle. For the configuration of missing fillet, a maximum plastic shear

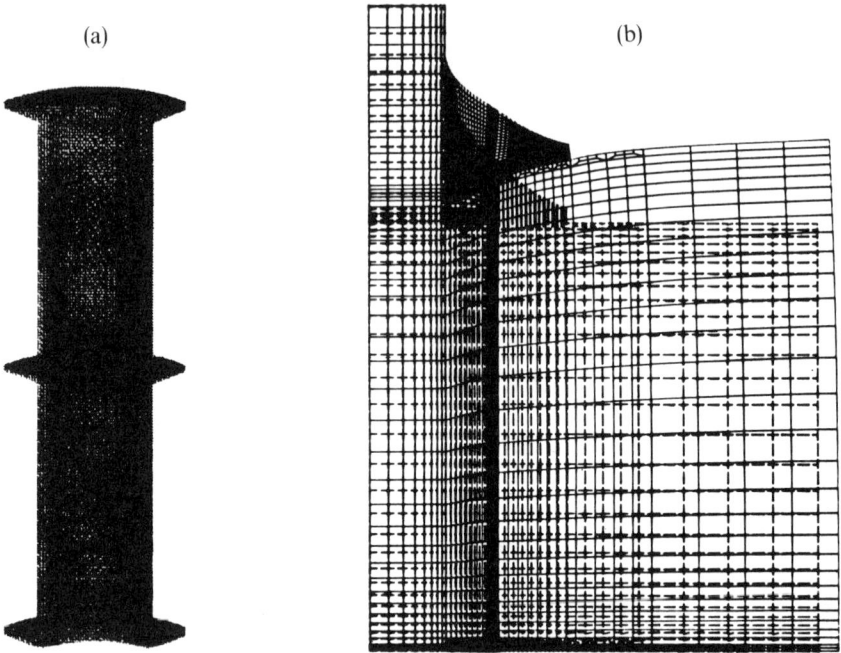

FIGURE 13.7. Finite element analysis. (a) Solid model of PTH copper; (b) displacement of the PGA assembly (from Lau et al. [20]).

strain $\gamma_p = 0.021$ was found, while for a full fillet, $\gamma_p = 0.05$ at the inside corner with the pin. If solder was not present ("cracked off") from the top half of the barrel, then the maximum γ_p was 0.019. Lau et al. also estimated the fatigue life of all three types of joints based on Solomon's theory [22]. They did not base their fatigue calculation on the maximum plastic shear strain γ_p (which would be too conservative), but rather on a reasonable estimate of the crack size. Accordingly, the fatigue lives were 4500, 3400, and 1200 cycles, respectively, at $\Delta T = 100\,°C$. Figure 13.8 shows plastic shear strain contours γ_{rz}, in the vertical plane.

Barker et al. [23] analyzed transient heat conduction and the ensuing thermal stresses caused by wave soldering using finite element methods. They found sharper thermal gradients to exist without inner planes, tending to increase stresses in the copper barrel.

3.4. Copper Inner-plane Bending

The trend for barrel stress is often contrary to inner-plane (IP) bending stress. Barrel stress is greatest in the middle of the board; on copper inner-planes or pads, bending gets more severe for those located nearer the surface of the MIB, in the region where barrel-to-board load transfer has a higher rate. When the PTH is filled out (by solder and pin), IP bending gets more severe

FIGURE 13.8. Plastic shear strain in the transverse direction for solder joint of PGA (from Lau et al. [20]).

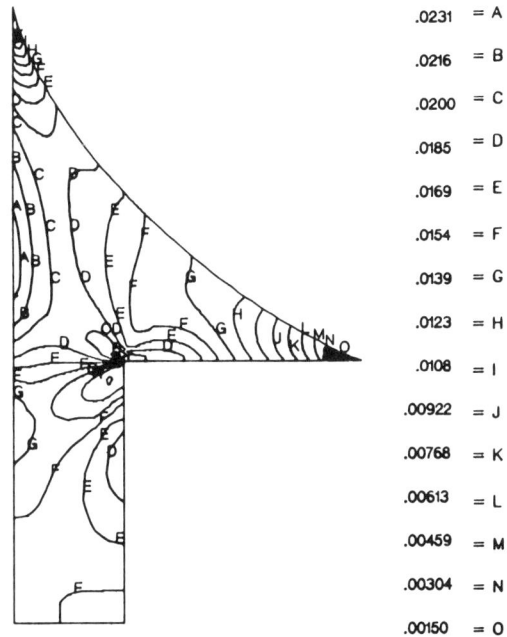

.0231	= A
.0216	= B
.0200	= C
.0185	= D
.0169	= E
.0154	= F
.0139	= G
.0123	= H
.0108	= I
.00922	= J
.00768	= K
.00613	= L
.00459	= M
.00304	= N
.00150	= O

since there is less barrel compliance to rotation. However, soldering itself imposes very severe flexure on the extreme IP. An increase of PTH spacing s is expected to induce more stress on long lands, but this stress converges to an asymptotic value as s gets large.

Mirman [13] modeled several pads as disks, each of variable thickness, attached to a barrel that had variable thickness between lands; he assumed elastic foundation action by the polymer in the inner-plane region. He treated the assembly in polar symmetry, and after substituting the boundary conditions, he obtained the pad loads as a double sum of Bessel functions.

To combat the separation of a hard and brittle land from the copper barrel during severe soldering operations, Wild found doubling the thickness of the extreme lands effective [8]. This minimizes the initial angle of the land and lowers the bending stress. The "etchback" (by chemical or plasma methods) of epoxy glass prior to plating a drilled hole improves the PTH-to-IP connection following plating; it can, however, introduce more stress in the barrel if more than 8 μm (0.3 mil) is etched away.

4. Experimental Methods

The module versus card mismatch has been simulated by Engel et al. [6] through mechanical means; a sliver of the module with a row of pins was sheared, subjecting PTHs to the desired displacement without a temperature effect. Thermal effects of the solder can be added by testing in an oven.

Fundamental experiments on the z-direction mismatch effect were originated by Wild [8, 24], modifying a hot stage metallograph (Fig. 13.3a). The unrestrained thermal deflection of G-10 epoxy glass MIB test specimens and PTHs was optically measured at low and elevated temperatures (Fig. 13.9). Wild used a strain gauge method as well, on button specimens of MIBs; for α measurements, two opposite strain gauges were applied along the barrel, eliminating flexural effects. For modulus measurements, the button specimens were adhered to test anvils, and a strain of 0.0002 applied. In four-hour test cycles of 1.5-mm MIBs between $-60°$ and 125 °C (a much larger ΔT than expected in practice), he induced hysteresis and permanent set. On the PTH, the set was negative (shrinkage), because of buckling of the barrel on the cooling side of the cycle. The MIB took a positive set. Barrel cracking was very dependent on temperature, more so above the T_g of the MIB; the cracks were short interrupted spiraling types. At larger PTH thicknesses (over 40 μm or 0.15 mil), IP cracking increased. Testing for electrical resistance growth, Wild found no electrical opens even after 600 cycles, long after cracking. Wild also used a thermomechanical analyzer (TMA) for strain measurements of PTHs.

Lee et al. [17] also used the hot plate experiment and, in addition, the LVDT displacement sensor, to measure barrel strain on sections of plated

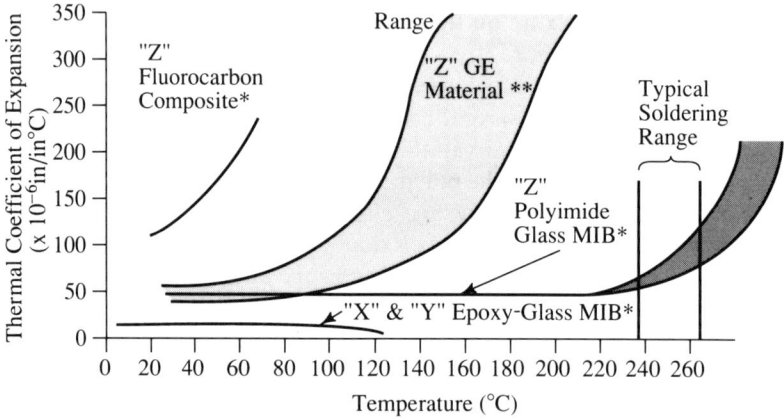

*Note the large expansion differences between the different MIB laminates and stressing directions X,Y, and Z

**General-purpose epoxy-resin-glass fabric

FIGURE 13.9. Board thermal expansion (from Wild [24]).

FIGURE 13.10. PTH peel test. (a) Schematic; (b) typical PTH peel strength data for a 4.6-mm-thick board (from Lee et al. [17], © 1984 International Business Machines Corporation; reprinted with permission).

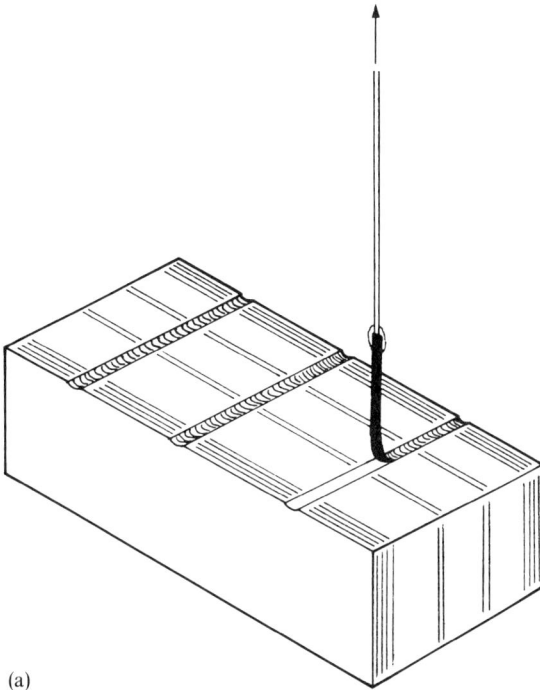

(a)

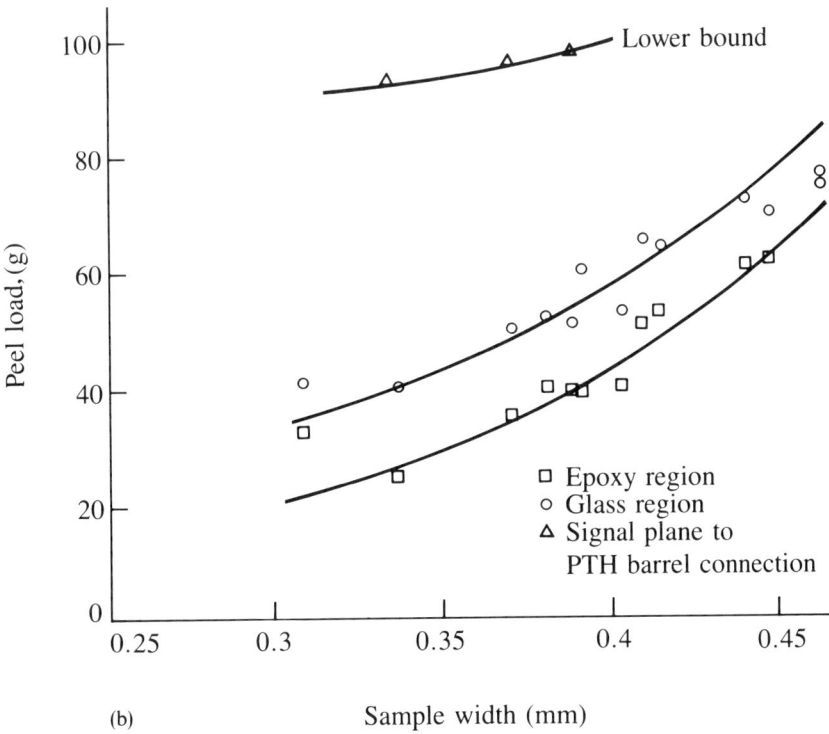

(b)

Sample width (mm)

holes. Studying the bond strength of a PTH, they sought to replace the conventional but unreliable PTH pull test. In the latter, the barrel is pulled out of the board by applying a tensile force to a soldered-in pin. This test does not properly measure PTH adhesion; rather, it depends a lot on hole roughness, which causes scatter in test results.

The PTH peel test of Lee et al. [17] employed 90-degree peel of a strip of the PTH (Fig. 13.10) at slow speeds, such as $v = 0.125$ mm/min. The peel strength was noted to rise slightly with speed. They found good adhesion of their samples, and sudden increases of peel strength at crossing IP connections to the barrel; when glass fibers from the MIBs were crossed, the peel force would also shoot up. Adhesion is not identical to peel strength [25], but some correspondence was established by finite element calculations.

Vecchio and Hertzberg [15] investigated the ductility of electroless copper used as a PTH barrel. They subjected three types of foil (A and B: uncontaminated; and C: intentionally made impure) to the ASTM E345-81 standard tensile test for thin ($t < 0.15$ mm or 0.006 in.) foils. Their result showed only 20% less R_A (percent reduction in area, a measure of strength) in type C than in types A and B, while great scatter in yield stress and elongation values was noted across types A, B, and C. Ductility was found quite satisfactory in electroless copper; it was the rough interface (characterized by sharp 10-µm-deep features) between epoxy glass and copper barrel that was shown to cause inferior strength of the latter. Special drilled rod specimens were made of epoxy and chopped fiber glass plated on the ends and in the drilled hole; these simulated PTH structures and could be subjected to fully reversed test cycles at various strain levels. The intent was to get information of the ongoing delamination and cracking process, by monitoring acoustic emission (AE). In addition, electrical resistance (ER) measurements were made [26], which could detect cracking but not delamination. Using a 2000 µΩ failure criterion, and three discrete nominal copper barrel strains ($\Delta\varepsilon = 0.01$, 0.023, and 0.035%), they derived a Coffin-Manson type failure criterion:

$$\Delta\varepsilon/2 = 0.4(2N_f)^{-0.45} \tag{13.16}$$

where N_f denotes the mean cycles to failure.

Good correlation was found between ER and AE events. The best way to improve fatigue life was found by reduction of the barrel-to-board interface surface roughness.

5. Vias

Programmed vias are little copper buckets bridging two parallel signal planes. Finite element modeling [18] (Fig. 13.11) demonstrated that thermal stresses on the bottom of the bucket switch sign (tension to compression or vice versa) from the corner to the midsection; thus a reliable via requires good adhesion of its bottom plane to the inner plane (IP). In order to test for this

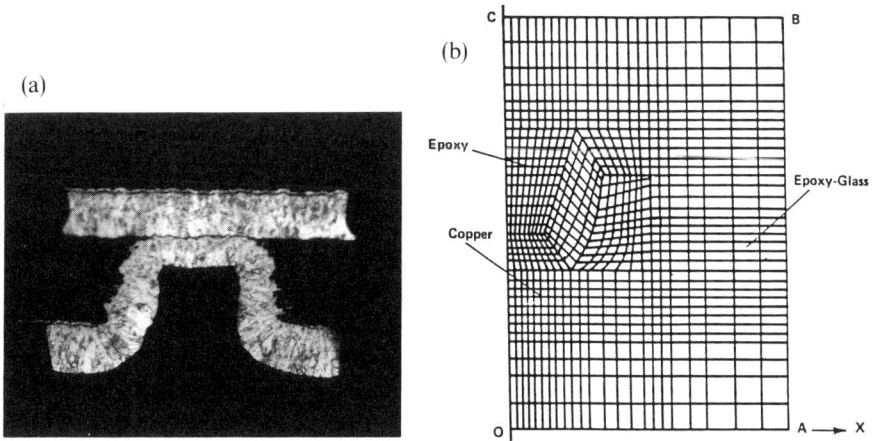

FIGURE 13.11. Programmed via. (a) Photo of via; (b) finite element (ANSYS) plot (from Chen et al. [18]).

adhesion [17], vias with a section of IP to which they were plated were carefully etched out of the MIB and embedded in a molding compound. The IP was then adhered to a pin, and a tensile force applied for a pull test of the assembly. This test proved reproducible, with less than 10% standard deviation, and adaptable to evaluate the quality of various via bonding processes.

6. Conclusions

Plated-through holes have several major thermomechanical failure mechanisms. The module-to-card mismatch induces longitudinal cracks in the tubular solder sleeve joining a pin to the PTH. More dangerous is the z-direction mismatch of board to PTH barrel. This tends to give rise to copper barrel tension and compression, IP bending, and barrel delamination. Vias must possess good adhesion in their bond to inner planes.

7. Exercises and Questions

1. List the potential failure mechanisms of a PTH.
2. Consult the thermal analysis of bonded layers (e.g., Reference [27]) to discuss its similarities with respect to the z-directional mismatch problem of a PTH.
3. What are some of the important finite element modeling features for analyzing (a) a PTH and (b) a programmed via.
4. What experimental facilities are useful for PTH analysis?

References

1. Tummala, R.R., and Rymaszewski, E.J. (1989), *Microelectronics Packaging Handbook*, Van Nostrand, New York.
2. Seraphim, D.P. 1982, "A New Set of Printed-Circuit Technologies for the IBM 3081 Processor Unit," *IBM J. Res. Dev.*, **26**(1), 37–44.
3. Niu, T.M., and Chen, P.C. (1985), "Finite Element Stress Analysis Versus Experiments of Drills Under Bending," *Proc. SEM Spring Conf.*, Las Vegas, Nev.
4. Ramirez, C.N., and Thornhill, R.J. (1992), "Drill wear monitoring in Circuit Board Manufacturing using Drilling Forces and their Spectra," *ASME J. Elec. Packag.*, **114**(3), 342–348.
5. Nowak, T., and Pryputniewicz R.J. (1992), "Theoretical and Experimental Investigation of Laser Drilling in a Partially Transparent Medium," *ASME J. Elec. Packag.*, **114**(1), 71–80.
6. Engel, P.A., Lim, C.K., Toda, M.D., and Gjone, R. (1984), "Thermal Stress Analysis of Soldered Pin Connectors for Complex Electronics Modules," Computers in Mech. Eng., **2**(6), 59–69.
7. Engel, P.A., Toda, M.D., and Trivedi, A.K. (1983), "Design Guide for Solder Cracking in Module-Card Assemblies," IBM Tech. Rep. 01.2678, Endicott, N.Y.
8. Wild, R.N. (1969), "Mechanical Properties of/and Thermal Effects on G-10 Multilayer Interconnection Boards (MIBs)," IBM No. 69-825-2367, Owego, N.Y.
9. Menichello, J.M. (1969), "NFT-F $10'' \times 10''$ Board Analytical Study," IBM No. 69-561-017, Owego, N.Y.
10. Oien, M. (1976), "A Simple Model for the Thermo-Mechnical Deformations of Plated-Through-Holes in Multilayer Printed Wiring Boards," *Proc. 14th IEEE Reliability Physics Symp.*, pp. 121–128.
11. Oien, M. (1976), "Methods for Evaluating Plated-Through-Hole Reliability," *Proc. 14th IEEE Reliability Physics Symp.*, pp. 129–131.
12. Baker, E. (1972), "Some Effects of Temperature on Material Properties and Device Reliability," *IEEE Trans.*, PHP-8(4), 4–14.
13. Mirman, B.A. (1988), "Mathematical Model of a Plated-Through-Hole Under a Load Induced by Thermal Mismatch," *IEEE Trans.*, **CHMT-11**(4), 506–511.
14. Conway, H.D., and Farnham, K.A. (1968), "The Shrink Fit of a Flexible Sleeve on a Shaft," *Intl. J. Mech. Sci.*, **10**, 757–764.
15. Vecchio, K.S., and Hertzberg, R.W. (1986), "Analysis of Long Term Reliability of Plated-Through-Holes in Multilayer Interconnection Boards. Part A: Stress Analyses and Material Characterization," *Microelec. Reliab.*, **26**(4), 715–732; "Part B: Fatigue Results and Fracture Mechanics," *ibid.*, 733–751.
16. Goel, R.P., and Guancial, E. (1980), "Stress Distributions Around an Interference-Fit Pin Connection in a Plated-Through-Hole," *IEEE Trans.*, **CHMT-3** (3), 392–402.
17. Lee, L.C., Darekar, V.S., and Lim, C.K. (1984), "Micromechanics of Multilayer Printed Circuit Boards," *IBM J. Res. Dev.*, **28**(6), 711–718.
18. Chen, W.T., Lee, L.C., Lim, C.K., and Seraphim, D.P. (1985), "Mechanical Modeling of Printed Circuit Boards," *Circuit World*, **11**(3), 68–72.
19. Engel, P.A., and Lee, L.C. (1986) "Surface Solder Stress Design for Thermal Loading of Modules," *Proc. NEPCON East Conf.*, Boston, Mass., pp. 263–270.
20. Lau, J., Subrahmanyan, R., Rice, D., Erasmus, S. and Li, C. (1991), "Fatigue Analysis of a Ceramic Pin Grid Array Soldered to an Orthotropic Epoxy Substrate," *ASME J. Elec. Packag.*, **113**(2), 138–148.

21. Bhandarkar, S., Dasgupta, A., Barker, D., Pecht, M., and Engelmaier, W. (1992), "Influence of Selected Design Variables on Thermo-mechanical Stress Distributions in Plated-Through-Hole Structures," *ASME J. Elec. Packag.*, **114**(1), 8–13.

22. Solomon, H.D., 1989, "Strain-Life Behavior in 60/40 Solder," *ASME J. Elec. Packag.*, **111**(2), 75–82.

23. Barker, D., Pecht, M., Dasgupta, A., and Naqvi, S. (1991), "Transient Thermal Stress Analysis of a Plated Through Hole Subjected to Wave Soldering," *ASME J. Elec. Packag.*, **113**(2), 149–157.

24. Wild, R.N. (1977), "Thermal Characterization of Multilayer Interconnection Boards – Phase II," IPC Conf., Orlando, Fla., IBM No. 77TPA0057, Owego, N.Y.

25. Kim, K.S. (1985), "Elasto-Plastic Analysis of the Peel Test," *T&AM Report No. 472*, Univ. of Illinois at Urbana-Champaign.

26. Rudy, D.A. (1976) "The Detection of Barrel Cracks in Plated Through Holes Using Four Point Resistance Measurements," *14th Proc. IEEE Reliability Physics Symp.*, pp. 135–140.

27. Mirman, B. (1992), "Interlaminar Stresses in Layered Beams," *ASME J. Elec. Packag.*, **114**(4), 389–396.

Chapter 14

Assembly of Cards and Boards

1. Physical Description

As an example for the circuit card and board assembly of a computer, in most of this chapter we shall describe the IBM 9370 card enclosure system [1] and its structural analysis features. The length, height, and depth $(L \times H \times D)$ dimensions of the assembly are $38 \times 30 \times 27$ cm ($15 \times 12 \times 11$ in.). The major structural elements (Fig. 14.1) are 1) the frame, 2) planar boards, 3) card-to-board connectors, and 4) the cards.

The frame, of horizontal U-shape (Fig. 14.2), is a one-piece molded structure, open from front and back. An earlier design of aluminium was replaced by polyester. The planar boards are mounted one on top and one on the bottom of the frame by molded-in threaded inserts. Planars are stiffened around their periphery and by steel bars along the span.

The circuit cards are connected into the unit by zero insertion force (ZIF) connectors fastened to the planars in as many as 23 parallel planes. After insertion "at zero force" into the slot provided, the connector actuator above and the one below, are hand-operated one after the other; the mechanism pushes, while wiping, the tips of tiny ZIF spring connectors against gold- and/or palladium-plated tabs on the card (Fig. 14.3). Sixty-four parallel ZIF connectors in one or two parallel rows on both sides of the card may add up to 256 contacts on top and as many on the bottom.

At actuation, each contact exerts a normal contact force n of approximately 1.75 N (0.39 lb); this generates an appreciable friction force f upon the card tab (Fig. 14.4). These friction forces act in concert upon the planars, tending to deflect them apart. The planar deflection w would detract from the connector wipe u while the latter is vital to assure clean metal-to-metal contacts for electrical conduction. Since providing a minimum of 0.25 mm (10 mil) contact wipe upon actuation of any card was a firm design goal, planar stiffness was a primary requirement.

The laminated copper and epoxy glass circuit cards have a 25×17.5-mm (10×7-in.) size in the $H \times D$ directions when assembled. Card covers are provided both for protection against direct airflow from the computer's fan

FIGURE 14.1. Card enclosure system of the IBM 9370 computer.

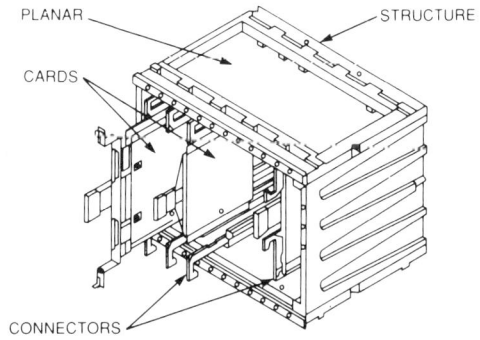

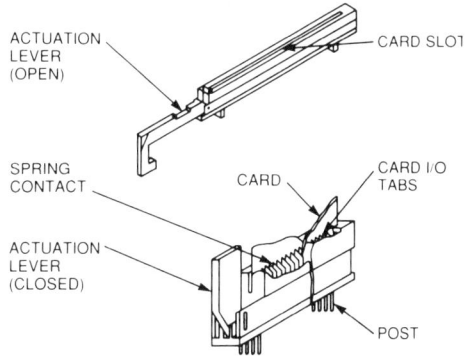

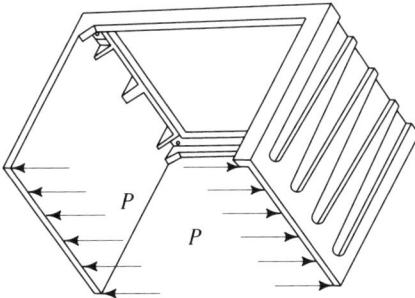

(a)

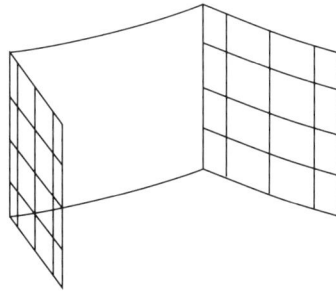

(b)

FIGURE 14.2. Frame stiffness perpendicular to the enclosure wings. The spring rate is 13 N/mm (75 lb/in.). (a) Test scheme; (b) NASTRAN plot.

FIGURE 14.3. Cross-sectional view of ZIF connector system.

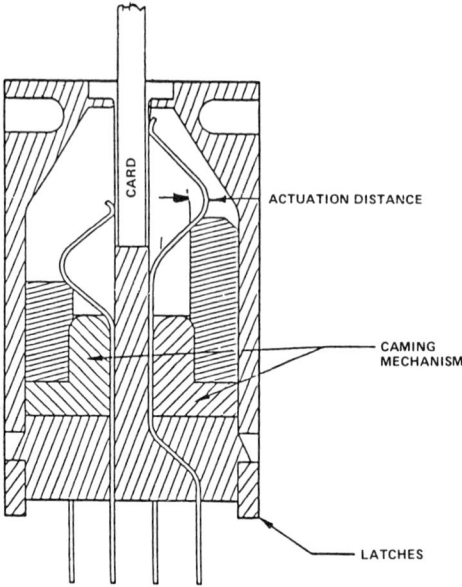

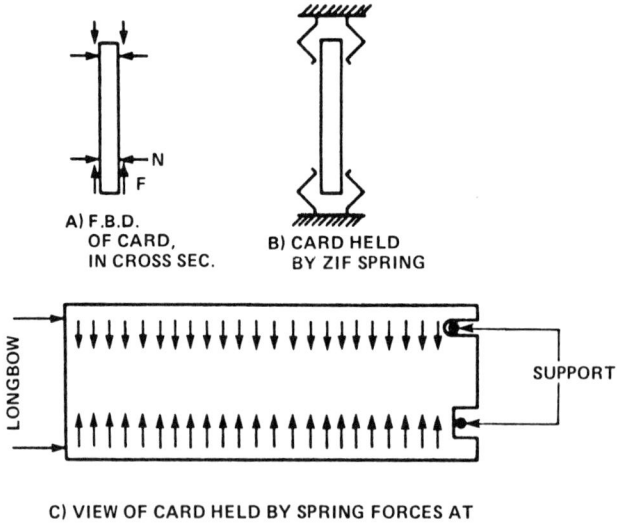

CARD

ACTUATION DISTANCE

CAMING MECHANISM

LATCHES

A) F.B.D. OF CARD, IN CROSS SEC.

B) CARD HELD BY ZIF SPRING

LONGBOW

SUPPORT

C) VIEW OF CARD HELD BY SPRING FORCES AT

FIGURE 14.4. Schematic of the actuation of parallel ZIF spring versus circuit card tabs. (a) F.B.D. of card, in cross section; (b) card held by ZIF spring; (c) view of card held by spring forces at the tabs, and the longbow, pressing the card against supports on the right-hand side (from Engel et al. [3]).

located in the back and to stiffen the card boundaries against bending during
actuation.

2. ZIF Connector Actuation

The free-body diagram of the card during contact actuation is shown in
Fig. 14.4. During actuation, normal forces n and friction forces $f = \mu n$ are
induced on each card tab. Both the tabs and the tiny spherical contact bumps
(these "Hertz dots" have a radius of curvature $R = 0.5$ mm or 20 mil and are
cold-formed on the spring tips) are palladium- or gold-plated multilayer
structures. The tabs are boundary-lubricated [2] to reduce the friction
coefficient μ from a potential 0.6 or, perhaps, higher, to around 0.2–0.3. This
is important for the load on the planar. With lubrication, the concerted
friction forces may exert an $F = \sum f = 0.3 \times 256 \times 1.75 = 134$ Newton (30 lb)
force on the card ends, i.e., on the planars on each side.

The actuation mechanism (Fig. 14.3), the spring stiffness before and during
wipe (Fig. 14.5), and the magnitude of wipe u expected for varying μ are
described in Ref. [3]. Actually, u was not found to be very sensitive to μ, and
the possibility of adherence instead of slip during actuation was ruled out for
real conditions. Two types of analysis were carried out for the ZIF springs.
On one hand, a three-segment cantilever structure was modeled by beam
theory. Another finite element (ANSYS) analysis included large-displacement
friction-combined treatment, where the contact features were represented by
gap elements. A Monte Carlo dimensional tolerance analysis was added, in
view of the sensitive geometrical parameters of the springs.

As illustrated by Fig. 14.4, actuation of the ZIF connectors compresses the
card in its plane by a force F as the connectors slip. Since the actuation
mechanism is attached to the planar, an equal and opposite force F tends to
pry open the enclosure simultaneously, along the card width D. When the

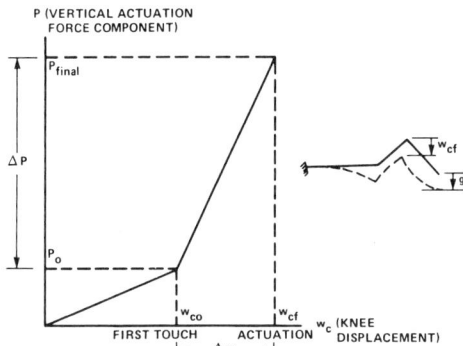

FIGURE 14.5. Graphic relationship between P (vertical component of the actuation
force) and w_c (transverse knee displacement) (from Engel et al. [3]).

card is deactuated for removal, the friction force is less, while slipping under
tension. Figures 14.6a and 14.6b show a simple model of the spring contact
[4]. It includes an extensional spring K and a rotational spring K_ϕ at the
connector spring knee. P_1 is the actuating force, assumed to have only
a transverse component, for simplicity. Compression slip produces the tangential force F_c, and tension slip would require a tensile force F_t:

$$F_{c,t} = \frac{P_1}{1/\mu \mp \dfrac{KL^2 \sin 2\beta}{2(K_\phi + KL^2 \sin 2\beta)}}. \qquad (14.1)$$

From Eq. (14.1), assuming $K_\phi \ll KL^2$, we get the tension/compression slip
ratio

$$\zeta = F_t/F_c = (1 - \mu \cot \beta)/(1 + \mu \cot \beta). \qquad (14.2)$$

For $\beta = 45$ degrees, and $\mu = 0.2$, we get $\zeta = 2/3$, which was experimentally
proven by the Instron load test of a group of 16 springs acting parallel on
a card segment (Fig. 14.7).

During the actuation of a card, the force f of impending friction arises at
each tab contact. The total force on a card $F = \sum f$ is due to compression
slip; it compresses the card while its equal and opposite reaction is prying
open the card enclosure box. Any tendency of the box to return to its original
shape would encounter more frictional resistance in compression slip, i.e.,
more work against the frictional resistance force would have to be done.

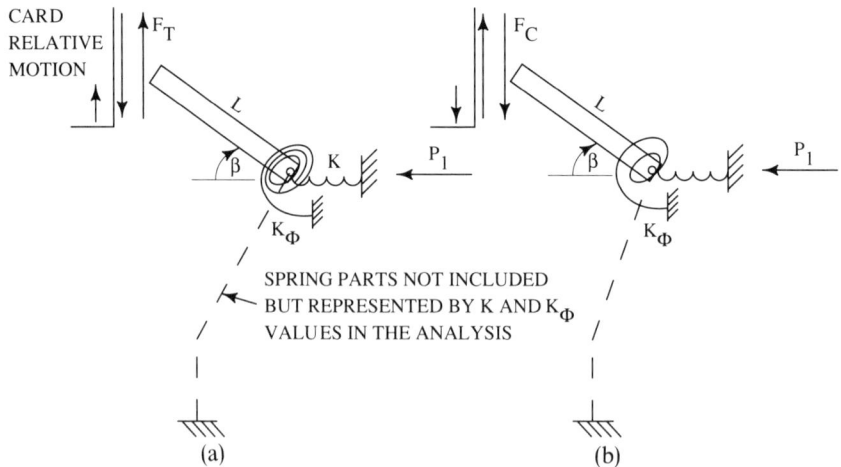

FIGURE 14.6. Idealized model of two possible kinds of slip of connector spring versus
tab. (a) Tension slip; (b) compression slip.

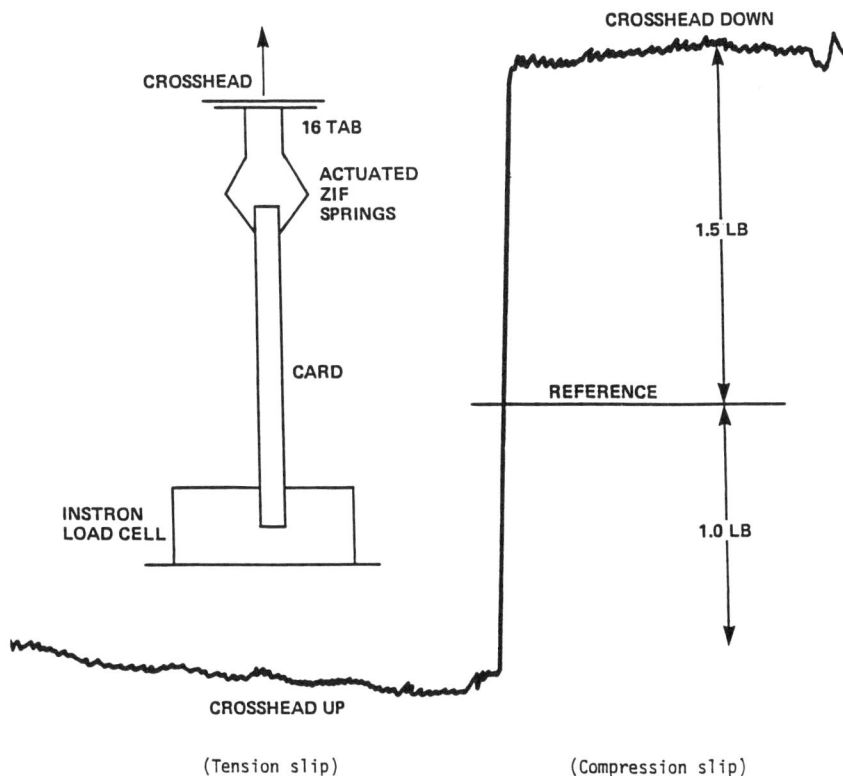

FIGURE 14.7. Experimental determination of tension and compression slip force in the Instron tester. 1 LB = 4.45 N.

Therefore the actuation force F tends to stay on, as we shall see, at least until a neighboring card is actuated.

As a matter of fact, as consecutive bays (as card locations will be called) are actuated ("plugged") (Fig. 14.8), the prior plugged cards tend to lose some of the actuation force induced in them. The explanation is simple: actuating Bay $i + 1$, we pry apart bay i as well, lessening the frictional compression there by a force ΔT_i. The incremental deformation of the prior-plugged bay(s) is not taking place under impending friction, in general. Rather, the prior-plugged joint behaves as if adhering rigidly, hence the incremental tensile force ΔT_i induced in the prior-plugged joint i during actuation of $i + 1$. The current force compressing any bay will then be the algebraic sum of the original actuation force F and the incremental forces $\sum \Delta T$ accrued by the plugging of further bays. Figure 14.9 shows an example: the sequence of the actuation of three consecutive bays. First, Bay 9 was actuated, followed by Bay 8 and then Bay 11.

For the comprehensive analysis of the state of compression of any bay, based on an arbitrary schedule of consecutive plugging events [4], two sets of

Computed Card Forces
(Due to Consecutive Plugging) by NASTRAN

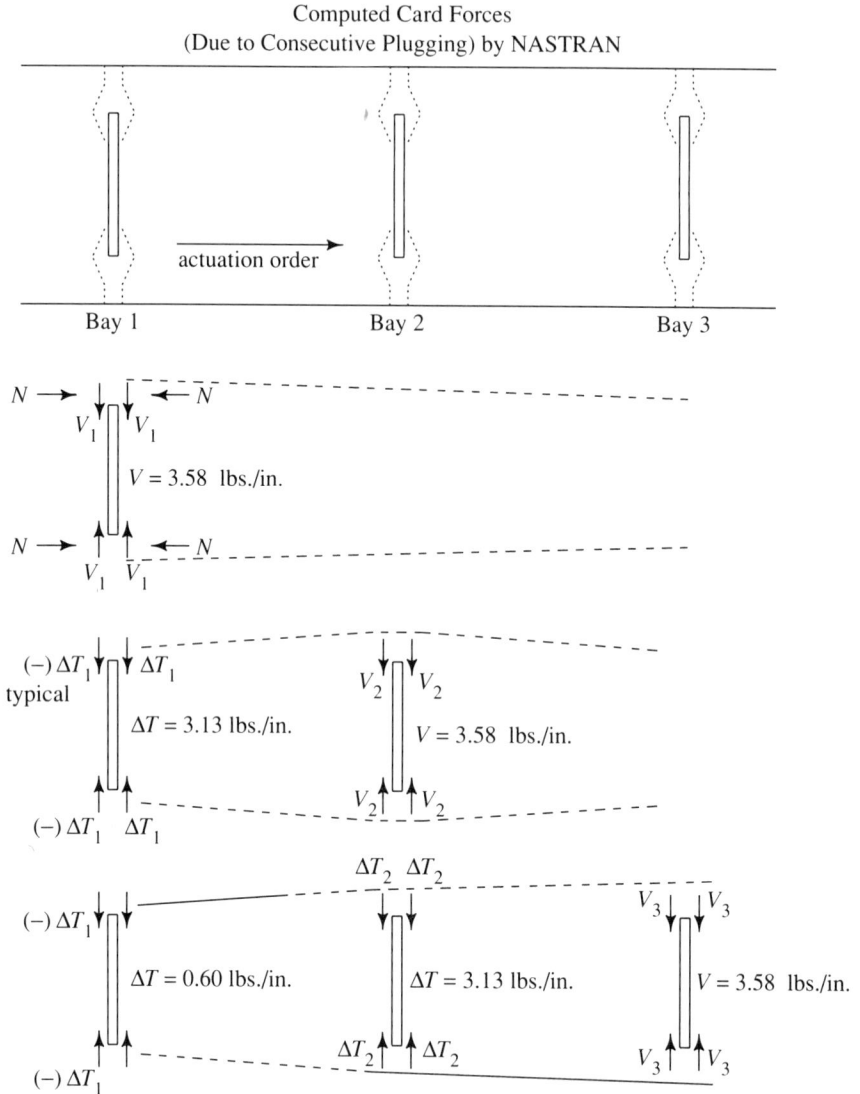

FIGURE 14.8. Example for a NASTRAN calculation of card forces, relaxed by consecutive plugging of new cards. Note V plugging forces and ΔT release forces, shown per unit width of card (from Engel et al. [1]). 1 lb./in. = 0.175 N/mm.

data at each bay i needed first to be determined. The plugging force F_i is a function of the number of spring connectors in that particular bay. Another characteristic quantity, the spring constant k_i of a bay, is defined as the force F_i necessary to pry the bay open to the extent of a unit deflection $w_i = 1$, under condition of adhering (nonslipping) spring contacts. This k_i can be calculated by finite element methods.

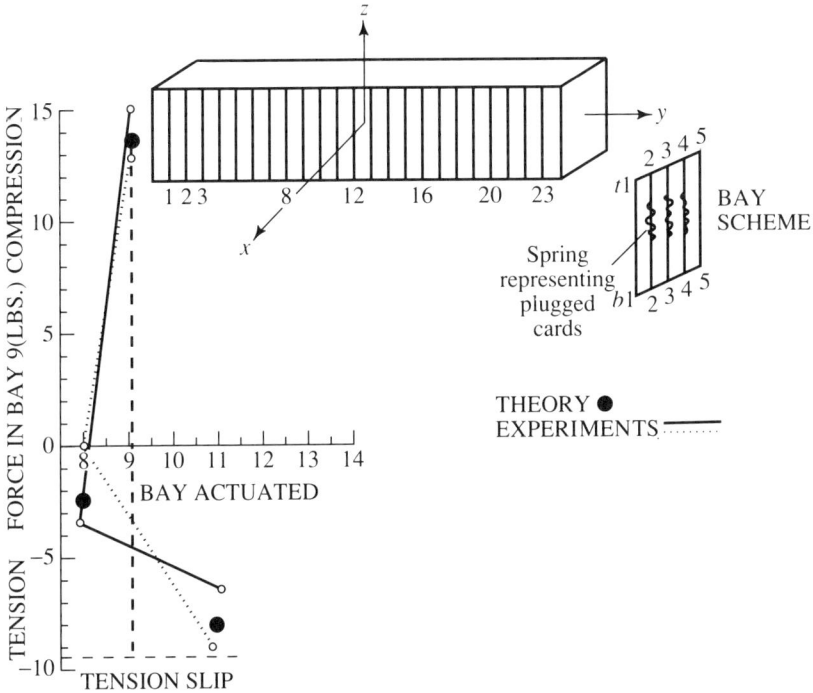

FIGURE 14.9. Card force in Bay 9 due to consecutive plugging of three bays: 9, 8, and 11. 1 lb. = 4.45 N.

A set of influence functions w_{ij} was computed. Such a set, for the displacement w_{ij} ($i = 12$), due to plugging another bay j while one single bay (marked 4, 8, 9, 10, 11) has already been prior-plugged, is shown in Fig. 14.10. The following assumptions were made. 1) After a card has been plugged, it never slips during consecutive pluggings at other bays; 2) After a card has been plugged, it preserves the force induced during its plugging until another bay gets plugged; 3) After a card has been plugged, the plugging of another bay

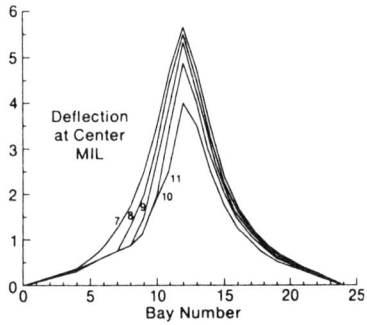

FIGURE 14.10. Influence functions for displacement at bay 12 when an $F = 107$ N (24 lb) insertion force is applied at any other bay shown in the abscissa. 1 mil = 39.4 μm.

may induce in it a force of opposite sign; 4) Maxwell's reciprocal deflection law $w_{ij} = w_{ji}$ is valid.

Based on these assumptions, a bay j that has been "preplugged" will behave as a structural support, the card and connecting springs adding rigidity to the box. When the known compressive force F_i of a new plugging is added at bay i, the influence coefficient w_{ji} allows computation of the incremental tension ΔT_j at bay j from the longitudinal stiffness k_j of the parallel springs at j. Thus

$$\Delta T_j = k_j \cdot w_{ji} \cdot F_i . \tag{14.3}$$

Note that the influence coefficients w_{ij} must be modified when other cards have already been actuated, stiffening the box. The preceding analysis was verified experimentally, by strain gauging of certain cards and performing arbitrary plugging sequences. In Fig. 14.9, the analytically achieved values are represented by dots, and the experimentally measured sequences by connected lines. For the study, cards of various stiffnesses were used. In bays 2, 5, 8, 11, 14, 17, and 20 there were 256 ZIF connectors on both top and bottom, whereas in the rest of the bays (3, 4, 6, 7, etc.) only half as many, 128 ZIF springs, were actuated. This meant different plugging forces at various bays as well, e.g., $F_2 = 2F_3$.

The k values of the central bays were very close to one another, since the prying force is only a weak function of the longitudinal position along the box structure. By the same token, a group of influence functions (such as Fig. 14.10) is likely to be very similar whether the observed bay is 12 (the middle bay) or any other neighboring bay, say i. Thus the numbers 4, 8, 9, 10, and 11 attached to the curves of Fig. 14.10 could be shifted by $12 - i$, and an approximation to the shifted group of influence functions obtained.

In Fig. 14.9, first Bay 9 was actuated, followed by Bay 8 and Bay 11. It is seen that the original compressive force F_9 is turned into a net tension force in two incremental steps, adding to it $\Delta T_{9,8}$ and $\Delta T_{9,11}$.

3. Connector Contact

The spring connector's Hertz dot versus card tab contact is a crucial part of the computer, because over 100,000 of them must work without a glitch through many potential sequences of actuation over years of use. The contacting surfaces of both the spring and the tab are noble or seminoble platings (gold, palladium, or their alloys) deposited on Ni plating on a Cu or Be-Cu base (Fig. 14.11). Electrical conduction is the main function of the surface films. They are required to have sufficient ductility to deform plastically, without cracking, under the normal contact load n. Such plastic deformations ensure a larger real contact area [5], which increases conductance. Maintenance of a wiping distance u on pin mating is also important, to keep away oxides, corrosion and contamination.

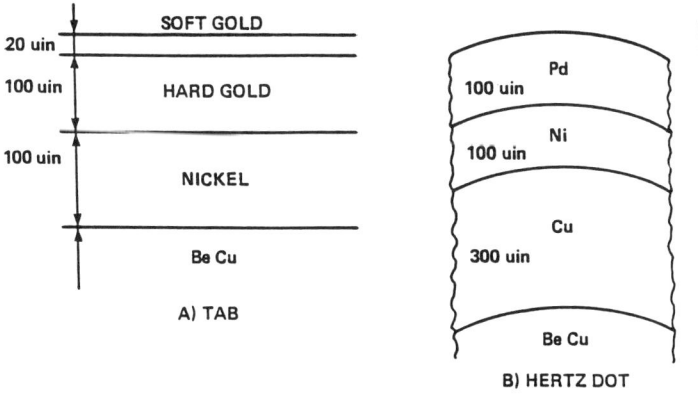

FIGURE 14.11. Sketch of platings for contact partners in the ZIF connector system. (a) Tab; (b) Hertz dot. 100 μin. = 2.54 μm.

Normal force N reduces the constriction resistance R (Fig. 14.12) comprised of plastically deforming asperity contacts on the interface:

$$R = CH/N \qquad (14.4)$$

where C is a constant proportional to the resistivity and H is the hardness of the composite plating. In the meantime, the wear W of the surface film is

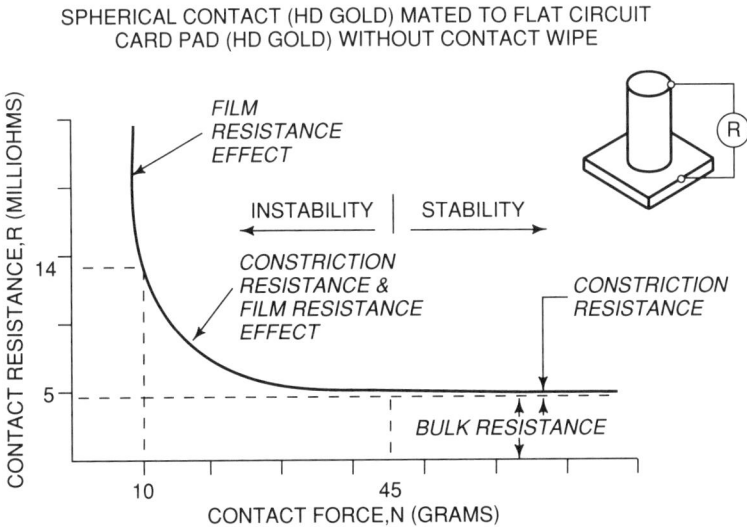

FIGURE 14.12. Contact resistance as a function of normal contact force.

proportional to N; the wear relationship for customary abrasive or adhesive wear mechanisms [6] is

$$W = KNx/H \tag{14.5}$$

where x is the rubbing distance, W is the worn volume, and K is a wear constant for the pair of surfaces. Another aspect entering connector design is the friction force F of plugging or unplugging the connector system, which may add up to a formidable structural load; it is related to N by the Coulomb relationship:

$$F = \mu N . \tag{14.6}$$

It is desirable for engineering considerations to minimize F and W while maximizing N. These severely conflicting requirements may not necessarily be optimized, but a satisfactory compromise for a given design must be reached. Such a compromise may be based on setting a realistic limiting number for the plugging-unplugging operations, dictating the total sliding distance x. With that, and perhaps reducing μ by lubrication, a value N should be found to get the design past the "knee" of the R versus N curve. The usual specification of desirable contact pressure is by prescribing the Hertz pressure [7] for a given contact. Since contacts deform plastically as a matter of course, application of Hertz's elastic contact stress formulas (Table 14.1) is a mere measure of the actual pressure.

Various aspects of the connector design will now be surveyed.

3.1. Contact Forces and Friction Regimes

Figure 14.11 shows one scheme of plated materials on the beryllium copper substrate of both Hertz dot and tab. Be-Cu has the modulus of Cu but a high yield strength. However, even the nominal normal force $n = 1.75$ N, under lubricated conditions, may cause a stubbing phenomenon (a fewfold increase of friction) if the incidence β of the spring (Fig. 14.6) is too shallow. At $\beta < 21$ degrees for the dry and 16 degrees for the lubricated conditions, with a rigidly held spring stem, such a condition could be induced for some metallurgies. The experimental measurement of connector force and wipe using a piezoelectric sensor is described in Ref. [8].

3.2. Electrical Contact Resistance

In an experimental study of factors inducing or affecting electrical contact resistance [9], the following parameters were varied: 1) Hertz dot diameter, 2) lubricant (clean versus Stauffer's CL 920), 3) contamination (e.g., dust or artificial perspiration), 4) rate of actuation (250–800 mm/s), 5) wipe (0–0.75 mm), 6) contact force (50–200 gram), and 7) accelerated aging of card surfaces. The most significant effect was contamination, followed by force and wipe (Fig. 14.13). Normal force, while necessary for good conductance [5], must be controlled to reduce wear.

TABLE 14.1. Spherical contact formulas, by Hertz theory (from Engel [7]).

Contact quantities: P, α, a, q_{max} (Normal force, elastic approach, contact radius, maximum pressure)

Geometric constants: R_1, R_2; $\beta \equiv \dfrac{1}{2}\left(\dfrac{1}{R_1} + \dfrac{1}{R_2}\right)$, combined curvature

Material constants: E_1, v_1, E_2, v_2; $E_r \equiv \left(\dfrac{1 - v_1^2}{\pi E_1} + \dfrac{1 - v_2^2}{\pi E_2}\right)^{-1}$, reduced modulus

Exponent table:

	P	α	a	q_{max}
$P \sim$	1	3/2	3	3
$\alpha \sim$	2/3	1	2	2
$a \sim$	1/3	2/3	1	1
$q_{max} \sim$	1/3	2/3	1	1

e.g., $\alpha \sim P^{2/3}$

Relations:

$P = P,$

$\alpha = \left(\dfrac{9\pi^2 \beta}{8E_r^2}\right)^{1/3} P^{2/3},$

$\alpha = \left(\dfrac{3\pi}{8\beta E_r}\right)^{1/3} P^{1/3},$

$q_{max} = \dfrac{6}{\pi}\left(\dfrac{\beta E_r}{3\pi}\right)^{2/3} P^{1/3},$

$P = \dfrac{8\beta E_r}{3\pi} a^3,$

$\alpha = 2\beta a^2,$

$a = a,$

$q_{max} = \dfrac{4\beta E_r}{\pi^2} a,$

$P = \sqrt{\dfrac{8}{\beta}\dfrac{E_r}{3\pi}}\, \alpha^{3/2}$

$\alpha = \alpha$

$\alpha = \dfrac{1}{\sqrt{2\beta}} \alpha^{1/2}$

$q_{max} = \dfrac{\sqrt{2\beta}\, E_r}{\pi^2} \alpha^{1/2}$

$P = \left(\dfrac{\pi}{6}\right)^3 \left(\dfrac{3\pi}{\beta E_r}\right)^2 q_{max}^3$

$\alpha = \dfrac{\pi^4}{8\beta E_r^2} q_{max}^2$

$a = \dfrac{\pi^2}{4\beta E_r} q_{max}$

$q_{max} = q_{max}$

3.3. Plating Ductility

In the absence of sufficient ductility, the top plating may develop cracks under contact pressure; this is because a radial tension prevails in the outer regions of spherical contacts [7]. Once cracked, the film admits corrosive agents of the environment, leading to eventual electrical failure.

For quality control of essentially Pd-plated ZIF springs, a ductility check by a sharp conical diamond indenter was devised [10]. Indenter tips were

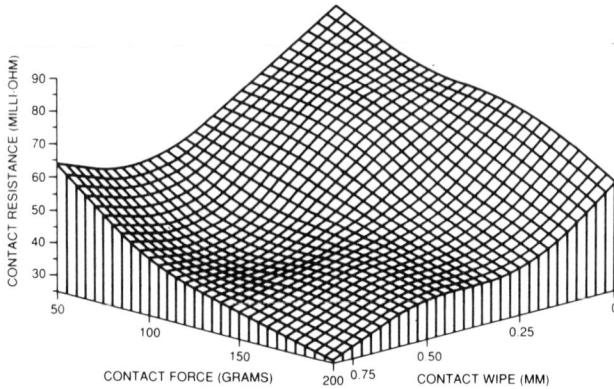

FIGURE 14.13. Three-dimensional plot of contact resistance versus contact force and contact wipe; the card surfaces have been contaminated (from Brodsky [9]).

varied: the cone angle α was between 60 and 90 degrees and the spherical tip radius R was between 25 and 75 μm (1–3 mil). The indentation crater radius a (observed following the indentation) corresponded to an average strain ε (Fig. 14.14), calculated from the extension of an original surface segment of length $2a$ into an arc $2s$:

$$\varepsilon = (s - a)/a. \tag{14.7}$$

This definition neglects friction and adherence of the indented material. Subjected to the cone indentation test, the larger $\varepsilon = \varepsilon_0$ the material would support without cracking, the more ductility it would prove; hence ε_0 was adopted as a measure of ductility. The radius a can be converted into force P, since the hardness $q = P/\pi a^2$ is nearly constant in the force range (> 10 N) of the test. Thus ε_0 versus P curves can be obtained; a certain indentation force P may be tested for proving a ductility goal ε_0.

In the ductility test, slowly pressing the indenter into the spring surface having the same metallurgy as the Hertz dot, the force P was raised until cracks (if any) could be noticed by SEM; a through-crack would expose the Ni underplating to an X-ray check (EDEX). On the empirical basis of desirable connector performance, an acceptable ductility threshold $\varepsilon_0 = 0.6$ was then specified; at least this amount of ductility was available in the absence of a through-crack, upon the application of a limit load, $P = 22.5$ N (5 lb) with an indenter having $\alpha = 60$ degrees and $R = 25$ μm.

3.4. Friction and Wear (Tribology)

While lubrication is crucial for reducing system loads, it can only be effective with a surface of limited roughness [11]. Surface texture is also important for electrical conductivity.

FIGURE 14.14. Indentation test
to measure plating ductility.
(a) Indentation geometry;
(b) strain versus indenting
force.

(a)

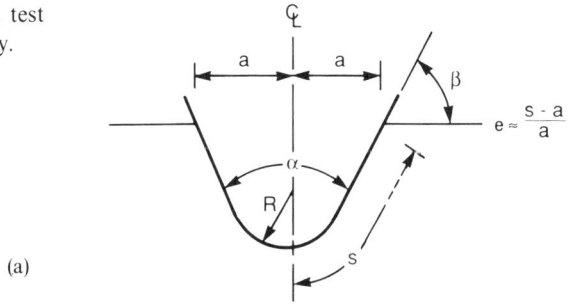

THE RADIAL STRAIN FOR $a > R \sin \beta$ is:

$$e = \frac{R(\beta \cos \beta - \sin \beta) + a}{a \cos \beta}$$

WHERE $\beta = \pi/2 - \alpha/2$; AND FOR $a < R \sin \beta$:

$$e = \frac{\sin^{-1}(a/R)}{a/R} - 1$$

(b)

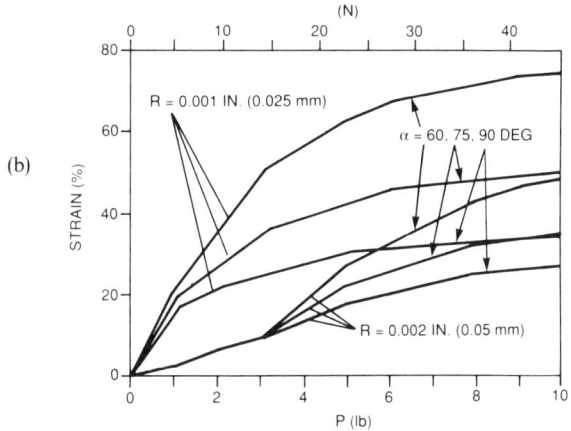

A matrix of systematic experiments on wear behavior concluded the desirability of eliminating soft gold films in favor of hard gold films, when gold was used instead of Pd in ZIF contacts [12]. The wear-resistant role of various plating layers was ranked. For the wear of electrical contacts, several engineering studies and formulas have been produced [13]. It must be recognized that one of the crucial effects of circuit card vibration is the aspect of fretting wear it causes in the contacts [14]: see Example 1 in Section 4.

The Vickers hardness test is an important surface check; for a non-homogeneous material, however, it reflects a great deal of the substrate properties. Reference [15] interprets the superficial Vickers microhardness reading of multilayer plating sandwiches, like those of Fig. 14.11, for given data on the intrinsic hardnesses and thicknesses of the constituent platings.

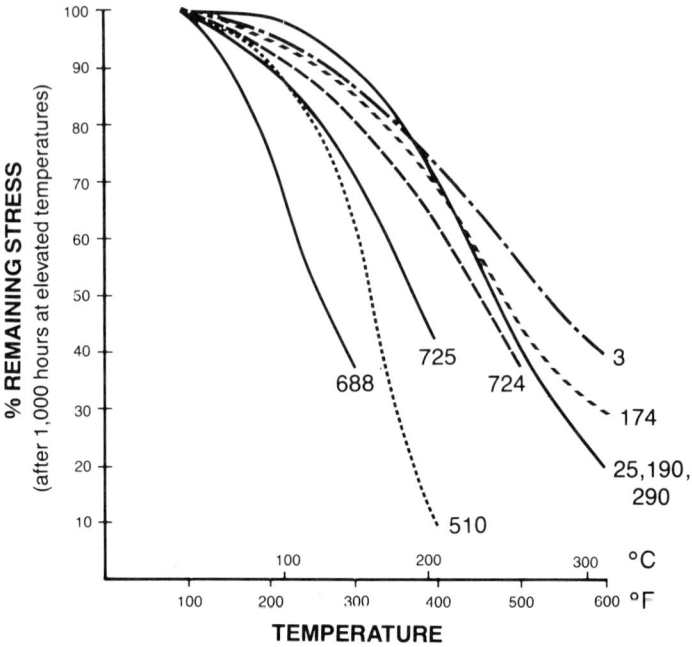

FIGURE 14.15. Effect of temperature on stress relaxation on various alloys used as electrical contacts (from [17]).

3.5. Contact Temperature

Temperature also plays a decisive role in the working of a connector. N will decay to some extent due to stress relaxation [16], and higher temperatures increase this tendency. Figure 14.15 shows the time decay of normal stress for various connector materials, among which Be-Cu is an excellent performer [17]. As N is reduced, R grows; thus the ohmic heat Q generated according to the $Q = IR^2$ law (where I is the current) increases, causing, in turn, more stress-relaxation tendency [18]. This phenomenon may induce instability, resulting in functional failure of the connector system.

4. Dynamic Response

Stationary random vibration response of the total card enclosure structure of Fig. 14.1 was obtained by the modal analysis technique (Chapter 12, Section 2). Figure 14.16a shows the location of accelerometers. The box was subjected to vibration in the x- and z-directions, both lightly loaded (one middle card actuated) and normally loaded (with 10 cards), with each card weighing 8 N (1.8 lbs). In general, the lightly loaded package produced similar but slightly higher responses.

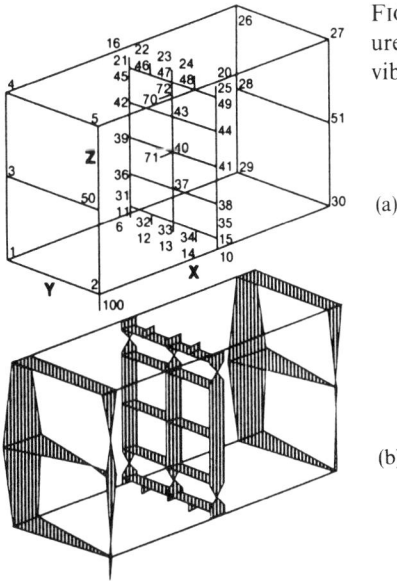

FIGURE 14.16. Modal analysis for card enclosure. (a) Location of accelerometers; (b) second vibration mode shape.

(a)

(b)

For a horizontal (x-direction) 15-g 11-ms half-sine shock, the largest displacement was 2.5 mm. For a steady sinusoidal vibration of 2 g amplitude at the second natural vibration frequency of 109 Hz (Fig. 14.16b), the relative displacement amplitude of point 3 with respect to the mounting point 2 was 1.5 mm. For the vertical shock (z-direction), the maximum deflection, at point 4, was 3.3 mm; the mounting points moved 1.8 mm.

The magnification at resonance Q (at point 3) and natural frequency f_1 for the normally loaded frame were 24.5 and 62.3 Hz, respectively. For the lightly loaded frame the corresponding values were 26.4 and 73 Hz.

The response of actuated cards was monitored by strain gauges and accelerometers while the box was mounted on the shaketable and vibrated in the x- and z-directions. The frequency region from 10 to 300 Hz was swept with sinusoidal vibration; the fundamental natural frequencies, accelerations, and strain responses ε_y and ε_z are shown in Table 14.2.

Dynamic flexure (Fig. 14.17) poses a threat to the external conductors, such as metallized lines of a card or board; an amplitude w on a simple span L causes a bending strain $\varepsilon = 4hw_0/L^2$ in the extreme fiber. In the meantime, the rubbing Δu per cycle of an electrical contact like that of the ZIF spring should also be checked since the volume of wear is proportional to the total tangential excursion u by Eq. (14.5). On a simple span, from the shortening of the card having an approximately sinusoidal deflection $w = w_0 \cdot \sin \pi x/L$, we have

$$\Delta u = \int_0^{L/2} (ds - dx) \approx 1/2 \int_0^{L/2} w'^2 \, dx \;,$$

TABLE 14.2. Vibration response of cards in card enclosure system.

Fundamental natural frequency, Hz	$10^{-6} \times$ strain ($\mu\varepsilon$)		Acceleration		Magnification $Q = \dfrac{\text{Input}}{\text{Output}}$
	z	x	g (input)	g (output)	
Without Cover					
40	360	160	0.55	4.9	8.9
40	125	18	0.20	1.8	9.0
40	420	65	1.00	5.0	5.0
With Cover					
47	260	42	0.55	5.1	9.3
47	70	70	0.20	1.6	8.0
47	360	44	1.00	7	7.0

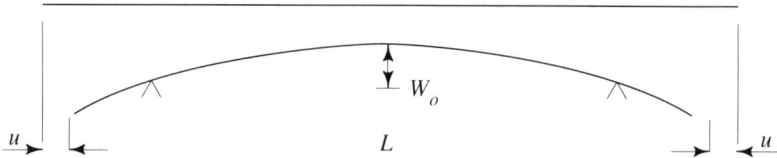

FIGURE 14.17. Flexure of a beam, showing bow w_0 and end shortening u.

resulting in a rubbing distance per quarter cycle of vibration:

$$\Delta u = 2.47\, w_0^2 / L \; . \tag{14.8}$$

Example 1

A ZIF-supported FR-4 card $L = 216$ mm of thickness $h = 1.27$ mm undergoes forced sinusoidal vibration at resonance with a 0.1-g input; it has a magnification at resonance ($f_1 = 33$ Hz): $Q = 6$. We shall find the dynamic strain ε of a line at midspan, the contact slip Δu, and estimate the wear of the contact, assuming that the tab and Hertz dot are of the same material, the top film, hard gold, having hardness $H = 1960$ MPa and wear constant at full sliding $K = 3.7 \times 10^{-4}$.

Assuming "simple supports," the displacement amplitude for a plate rigidity $D = 2328$ N . mm and unit weight 22.2×10^{-6} N/mm^3 is

$$w_0 \cong \frac{(5)(1.27)(22.2 \times 10^{-6})(216^4)(0.1)(6)}{(384)(2328)} = 0.206 \text{ mm} \; .$$

Thus the maximum strain in the extreme card fiber is, from beam theory:

$$\varepsilon_{max} = \frac{(4)(1.27)(0.206)}{216^2} = 22.4 \times 10^{-6},$$

and by Eq. (14.8) there results

$$\Delta u = \frac{(2.47)(0.206^2)}{216} = 0.000485 \text{ mm} .$$

The spring contact area is approximately $A = N/H = 1.75/1960 = 0.000893 \text{ mm}^2$. The contact radius is then $r = (0.000893/\pi)^{1/2} = 0.0169 \text{ mm}$. Since the fretting motion has an amplitude $\Delta u \ll r$, the sliding wear constant K should be proportionally reduced to a factor of $\Delta u/r$, by interpretation of the wear theory of Ohmae and Tsukizoe [19]; thus $K = 3.7 \times 10^{-4} \times 0.000485/0.0169 = 1.06 \times 10^{-5}$. The average wear depth $\Delta W/A$ per vibration cycle (4 times Δu) is, from Eq. (14.5), $\Delta h_w = 4\Delta u K$, that is, for n cycles of vibration the depth of wear expected is

$$h_w = 4nK \Delta u = 1.94 \times 10^{-8} n ;$$

For one hour of vibration at the natural frequency 33 Hz, $n = 33 \times 60^2 = 118{,}800$, so that the wear depth becomes $h_w = 1.94 \times 10^{-8} \times 118{,}800 = 0.0023 \text{ mm}$ (91 μin.). This relatively large amount of wear may force us to stiffen or isolate the card; hence the card covers are beneficial to reduce cyclic slip and wear.

5. Thermal Stress

The card enclosure is basically the assembly of three materials: the polyester frame, epoxy glass planars and cards, and steel planar stiffeners. Thus during computer operation, thermal mismatch is likely to cause stresses between the members. The room temperature CTE of a 22% glass-filled polyester is $\alpha = 22 \times 10^{-6}/°C$, its modulus $E = 9.6 \text{ GPa}$ (1.39 Mpsi), yield strength 47 MPa (6810 psi), and ultimate strength 98 MPa (14,200 psi). These properties change with temperature, and at $80 °C$ a dramatic increase in the CTE and a decline in E are evident; e.g., the CTE becomes $63 \times 10^{-6}/°C$.

The shape of the structure under a temperature rise $\Delta T = 100 °C$ is shown in Fig. 14.18. The sequence shows (a) a cardless box, (b) one central card, (c) five cards, and (d) three cards unsymmetrically located. The finite element NASTRAN plots were computed assuming no slipping on the actuated cards, so the elastic in-plane compliance of the cards is that of the spring connector forces in series with the card. The calculated card forces F_c were compared with the force F of impending friction on the ZIF connectors, to find whether the adhering contact would give way to tensile slip. Assuming constant CTEs and constant moduli, a critical temperature change of 14 °C was calculated, at which slipping may start. This number proved overly conservative, understandably so in view of the temperature-dependent properties of the materials. No slip was experimentally measured under operating thermal conditions [4].

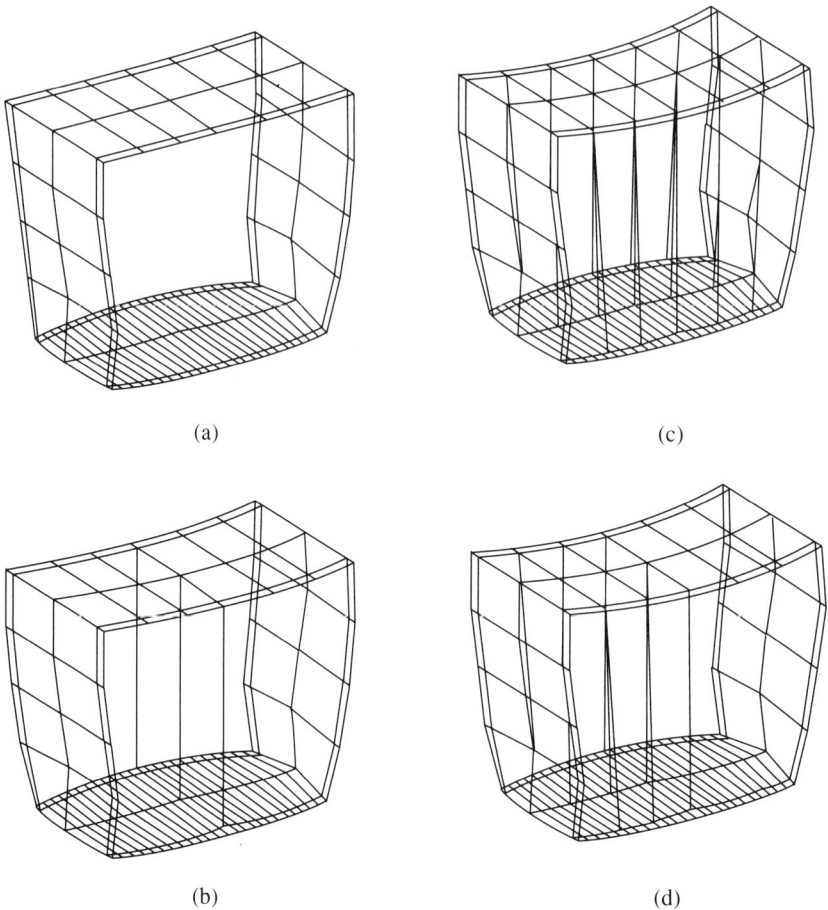

FIGURE 14.18. Card enclosure frame under uniform temperature change. (a) cardless box; (b) one central card inserted; (c) five cards; (d) three cards unsymmetrically located.

6. A Combined Permanent and Separable Connector System

In general, electrical connectors are either separable or permanent. The ZIF connectors discussed earlier, attaching circuit cards to the heavier circuit boards, are of the first kind, allowing unplugging and replugging as the need arises.

FIGURE 14.19. Card-to-board connector. (a) Section of connector system showing board, housing, and card; (b) spring connector to be inserted in plastic housing; (c) forces arising at insertion of pin.

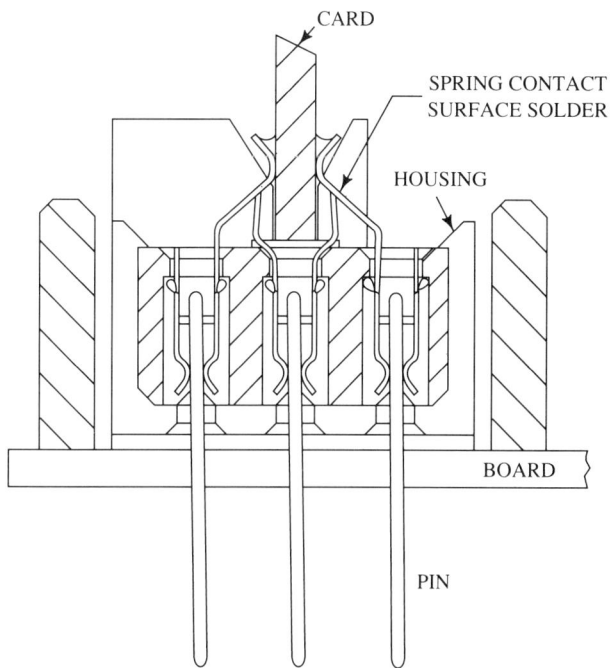

CARD

SPRING CONTACT
SURFACE SOLDER

HOUSING

BOARD

PIN

(a)

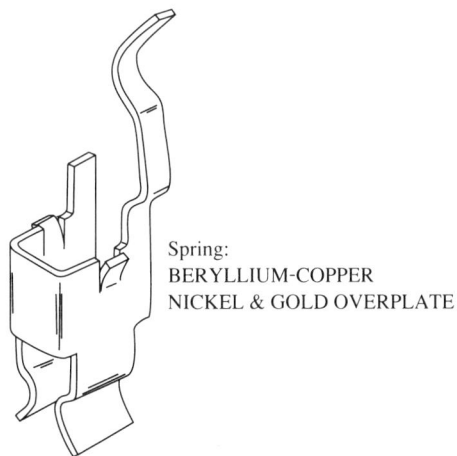

Spring:
BERYLLIUM-COPPER
NICKEL & GOLD OVERPLATE

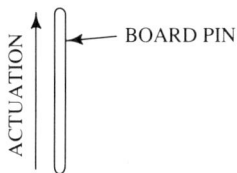

ACTUATION

BOARD PIN

(b)

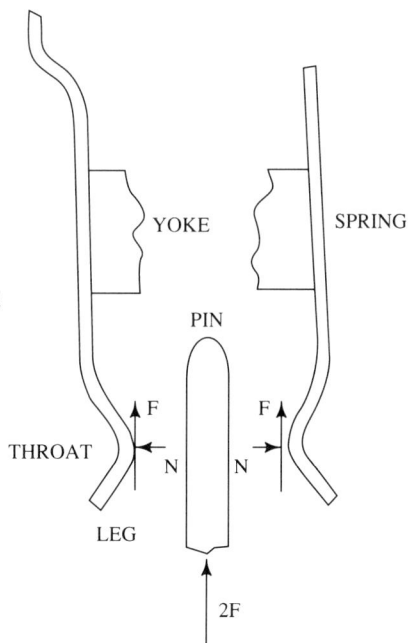

YOKE

SPRING

PIN

F   F

THROAT   N   N

LEG

2F

(c)

A card-to-board connector device having both separable and permanent features is illustrated in Fig. 14.19. It relies on an extruded Be-Cu spring connector that is inserted into the holes of a plastic housing. The housing with its holes is positioned underneath the card and parallel to its edge. Retention of each spring is maintained by its lances, which press it against the inner wall of the housing hole. The tail of the spring stands out upward and is soldered to a conductive tab surface on the card. The contact gap below makes a flexible socket into which a connector pin soldered into and protruding from the board can be pressed to establish the card-to-board connection. The contacting surfaces of both the pin and the spring can be made of noble or seminoble platings deposited on the Be-Cu surface.

7. Conclusions

The IBM 9370 computer card enclosure was discussed and analyzed as an assembly of frame, planar boards, ZIF electrical connectors, and cards. The design criteria were centered on maintaining great stiffness during actuation of the cards and for dynamic (vibration and shock) loading. Friction, lubrication, and wear of electrical connector contacts were crucial considerations.

8. Exercises and Questions

1. Calculate the wipe (i.e., the tangential displacement u) of a ZIF connector such as shown in Fig. 14.3 if the actuator force applied at the knee has a transverse component $P = 1$ N and a tangential component $T = 0.5$ N. Let the length of the Be-Cu ZIF spring be 2 cm, its thickness and width 0.25 mm, and scale the approximate geometry from Fig. 14.3.
2. Derive Eq. (14.1).
3. Based on the development of Eq. (14.3) and Fig. 14.9, estimate the card force in bay 12, if the actuating sequence is (a) 12, 8, 10; (b) 12, 10, 8. Assume that k is the same for each bay, and the same as applied toward constructing the graphs in Fig. 14.9.
4. Based on the data for the wearing of a Hertz dot in Example 1 ($K = 1.06 \times 10^{-5}$), find the depth of wear upon 30 actuations, each stroke length being 0.25 mm.

References

1. Engel, P.A., Toda, M.D., and Covert, D. (1988), "Card Enclosure Design Doubles I/O Connections," *Connection Technol.*, Sept., pp. 35–40.
2. Hsue, E.Y., and Bayer, R.G. (1989), "Metallurgical Study and Tribological Properties of Edge Card Connector Spring/Tab Interface," *IEEE Trans.*, **CHMT-12**(2), 206–214.

3. Engel, P.A., Nemier, S.E., and Toda, M.D. (1989), "Stress and Tolerance Analysis for Zero Insertion Force (ZIF) Connector," *ASME J. Elec. Packag.*, **111**(1), 9–13.

4. Engel, P.A., et al. (1985), Stress Analysis for the Corinthian Second Level Package," IBM Endicott Tech. Rep. 01.A064.

5. Holm, R. (1967), *Electrical Contacts*, 4th ed., Springer-Verlag, New York.

6. Rabinowicz, E. (1965), *Friction and Wear of Materials*, Wiley, New York.

7. Engel, P.A. (1976), *Impact Wear of Materials*, Elsevier, New York.

8. Chikazawa, T., Lim, C.K., Luu, H.V., Toda, M.D., and Vogelmann, J.T. 1990, "Sensor Technology for ZIF Connectors," *ASME J. Elec. Packag.*, **112**(3), 187–191.

9. Brodsky, W.L. (1987), "Testing of Design Parameters for Zero Insertion Force Connector," *Proc. 37th ECC Conf.*, pp. 32–40.

10. Engel, P.A., and Questad, D.L. (1990), "Indentation Method to Measure Plating Ductility," *ASME J. Elec. Packag.*, **112**(3), 272–277.

11. Trzeciak, M.J. (1971), "Studying Contact Tab Lubrication by Scanning Electron Microscopy," *Proc. Holm Conf. on Electrical Contacts*, pp. 145–156.

12. Bayer, R.G., and Roshon, D.D. (1965), "Some Design Considerations for Low Voltage Contacts," *Microelec. and Reliab.*, **4**, 131.

13. Engel, P., Talke, F., Bayer, R., Chai, S., Martin, J., Adams, C., and Lee, F. (1978), "Review of Wear Problems in the Computer Industry," *ASME J. Lub. Tech.*, **100**(2), 189–198.

14. Antler, M. (1985), "Electrical Effects of Fretting Contact Materials: A Review," *Wear*, **106**(1), 5–33.

15. Engel, P.A., Chitsaz, A.R., and Hsue, E.Y. (1992), "Interpretation of Superficial Hardness for Multilayer Platings," *Thin Solid Films*, **207**, 144–152.

16. Pope, R.A., and Schoenbauer, D.J. (1987), "Temperature Rise and Its Importance to Connector Users," *Proc. IEEE Holm Conf.*, pp. 24–31.

17. *Connector Design Guide* (1988), Brush Wellman Inc. Alloy Div., Cleveland, Ohio.

18. Kear, F.W. (1983), "Failure Modes in Printed Circuit Boards," *PC Fab.* August, pp. 73–80.

19. Ohmae, N., and Tsukizoe, T. (1974), *Wear*, **27**, 281–294.

Author Index

Subject Index

ABAQUS, 34, 211, 251
Acceleration transform, 222
Accelerometer, 228
Actuation of cards, 262–280
Adhesion, 250, 257
Airy stress function, 97
Alloy 42, 47
Amzirc, 47, 48, 224, 240
ANSYS, 34, 141, 250, 257, 263
Anticlastic curvature, 8
Arrhenius function, 217
Assembly, 260
ASTM, 236, 256
Avionics specification, 227
Axisymmetrical plate deformation, 10

Barrel-to-board mismatch, 246
Beams, 1
   coupled, 127
   on elastic foundation, 2, 78, 80, 119
Bessel functions, 11
Bimaterial rod, plate, 12, 67
Birefringent materials, 18
Bond stress, 14
Boundary lubrication, 263
Building block approach, 136, 150, 152

CAEDS, 34
Capacitance measurement, 17
Card enclosure system, 260
Ceramic substrate, 17
Circuit card and board, 17, 35

Boundary conditions, 226
   drilling, 244
   flexural vibration, 226
   manufacturing, 43
   multilayer (MIB), 41, 44, 243–255
Coffin Manson Equation, 256
Collocation, 99
Connector spring, 260–280
Constitutive equation, 72
Contact force, 18, 260–280
Contact temperature, 274
Copper-Invar-Copper, 42
Cornell University, 17
Coulomb's law, 270
Creep, 217
CTE, 12, 41, 45, 46, 61, 204, 277

Damage, 238
   boundary curve, 234
Degree of freedom (dof), 3, 22
Delamination, 246
Dielectric constant, 244
Direct stiffness, 22
Drop test, 236
Ductility, 250, 271
Dwell time, 220
Dynamic
   environment, 227
   measurement, 17, 227
   response, 226, 274
   vibration absorber, 234
DYNA2, DYNA3, 34